T0178192

Foundations of Multi-Paradigm Modelling for
Cyber-Physical Systems

Paulo Carreira • Vasco Amaral • Hans Vangheluwe
Editors

Foundations of Multi-Paradigm Modelling for Cyber-Physical Systems

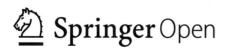

EUROPEAN COOPERATION
IN SCIENCE & TECHNOLOGY

Editors
Paulo Carreira
INESC-ID & Instituto Superior Técnico
Universidade de Lisboa
Lisboa, Portugal

Vasco Amaral
NOVA LINCS & Departamento de Informática
Faculdade de Ciências e Tecnologia
Universidade NOVA de Lisboa
Caparica, Portugal

Hans Vangheluwe
Department of Computer Science
University of Antwerp - Flanders Make
Antwerpen, Belgium

*In memory of our dearest friend and colleague Bernhard Schätz
(1965–2017), an extraordinary scientist and one of the founders of
the MPM4CPS network. He will be forever remembered for his
remarkable work ethic—an inspiration for everyone around him.
His enthusiasm and his always positive attitude are greatly missed.*

Foreword

The explosion of desktop computing power that started in the Eighties made computational modeling available to engineers and scientists in exponentially increasing numbers. A key impediment to unlocking the potential, however, was the necessity to formulate models in programming languages, for instance, FORTRAN and Algol. Whereas organizations such as the national laboratories were able to employ software engineers who could program mathematical models, desktop users did not have access to like resources.

As a natural course, more accessible languages for computational modeling were developed that aligned with modeling practices and formalisms in various fields. To list a few examples, algorithmic array-based languages such as MATLAB® proved exceedingly useful in Signal Processing. In Controls, models were often preferred in the form of block diagrams. In Communications, protocols would be modeled well with Petri nets. Physics would be most conviniently represented by noncausal models such as bond graphs. Moreover, not any less important, intuitive environments for modeling were developed, oftentimes building on graphical user interface (GUI) technology that emerged.

While the democratization of computational modeling that desktop computing brought about was initially providing simulation as an aid in the work by engineers and scientists, in due time the computational artifacts became of additional value. Not only to the point where modeling and computational simulation became a critical step in individual analysis but also by an increasing value proposition in sharing models across teams, departments, and even organizations.

It is around the turn of the millennium when this trend of models as first-class artifacts in engineered system design starts to take shape. While I was employed by the DLR Oberpfaffenhofen, Hans Vangheluwe and I meet during one of the Modelica language design meetings and our discussions start to uncover the necessary research fundamental to addressing the growing need for model sharing and reuse. The magnificent challenge before us is in connecting, combining, and integrating models with different syntax (e.g., textual, different graphical elements, different types of relations), different semantics (e.g., imperative vs. declarative, data flow vs. control flow, time-based vs. untimed), different contexts (individual vs. team based, critical design artifact vs. one-off study, debugging vs. production), different perspectives (e.g., functional vs. performance, structure vs. behavior), different purposes (e.g., analysis vs. synthesis, safety vs. creativity), different processes (from mathematics to computation vs. computational semantics, first principle modeling vs. data-driven modeling), and different workflows (e.g.. top-down vs. bottom-up, continuous integration vs. periodic sharing, continuous refinement vs. back-to-back testing).

One day, the director of my group at the DLR Oberpfaffenhofen, Georg Gruebel and I are considering a special session on the topic at the IEEE Computer Aided Control System Design symposium (CACSD) in 2000. Our considerations of a title do not result in anything befitting and he leaves for home. Only to return to my office after about fifteen minutes with a name for the budding field that was developing: Computer Automated Multiparadigm Modeling (CAMPaM).

The CACSD 2000 initiative was successful enough that it required two conference sessions. We followed that up with special issues of SCS SIMULATION and IEEE Transactions on Control System Technology. In 2004 we started a Bellairs series of workshops on CAMPaM that is active to this day.

Over time, the computer automated ambition proved to be overly optimistic and we tempered our enthusiasm by extending the scope to multiparadigm modeling (MPM) in the large, including computer-aided approaches. Correspondingly, a series of MPM workshops were organized at the ACM/IEEE International Conference on Model Driven Engineering Languages and Systems (MODELS)from 2010 through 2015.

Today, the prescience of our observation at the turn of the millennium is reflected by the critical need for models in the emerging proliferation of Cyber-Physical Systems (CPS). CPS embed computation and communication as central to their physical operation. Most prominently, this enables a degree of flexibility or openness that is distinctly different from the rigidly singular operation of computation in the embedded systems paradigm that preceded CPS. CPS are open in a vertical sense such that compute platforms in hardware and software may change independent of the logic that the computations implement. CPS are open in a horizontal sense such that systems connect with other systems to achieve overall goals as a result of behavior that emerges post deployment. And it is precisely this trend toward ever more openness that necessitates an increasing use of models across paradigms, not limited to system design but just as much during operation (at run time).

And so the compilation of this textbook on Foundations of Multi-Paradigm Modelling for Cyber-Physical Systems is exceedingly timely. Much of the research efforts have solidified over the years while the underlying needs for multiparadigm modeling by the practitioner are still growing. This books provides an excellent collection of content that focuses on key aspects of CPS, ranging from modeling the physics, the computation and communication, control, performance characteristics, and system architecture. As an essential textbook on mutiparadigm modeling for Cyber-Physical Systems, all of these aspects are layered on top of foundations of multiparadigm modeling.

Over the years it has been a great pleasure working with all of the authors and editors of this collection. I am excited to see the field of CAMPaM coming into maturity as signified by this book. Many thanks to all involved for their time and effort to make this happen!

Natick, MA, USA, May 2019

Pieter J. Mosterman
Chief Research Scientist and Director of the
MathWorks Advanced Research & Technology Office

Preface

Modelling and analysis are essential activities in the development of Cyber-Physical Systems. The inherent cross-disciplinary nature of CPS requires distinct modelling techniques related to different disciplines to be employed. At the same time, to enable communication between all specialities, common background knowledge of formalisms is expected.

An important realization is that distinct models need to be woven together consistently to form a complete representation of a system to enable, among other global aspects, performance analysis, exhaustive simulation and verification, hardware-in-the-loop simulation, determining the best overall parameters for the system, prototyping, and implementation. Conceivably, a unifying formalism would represent the connections between the mentioned distinct models to enable reasoning about them. Likewise, recent advances show that coupling disciplines by using co-simulation will allow them to cooperate without enforcing new tools or design methods.

As first, CPS is a relatively new field, with no established standards and with different actors, each of which adopts a slightly different definition and a different approach for the topic. While multi-paradigm modelling embraces a vast spectrum of distinct topics and has a very elastic and generic (informal) definition, the result is that existing tools were created from different directions, as the field spans from very theoretical to application-oriented approaches.

The goal of this book volume is to serve as a reference text that compiles, in a coherent manner, well-established knowledge around fundamentals and formalisms for modelling of CPS. The book focuses on state-of-the-art solid knowledge and practice, or fundamentals (not research) and strives to provide a homogeneous treatment of the cross-disciplinary nature of CPS modelling. It can be used as a bridge for anyone entering CPS from any of the related areas of engineering (e.g., electric engineering, mechanics, physics, computer science). Though chiefly intended for master and post-graduate level students in computer science and engineering, it can also serve as a reference text for practitioners at large.

Acknowledgements

This book was supported by the COST Action IC1404 Multi-Paradigm Modelling for Cyber-Physical Systems (MPM4CPS), COST is supported by the EU Framework Programme Horizon 2020.

Contents

1 Multi-Paradigm Modelling for Cyber-Physical Systems: Foundations 1
Paulo Carreira, Vasco Amaral, and Hans Vangheluwe
1.1 Introduction 1
1.2 Understanding Cyber-Physical Systems 2
    1.2.1 Systems and their models 3
    1.2.2 Types of systems 3
    1.2.3 What are Cyber-Physical Systems? 4
    1.2.4 Sources of Engineering Complexity 5
1.3 Modeling of a Cyber-Physical System 6
    1.3.1 What is a Model? 7
    1.3.2 Multiple Formalisms in CPS 8
1.4 Multi-Paradigm Modelling of CPS 8
    1.4.1 What is a Paradigm? 9
    1.4.2 The dimensions of Multi-Paradigm Modelling 9
1.5 A foundation for MPM4CPS 10
    1.5.1 Modelling physical components 10
    1.5.2 Joining the 'Physical' with the 'Cyber' 11
    1.5.3 Tooling support for MPM4CPS 13
1.6 Summary 13

2 Bond Graphs: A Unifying Framework for Modelling of Physical Systems 15
Jan F. Broenink
2.1 Introduction 15
2.2 Bond-Graph Examples 16
2.3 Foundation of Bond Graphs 19
    2.3.1 Starting Points 19
    2.3.2 Bonds and Ports 20
2.4 Bond-Graph Elements 20
    2.4.1 Storage Elements 21
    2.4.2 Resistors 22
    2.4.3 Sources 24
    2.4.4 Transformers and Gyrators 25
    2.4.5 Junctions 26
    2.4.6 Positive Orientation 26
    2.4.7 Duality and Dual Domains 28
2.5 Systematic Procedure to Derive a Bond-Graph Model 28
    2.5.1 The Eight Steps of the Systematic Procedure 28
    2.5.2 Illustration of the Systematic Procedure 29
2.6 Causal Analysis 33
    2.6.1 Causal Constraints 33
    2.6.2 Causal Analysis Procedure 34

	2.6.3	Model Insight via Causal Analysis	35
	2.6.4	Order of the Set State Equations	36
	2.6.5	Matrix Form of Linear Systems	37
2.7	Generation of Equations		38
2.8	Expansion to Block Diagrams		40
2.9	Simulation		41
2.10	Summary		43
2.11	Literature and Further Reading		43
2.12	Self-Assessment		43

3 Modelica: Equation-Based, Object-Oriented Modelling of Physical Systems ... 45
Peter Fritzson

3.1	Introduction		45
3.2	Getting Started with Modelica		46
	3.2.1	Variables and Predefined Types	49
	3.2.2	Comments	50
	3.2.3	Constants	51
	3.2.4	Variability	52
	3.2.5	Default Start Values	52
3.3	Object-Oriented Mathematical Modelling		52
	3.3.1	Classes and Instances	53
	3.3.2	Inheritance	56
	3.3.3	Generic Classes	57
3.4	Equations		59
	3.4.1	Repetitive Equation Structures	60
	3.4.2	Partial Differential Equations	61
3.5	Acausal Physical modelling		61
3.6	The Modelica Software Component Model		63
	3.6.1	Components	64
	3.6.2	Connection Diagrams	64
	3.6.3	Connectors and Connector Classes	65
	3.6.4	Partial Classes	68
	3.6.5	Reuse of Partial Classes	69
3.7	Component Library Design and Use		70
	3.7.1	The Simple Circuit Model	72
	3.7.2	Arrays	73
3.8	Algorithmic Constructs		75
	3.8.1	Algorithm Sections and Assignment Statements	75
	3.8.2	Statements	76
	3.8.3	Functions	76
	3.8.4	Operator Overloading and Complex Numbers	78
	3.8.5	External Functions	79
	3.8.6	Algorithms Viewed as Functions	80
3.9	Discrete Event and Hybrid modelling		80
3.10	Modularity Facilites		84
	3.10.1	Packages	84
	3.10.2	Annotations	86
	3.10.3	Naming Conventions	86
3.11	Modelica Standard Library		87
3.12	Implementation and Execution of Modelica		88
	3.12.1	Hand Translation of the Simple Circuit Model	90
	3.12.2	Transformation to State Space Form	91
	3.12.3	Solution Method	92
3.13	Tool Interoperability through Functional Mockup Interface		94
3.14	Summary		94

    3.15   Literature and Further Reading 95
    3.16   Self-assessment 95

4   Causal-Block Diagrams: A Family of Languages for Causal Modelling of Cyber-Physical Systems 97
    Cláudio Gomes, Joachim Denil, and Hans Vangheluwe
    4.1   Introduction 97
    4.2   Background 99
          4.2.1   Models of Physical Systems 99
          4.2.2   Discrete Time Models 100
    4.3   Algebraic Causal Block Diagrams 101
          4.3.1   Syntax 102
          4.3.2   Semantics 103
    4.4   Discrete-time CBDs 109
          4.4.1   Syntax 109
          4.4.2   Semantics 109
    4.5   Continuous-time CBDs 112
          4.5.1   Syntax 112
          4.5.2   Semantics 112
    4.6   Advanced Concepts and Extensions 115
          4.6.1   Approximation Error... 116
          4.6.2   Other Numerical Methods 117
          4.6.3   Adaptive-Step Size 120
          4.6.4   Logic Blocks 121
    4.7   Global Error Euler Method 122
    4.8   Summary 124
    4.9   Literature and Further Reading 124
    4.10 Self Assessment 124

5   DEVS: Discrete-Event Modelling and Simulation for Performance Analysis of Resource-Constrained Systems 127
    Yentl Van Tendeloo and Hans Vangheluwe
    5.1   Introduction 127
    5.2   Atomic DEVS models 128
          5.2.1   Autonomous Model.. 128
          5.2.2   Autonomous Model with Output 130
          5.2.3   Interruptable Model 132
    5.3   Coupled DEVS Models 135
          5.3.1   Basic Coupling 135
          5.3.2   Input and Output 136
          5.3.3   Tie-breaking.. 136
          5.3.4   Translation Functions 137
          5.3.5   Closure Under Coupling 137
    5.4   The DEVS Abstract Simulator 139
    5.5   Application to Queueing Systems 141
          5.5.1   Problem Description 142
          5.5.2   Description in DEVS 143
          5.5.3   Performance Analysis 150
    5.6   DEVS Variants 150
          5.6.1   Parallel DEVS 150
          5.6.2   Dynamic Structure DEVS 152
          5.6.3   Cell-DEVS 152
          5.6.4   Other Variants 152
    5.7   Summary 153
    5.8   Literature and Further Reading 153
    5.9   Self-Assessment 153

**6   Statecharts: A Formalism to Model, Simulate and Synthesize Reactive and Autonomous
      Timed Systems** 155
      Simon Van Mierlo and Hans Vangheluwe
      6.1   Introduction 155
      6.2   Background 156
             6.2.1   Running Example 156
             6.2.2   Discrete-Event Abstraction 158
             6.2.3   Process 160
      6.3   Modelling with Statecharts 160
             6.3.1   States and Transitions 160
             6.3.2   Composite States 162
             6.3.3   Orthogonal Regions 163
             6.3.4   History 164
             6.3.5   Syntactic Sugar 166
             6.3.6   Full Statechart Model 166
      6.4   Detailed Semantics 166
      6.5   Testing Statecharts 168
      6.6   Deploying Statecharts 169
      6.7   Advanced Topics 171
             6.7.1   Semantic Variations 171
             6.7.2   Execution Platforms 173
             6.7.3   Dynamic Structure 174
      6.8   Summary 174
      6.9   Literature and Further Reading 175
      6.10   Self Assessment 176

**7   Petri Nets: A Formal Language to Specify and Verify Concurrent Non-Deterministic Event
      Systems** 177
      Didier Buchs, Stefan Klikovits, and Alban Linard
      7.1   Introduction 177
      7.2   Modelling Concurrency 178
             7.2.1   Petri Nets 179
             7.2.2   Common Petri net patterns 180
             7.2.3   Formal syntax and semantics 182
             7.2.4   Deduction Based on Rules 184
             7.2.5   Reachability Graph 185
             7.2.6   Monotony 186
      7.3   Properties of Petri Nets 187
             7.3.1   Marking Properties 188
             7.3.2   Sequence Properties 189
             7.3.3   Invariants 189
             7.3.4   Formal Definition of Invariants 190
             7.3.5   Computing P-invariants 191
             7.3.6   Using Invariants for Proving Properties 192
      7.4   Techniques for Model Checking 193
      7.5   Data manipulation in Petri nets 194
             7.5.1   Drone Controller 195
             7.5.2   Formalising High-level Petri nets 195
             7.5.3   Other High-level nets 197
      7.6   Combining Model Semantics and Simulation 198
             7.6.1   PNFMU Formalisation 199
             7.6.2   PNFMU Example 200
             7.6.3   PNFMU Composition 202
             7.6.4   Advanced Composition Mechanisms 202
      7.7   Tooling 203

7.7.1 Tools for Petri net Modelling and Verification 203
7.7.2 Evaluation of Model Checking Techniques 204
7.8 Summary 205
7.9 Literature and Further Reading 206
7.10 Self Assessment 207

8 AADL: A Language to Specify the Architecture of Cyber-Physical Systems 209
Dominique Blouin and Etienne Borde
8.1 Learning Objectives 209
8.2 Introduction 209
8.2.1 Increasing Systems Complexity and Unaffordable Development Costs 210
8.2.2 Mismatched Assumptions in Collaborative Engineering 211
8.2.3 System Architecture Virtual Integration: Integrate, Analyze then Build 212
8.2.4 Architecture-Centric Authoritative Source of Truth 212
8.2.5 Organisation of the Chapter 213
8.3 AADL Overview 214
8.4 CPS Running Example 215
8.5 The Development Process 217
8.6 Modelling the Line Follower Robot with AADL 217
8.6.1 System Overview 218
8.6.2 Operational Concepts 227
8.6.3 Environmental Assumptions 227
8.6.4 Functional Architecture 229
8.6.5 Physical Plant Model 233
8.6.6 Deployment 241
8.6.7 Analyses 250
8.6.8 Code Generation 255
8.7 Summary 257
8.8 Literature and Further Reading 257
8.9 Self Assessment 257

9 FTG+PM: Describing Engineering Processes in Multi-Paradigm Modelling 259
Moharram Challenger, Ken Vanherpen, Joachim Denil, and Hans Vangheluwe
9.1 Introduction 259
9.2 Model-based Systems Engineering 260
9.3 Development Lifecycle Models 261
9.3.1 Waterfall Model 261
9.3.2 V Model 262
9.3.3 Spiral Model 263
9.3.4 Agile Model 263
9.4 Modelling the Design Process 263
9.4.1 Rationale 263
9.4.2 What to model? 264
9.4.3 Reasoning about maturity 264
9.5 Activities 2.0 for modelling processes 265
9.6 The tool perspective: Formalism Transformation Graph 267
9.7 FTG+PM: Formalism Transformation Graph and Process Model 267
9.7.1 Reasoning about appropriateness of formalisms and heterogeneous modelling 268
9.7.2 Orchestrating Processes 269
9.8 Summary 270
9.9 Literature and Further Reading 270
9.10 Self-Assessment 270

Bibliography 273

List of Contributors

Albin Linard
Computer Science Department, University of Geneva, 7 route de Drize, Bât. A, Office 217, CH-1227 Carouge, Switzerland, e-mail: albinalban.linard@unige.ch

Cláudio Gomes
University of Antwerp, Campus Middelheim, Middelheimlaan 1, M.G.028, 2020 Antwerpen, Belgium, e-mail: claudio.goncalvesgomes@uantwerpen.be

Didier Buchs
Computer Science Department, University of Geneva, 7 route de Drize, Bât. A, Office 217, CH-1227 Carouge, Switzerland, e-mail: Didier.Buchs@unige.ch

Dominique Blouin
COMELEC / INFRES, Telecom ParisTech, 46 rue Barrault, 75013 Paris, France, e-mail: dominique.blouin@telecom-paristech.fr

Etienne Borde
COMELEC / INFRES, Telecom ParisTech, 46 rue Barrault, 75013 Paris, France, e-mail: etienne.borde@telecom-paristech.fr

Hans Vangheluwe
University of Antwerp, Campus Middelheim, Middelheimlaan 1, M.G.116 , 2020 Antwerpen, Belgium , e-mail: Hans.Vangheluwe@uantwerpen.be

Jan F. Broenink
University of Twente, Faculty EE-Math-CS, Department of Electrical Eng., Robotics and Mechatronics, Room Carre 3437, P.O. Box 217, NL-7500 AE Enschede, Netherlands, e-mail: j.f.broenink@utwente.nl

Joachim Denil
University of Antwerp, Campus Groenenborger,Groenenborgerlaan 171, G.U.135, 2020 Antwerpen, Belgium e-mail: Joachim.Denil@uantwerpen.be

Ken Vanherpen
University of Antwerp, Campus Groenenborger Groenenborgerlaan 171, G.U.123, 2020 Antwerpen , Belgium e-mail: ken.vanherpen@uantwerpen.be

Moharram Challenger
University of Antwerp, Campus Groenenborger Groenenborgerlaan 171, G.U.122, 2020 Antwerpen, Belgiume-mail: moharram.challenger@uantwerpen.be

Paulo Carreira
Instituto Superior Técnico, Universidade de Lisboa, and INESC-ID, Gab. 2N.5.17, DEI/IST Taguspark, Av. Prof Cavaco Silva, 2780-990, Porto Salvo, PORTUGAL e-mail: paulo.carreira@tecnico.ulisboa.pt

Peter Fritzson
Linköping University, 581 83 Linköping, Sweden e-mail: peter.fritzson@liu.se

Simon Van Mierlo
University of Antwerp, Campus Middelheim, Middelheimlaan 1, M.G.330 , 2020 Antwerpen, Belgium, e-mail: simon.vanmierlo@uantwerpen.be

Stefan Klikovits e-mail: stefan.klikovits@unige.ch
Computer Science Department, University of Geneva, 7 route de Drize, Bât. A, Office 217, CH-1227 Carouge, Switzerland,

Vasco Amaral
Departamento de Informática, Faculdade de Ciências e Tecnologia, Universidade NOVA de Lisboa,Quinta da Torre, 2829 -516 Caparica, Portugal e-mail: vasco.amaral@fct.unl.pt

Yentl Van Tendeloo
University of Antwerp, Campus Groenenborger,Groenenborgerlaan 171, G.U.135, 2020 Antwerpen, Belgium e-mail: yentl.vantendeloo@uantwerpen.be

Chapter 1
Multi-Paradigm Modelling for Cyber-Physical Systems: Foundations

Paulo Carreira, Vasco Amaral, and Hans Vangheluwe

Abstract Modeling and analysis of Cyber-Physical Systems (CPS) is an inherently multi-disciplinary endeavour. Anyone starting in this field will unavoidably face the need for a literature reference that delivers solid foundations. Although, in specific disciplines, many techniques are used already as a matter of standard practice, their fundamentals and application are typically far from practitioners of another area. Overall, practitioners tend to use the technique that they are most familiar with, disregarding others that would be adequate for the problem at hand. The inherent cross-disciplinary nature of CPS requires distinct modelling techniques to be employed, thus prompting for a common background formalism that enables communication between all specialities. However, to this date, no such single super-formalism exists to support the multiple dimensions of the design of a CPS. Indeed, to effectively design a CPS, engineers (in the role of modellers) either need to be versed in multiple formalisms, or a fundamentally new modelling approach has to emerge. Herein, we motivate Multi-Paradigm Modelling of CPS (MPM4CPS), introducing fundamental definitions and terminology regarding CPS modelling and Multi-Paradigm, and finally, laying the ground for the rest of the book.

1.1 Introduction

Cyber-Physical Systems (CPS) refer to systems that consist of *cyber* (as computerised implementations) and *physical* components [130]. The general idea is that the cyber and physical components influence each other in such way that the cyber is able to cause the physical component to change state, and that the change, in turn, will feed-back, resulting in a change of state on the cyber component.

Having emerged from earlier concepts, among other, in the fields of mechatronics, embedded systems, and cybernetics, literature gives the coining of the term 'Cyber-Physical System' (CPS) to Hellen Guille in 2006 [130]. CPS are often regarded as networks of multi-physical (mechanical, electrical, biochemical, etc) and computational (control, signal processing, logical inference, planning, etc) processes, often interacting with a highly uncertain and adverse environment, including human actors and other CPS.

Example application domains of CPS include energy conservation, environmental control, avionics, critical infrastructure control (electric power, water resources, and communications systems), high confidence medical devices and systems, traffic control and safety, advanced automotive systems, process control, distributed robotics (telepresence, telemedicine), manufacturing, and smart city engineering. The design of CPS is currently a driver

Paulo Carreira
Instituto Superior Técnico, Universidade de Lisbon, Portugal
e-mail: paulo.carreira@tecnico.ulisboa.pt

Vasco Amaral
FCT, Universidade NOVA de Lisboa, Portugal
e-mail: vma@fct.unl.pt

Hans Vangheluwe
McGill University, Canada
e-mail: hans.vangheluwe@uantwerp.be

© The Author(s) 2020
P. Carreira et al. (eds.), *Foundations of Multi-Paradigm Modelling for Cyber-Physical Systems*,
https://doi.org/10.1007/978-3-030-43946-0_1

for innovation across various industries, creating entirely new markets. More efficient and cheaper CPS will have a positive economic impact on any one of these applications areas.

CPS are notoriously complex to design and implement mostly because of their cross-discipline borders, leading to inter-domain interactions, in applications that are often safety-critical. Indeed, due to the nature of their application, failure or underperformance of CPS can have direct, measurable economic costs, can harm the environment or even directly affect humans. Expectably, engineering practice has, over the years, sought to address these concerns by improving the languages, frameworks, and tools used in the design and analysis of CPS. This effort led to the emergence and strong adoption of model-based design, in which systems are designed at a higher level of abstraction, and an implementation is then produced by automatic generation.

A striking aspect of CPS design is that it is inherently multi-disciplinary. One source of this multi-disciplinarity arises from the domain of the application itself, such as e.g., medical, biological, or aeronautical industries. Another source is the heterogeneous nature of CPS, which consists of computerised, electronic, and mechanic parts. To design a CPS, engineers from various disciplines need to explore system designs collaboratively, to agree, to allocate responsibilities to software and physical elements, and to analyse trade-offs between them.

Originating from the Modelling and Simulation Community, the term *Multi-Paradigm-Modelling* (MPM) finds its origin in 1996, when the EU ESPRIT Basic Research Working Group 8467 formulated a series of simulation policy guidelines [286] identifying the need for *"a multi-paradigm methodology to express model knowledge using a blend of different abstract representations rather than inventing some new super-paradigm"*, and later on proposing a methodology focusing on combining multiple formalisms [294]. Since 2004, during the yearly Computer Automated Multi-Paradigm Modelling (CAMPaM) Workshop series at McGill University's Bellairs Research Institute, many ideas surrounding MPM were developed. Since then, MPM became a well-recognised research field with a large body of research produced and published, in particular in the MPM Workshops co-located with MoDELS.

The recent COST Action IC1404 MPM4CPS (http://mpm4cps.eu) aimed at exploring how MPM can be employed to alleviate the engineering complexity surrounding the conception of CPS. Among other efforts, the scientific community gathered around this action surveyed existing languages, techniques, and tools commonly used for modelling CPS and organized them into an ontology [167].

This work identified the need come up with a theoretical foundation for MPM and identified useful language features. Among other, the most relevant can be summarised as follows: Closeness to the essential concepts that engineers use to reason about the behaviour of physical systems, in a computationally a-causal fashion (i.e., without the need to specify early on what are inputs and outputs); the ability to precisely describe computation, at different levels of detail of the time dimension; the ability to elegantly express concurrency, synchronisation and non-determinism and to reason about properties over all possible behaviours of a system; the ability to express modal, timed, reactive and autonomous behaviour and to synthesise code; suitability to model competition for shared resources, which leads to queueing, as a basis for quantitative performance analysis; suitability for easy and correct architectural composition; the ability to express workflows at a high level of abstraction and finally, the high-level feature of modularity, supporting re-use, compositional verification of properties and the integration of black-box components such as co-simulation units. The breath of the techniques is broad.

The results of the efforts held by the MPM community, mentioned above, culminated in a book that compiles in a coherent manner well-established knowledge around fundamentals and formalisms for modelling of CPSs with a particular focus on tools and techniques that are multi-paradigm.

1.2 Understanding Cyber-Physical Systems

We now turn to discussing which classes of systems can be considered a CPS, what are their properties, and the sources of engineering complexity worth solving using the so-called MPM approach. Before embarking in the discussion some preliminary concepts need to be made clear.

1.2.1 Systems and their models

The notion of *system* is a fundamental concept used in multiple disciplines. The term conceptualises a physical existence such as an ecosystem, an organism, a machine, or a purely *abstract existence*. The latter case refers to processes, rules (such as a socio-economic system), or mathematical models. Regardless, any conception of a system is understood *(i)* as a set of *components (ii)* identifiable as a *whole*, that *(iii) cooperate* (also said to *interact*) to *(iv)* perform a particular *function*. There is a distinction between the actual systems (physical or abstract) and the human understanding of them. The formalisation of this understanding, usually limited, is called a *model*. [308]

According to Systems Theory, a system can be understood in two ways. One way is the *black-box* approach that seeks to model the *external behaviour* of the system in terms of how it interacts with the environment. Here, the behaviour of the system is seen as the relationship between the evolution of the history of *manifestations* (also said *outputs*) and the history of *stimuli* (also said *inputs*). Another way to understand the system is the *white-box* approach, that seeks to understand its *internal structure* in terms of *components* and *connectors* through which interaction occurs.

Both the black-box and the white-box approach formalise knowledge about the behaviour of the system with *models*[1], which are simplified representations of the system.The utility of working representations is that they are cognitively effective means to reason about the actual system (or systems) they represent. By reasoning, we mean analysing properties of the system or predicting the future behaviour of the system. A system can thus have multiple models that cater to distinct requirements in terms of reasoning. Control Theory has historically taken the black-box approach where outputs are modelled in terms of inputs using differential equations; Computer Science has traditionally taken the white-box approach by modelling the internal structure of systems using object diagrams.

The driving idea behind understanding the internal structure is that components, bound through a certain arrangement of connections, display specific external behaviour. Moreover, the overall behaviour can be derived from *(i)* the known behaviour of the components and *(ii)* the characteristics of the connectors. Connections are abstract flows of information, or energy, that bind components through pre-designated points of interaction known as *ports*. Some component ports are *directional*, we can thus talk about *input ports (inputs)*, and *output ports (outputs)*. Other ports are *adirectional* in that they model an exchange and not necessarily a flow (consider a thermal coupler, for example).

A component can itself be seen as a system itself with its own components and connectors. The ability to continue composing systems from previously constructed ones is known as *hierarchical construction*. It is common also to model components with a *internal state* that reflects (partially and sometimes inaccurately) what the system knows about the surrounding environment and a *state-transition* mechanism typically referred to as *transition function* that is capable of producing new states from previous states upon receiving certain inputs and creating certain outputs.

1.2.2 Types of systems

Inputs, outputs and the state of the components are modelled in terms of *variables* that can take values from their corresponding support sets. These variables are said to be *continuous* or *discrete* if their support sets are, in a mathematical sense, dense or discrete, respectively. The behaviour of the system, as modelled in terms of internal state and outputs, evolves from a previous state according to some notion of time. Time can also be understood as continuous or discrete. In *continuous time*, it is possible to derive the new state and the new outputs for the system for an arbitrarily small time delta; whereas in *discrete time*, the new state and outputs can be derived only at predefined intervals, or upon the occurrence of a certain event.

Systems can be classified into distinct types depending on the nature of inputs, outputs, state variables, state transition function, and the notion of time. A well-accepted classification is as follows [308]:

- **Static vs Dynamic Systems.** A system is said to *static* if its output depends only on the present input. If the output of the system depends on the history of past inputs, then the system is said to be *dynamic*.

---

[1] The precise meaning of the term 'model' is discussed later in Section 1.3.1

- **Causal vs Acausal (Non-causal).** Whenever the output value of the system is independent of future values of input, the system is said to be a *causal system*; whenever the output values of the system depend on input values at any instant of time, the system is said to be a *acausal system*.
- **Linear vs Non-linear.** A system is said to be *linear* (respectively, *non-linear*) if changes on the output are proportional (respectively, not proportional) to the changes of the input.
- **Discrete State vs. Continuous State.** Those systems in which the state variable(s) change only at a discrete set of points in time are said to be *discrete state* systems; systems in which the state variable(s) change continuously over time (e.g., a water tank filling in), are said to be *continuous state* systems.
- **Discrete Time vs Continuous Time.** A system is said to be *discrete*, in contrast with *continuous*, if it has a countable number of states.
- **Time-Driven vs. Event-Driven.** A *time-driven* system changes state in response to a uniform physical time. While, a *event-driven* system changes state in reaction to the occurrence of asynchronous discrete events (not changing state between event occurrences).
- **Time-Variant vs. Time-Invariant.** A system in which certain state variables change with time causing the system to respond differently to the same input at different times is called *time-variant*; a system that yields the same output for a given input at distinct points in time is said to be *time-invariant*.

A dynamic system that exhibits both continuous and discrete time behaviour is said to be an *hybrid system*. The study of hybrid systems is very important as they arise often in the composition of discrete with continuous components typical of cyber-physical systems.

1.2.3 What are Cyber-Physical Systems?

It is assumed that the cyber component *controls* the physical component in the sense that the cyber component has some 'intelligence' or, at least, some strategy to drive the physical to reach a predefined observable goal. The converse is not true, i.e., the physical does not aim at driving the cyber to reach a certain predefined goal. This formulation puts CPS in the realm of Control Systems theory. Despite de comprehensive nature of Control Systems theory, CPS are not a particular case of Control Systems.

In a *control system* one component, known as the *controller* realises a control model that acts upon the physical environment component known as *plant* by means of a *control action*. The controller knows the desired value of a variable and the current value of that variable in the physical component. The controller then creates a sequence of control actions to correct (i.e., to minimise the distance over time) of a variable (measured from the physical system) to the desired value (the goal). This arrangement is known as *control system*. It is, therefore, reasonable to ask: *"what distinguishes a control system from a CPS?"* Besides the very idea of a cyber control component—a discrete computer algorithm controlling continuous physical phenomena—it seems that there is no single characteristic that of itself defines a CPS. However, it is well-accepted that cyber-physical systems consist of a large number of interacting components and display a number of recurring characteristics that distinguish them from classic control systems. In particular:

- **Extensive 'cyber' components** that encode complex control and supervisory control logic. Typically, they have multiple cyber sub-components that support complex action coordination and require the processing of very large amounts of historical data.
- **Very large scale of operation** outreaching several millions of elements (sometimes heterogeneous) involving an inherent complexity of hierarchies and interactions. Examples of those systems are Smart grids, Smart cities, Particle Physics Detectors, among others.
- **Hybrid discrete-continuous nature -** where a very large number of discrete components (especially 'cyber' components) are connected to physical components (continuous in nature) thus creating a hybrid system. Also, many of these components are quite heterogeneous with respect to their types (Section 1.2.2).
- **Integration with multiple external systems** by processing data and events in distinct formats from multiple systems with varying bandwidth and message delivery guarantees.
- **Highly networked and hierarchical** connecting many components typically through digital networks with distinct communication buses and protocols.
- **Adaptable** where the system must adjust their behaviour to patterns that could not be accounted for at design time.

- **Human in the loop** offering specific provision for Human actors to consume outputs and give inputs. Human actors are often modelled as components with specific behaviour requirements. and assumptions.

The pervasiveness of IoT with a large number of connected devices has created an upsurge of interest in CPS. However, the term is somewhat overused, and it has now become an umbrella for any system that interacts with the physical environment. For instance, at the current state of practice, a developer of a temperature logger might as well claim that his device is a CPS. This raises the question of *what is the minimum requirements for a system to be considered a CPS?* One key observation is that having identifiable cyber and physical components are not enough for a system to be considered cyber-physical; the cyber and physical components must influence each other[2]. Another reasonable question is whether this relationship must be of mutual influence, or whether, instead, the cyber component may not be influenced by the physical component. It is clear that since the cyber controls the physical, *the cyber must have the means to influence (act upon) the physical.* It follows, that a system where the cyber component does not influence the physical component is not a CPS.

CPS also builds on *embedded systems*, which are self-contained systems that incorporate elements of control logic and real-world interaction. An embedded system is typically a single device, while CPS include many constituent systems. Further, embedded systems are specifically designed to achieve a limited number of tasks, often with limited resources. A CPS, in contrast, operates at a much larger scale, potentially including many embedded systems or other CPS elements including human and socio-technical systems.

It becomes clear from the above that having a cyber and a physical component is not a suficient condition for system to be considered cyber-physical systems. The example of the temperature logger system, which only reads from the physical system, despite having a cyber and a physical component, is not a CPS.

1.2.4 Sources of Engineering Complexity

Cyber-physical systems are complex to build due to the inherent complexity of the problems they solve that consist of coordinating action to optimise multiple (possibly contradicting) goals in systems with vast numbers of sensors and actuators. The other source of complexity is, more of accidental nature, and has to do with the heterogeneity of the components.

One starting source of complexity is that cyber components are discrete in nature but must act in a time-driven fashion and, for that matter, they must reason about time. Concepts of *duration*, *deadlines*, and *simultaneity*, must be dealt with and modelled explicitly. Yet, programming languages abstract away the notion of time and provide little or no support for time[3]. Timing behaviour is achieved through dedicated timing hardware, interrupt control routines, and timer libraries—in the words of Eduard Lee, *"programmers have to step out outside of the programming abstraction to specify timing behaviour."* Instead of being explicitly modelled, timing behaviour is obfuscated and buried under the complexity of the orchestration of these mechanisms making the *correctness* of the composition between the cyber and physical components very hard to achieve. The problem is exacerbated as more components are added onto the system.

Not only discrete components often display incorrect timing behaviour but interfacing discrete and continuous components poses another unexpected engineering challenge. The composition of a deterministic model of the cyber with a deterministic model of the physical results in a non-deterministic model that is very difficult to analyse. To understand why to consider that it is impossible, in practice, to guarantee that the components that implement the two models will be perfectly aligned, concerning time. This applies especially to the mechanism that they use to communicate and synchronise because this mechanism operates according to certain assumptions (constraints) of time.

The actual implementation of CPS is largely component-based. These components are multi-vendor and multi-technology and, as a result, they often have different communication protocols, response timings, distinct tolerances and operating conditions. One side of the problem is that some of these constraints are not known or accounted for upfront. Another side of the problem is that while constructing the system, these constraints are not modelled and handled explicitly. A lot of accidental complexity arises when trying to assemble components with such variability, especially because typical CPS consists of a large number of components.

---

[2] In terms of modelling, this means that they must be bound through at least one connector

[3] A rigorous semantics of time is absent from standard computer languages

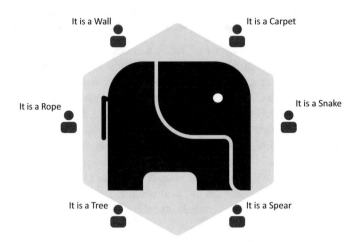

Fig. 1.1: Illustration of the tale of the 6 blind men. Each blind man, from his own perspective, understands the elephant as a diferent object.

CPS also have stringent requirements that cannot be relaxed. Besides the usual guarantees of functional correctness that apply to any engineered object, CPS are especially known for their large number of extra-functional (also known and 'non-functional') requirements that cover issues of reliability, performance, safety, security, among other. What is relevant to note is that extra-functional requirements are known to pose complex constraints to timing component design and to timing behaviour. Moreover, since these systems are critical, they often have to undergo strict qualification/testing processes. When changes to the system are needed, they are often discouraged due to the costs involved, making CPS difficult to adapt to changing requirements. There should be a reliable means to guarantee that certain aspects of a system remain untouched and therefore, do not need to be tested again.

Actual realisations of components sometimes interact in ways that were not designed upfront. Often physical components interfere due to electromagnetic or thermal interference. Emerging behaviour as of complex behaviour that arises from the interaction between components, which was not planned upfront. While developing CPS, it is important to have the means to explore alternative designs.

1.3 Modeling of a Cyber-Physical System

The engineering process of CPS requires distinct disciplines to be employed. Each discipline typically creates a design (also said to be a *view* or *understanding*) of the system for its own purposes in the form of a model. Models are created using abstractions, heuristics of decomposition, and tools of analysis typical of each discipline. Each discipline also brings along a body of knowledge that enables humans in charge of modelling the reality to critically assess the correctness and soundness of the model being produced. In this sense, the piecemeal approach of having distinct models organised by discipline is effective. Another motivation for having distinct models is that, even within the same discipline, they are required to answer distinct questions. Models also have to be produced with distinct levels of detail, and therefore, the modeller has to find the right balance and capture the right things creating a model adequate to the question being studied.

Modelling of a CPS system is inherently represented in multiple views of the system (most of the times following the principle of separation of concerns). As in the ancient tale of the six blind men (see Fig. 1.1), no single view (nor corresponding modelling formalism) can model all aspects of a system. Similarly, in CPS engineering, the results are models reflecting distinct views of the problem, expressed in multiple notations. One problem for the CPS engineer, as it is also for blind men, is *"how to integrate knowledge to form a more approximate model of the reality?"* Naturally, modelling of CPS systems calls for a trans-disciplinary approach that merges the different models into a unified abstraction of reality. An essential challenge of CPS is thus how

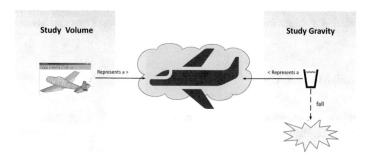

Fig. 1.2: Models of an airplane: The CAD model on the left to study volume; on the right side the model for studying gravity.

to conjoin abstractions of the various engineering disciplines and the models for physical processes including differential equations, stochastic processes, among others.

Currently, there is no standard design and modelling approach to integrate models produced by distinct disciplines of CPS. Indeed, the knowledge captured in models only comes together when assembling physical (prototype) implementations to evaluate some relevant properties of a given CPS. A more sophisticated approach is to avoid creating prototypical implementation altogether and (co-)simulate the models. This is a more generic approach as it enables verifying properties independently from the particularities of the implemented prototype.

Previously, we have mentioned the term *model* assuming some implicit notion of modelling based on abstraction, which is the process of removing details in the study of objects or systems in order to focus attention on details of higher importance. However, not all conceivable abstractions or representations of a system can be considered a model. Indeed, a more precise understanding of these concepts is required. Let us look at these in further detail.

1.3.1 What is a Model?

A model of a system is an abstraction (a representation) to make predictions or inferences [176] about a reality. More specifically, this reality is a system under study (SUS), whose governing rules and properties we want to understand, within the context of a given *experimental frame*. An experimental frame denotes the limited set of circumstances under which a system is to be observed or subject to experimentation.

According Stachowiak [256], three main properties should hold in a model. The first property is the *mapping feature*, which means that any model, to be called as such, should be faithful to or based on an original, that exists or simply be the formalisation of an idea to be realised at some point in the future. The second property is *reduction feature*. In this case, there is no model if it does not remove unnecessary detail, and select original properties useful to the purpose of the model in hands. The third is the *pragmatic feature*. In this last property, a model needs to be usable in place of an original with respect to some purpose. Other important, yet not fundamental characteristics of a model are *purposeful* [243], *understandable* and *cost-effective*.

To make the idea of models more concrete, let us illustrate these properties with a simple example. Consider the case of an aircraft where we are concentrating on the study of the single properties of mass and gravity. For this purpose, we can claim that a glass of water is a rough approximation of an aircraft (and it is chosen as a physical representation of the SUS). When it falls, it takes some time to reach the floor, and then it breaks. Can the glass be considered a model for the aircraft? For the purpose of evaluating the effects of gravity (our scope) we have *(i)* a mapping feature, as the object (glass) represents the original (the airplane); *(ii)* a reduction feature, as all the unnecessary details like shape, aerodynamics, architecture, among other, are removed, and *(iii)* it is pragmatic in terms that we can substitute the glass for the real airplane. Finally, it is *(iv)* purposeful, as it is meant to study gravity, *(v)* understandable, as it is a straightforward representation that everyone understands; and *(iv)* cost-effective, to study and substitute the real one during the fall it is several orders of magnitude cheaper and ethically acceptable. Another possible example of a model, for the same case study, can be the CAD drawing of the aeroplane for the goal of studying the volume of the machine, for purposes like studying how it fits in a hangar or what is the internal volume for the purpose of choosing the proper ventilation system.

1.3.2 Multiple Formalisms in CPS

As it became clear above, the engineering process of CPS results in a collection of models. Models are abstractions of a system (a reality) that has properties worth studying. It is well known as well that models, especially of distinct disciplines, are are expressed using correspondingly different modelling formalisms. A *modelling formalism* is a language that has formal syntax and semantics. The usual meaning of 'formal' is precisely and unambiguously defined, mathematically, in the form of Differential Equations, Finite State Automata, State Charts, Petri Nets, among others. Distinct formalisms exist because they are more concise and enable answering efficiently to distinct classes of questions. Indeed, no single formalism can be used to model all aspects of a system, as the formalism to be used depends on the nature of the problem to be solved.

Formalisms also enable the manipulation and re-writing of models, or parts of models, into other that are equivalent to the originals (for a given semantics), and whose realisations have more desirable properties. Such as, being less redundant, more compact, faster, or consuming less energy, for example.

In order to overcome the complexity of the problem, a common modelling practice, is to describe models of the same reality at different levels of abstraction (sometimes using correspondingly distinct formalisms.) Models expressed at distinct levels of abstraction are linked to one another through structure-preserving maps. Indeed, an overarching issue with distinct formalisms is merging models of the same system through these maps. There needs to be a notion of *consistency* among them.

These multiple formalisms are used to model a system interacting with its environment, its architecture and components, at different levels of detail, approximation and abstraction, and from different viewpoints, as well as the platforms the software components of the system will be deployed on. The integration of models produced according to distinct formalisms is achieved by mapping (compiling) the model into lower level *super formalisms* that integrate different domains such as Bond Graphs, or other formalisms to integrate discrete and continuous modelling constructs such as DEVS.

To support the design of CPS, not one single super-formalism, but rather of a multitude of modelling formalisms, chosen for their particular reasoning and analysis features need to be employed. These features make each of them most appropriate for a particular CPS design (sub-)goal. Pragmaticaly, the engineer (in the role of modeller) needs also to know the strategies (formalised as *processes*) to describe reality according to the formalism. The formalism together with these said processes and constraints form what is otherwise known and as a *modeling paradigm*—the object of study of Multi-Paradigm Modelling [286].

1.4 Multi-Paradigm Modelling of CPS

Multi-Paradigm Modelling (MPM) has been recognised as a powerful approach (a paradigm in its own right) that may be helpful in designing, as well as communicating and reasoning about CPS, which are notoriously complex because of their cross-discipline borders and inter-domain interactions.

To develop a CPS, project managers and engineers need to select the most appropriate development languages, software lifecycles and "interfaces" to specify the different views, components and their interactions of the system with as little "accidental complexity" [52] as possible. For example, when it is known that system/software requirements are likely to change frequently during the project's course, selecting an Agile development process may help to cope with evolution and change. If the system's behaviour requires that operations are triggered when data becomes available, similar to reactive systems, Data Flow languages may help to specify the most critical parts of the software behaviour in a precise way, making it amenable for timing analysis.

1.4.1 What is a Paradigm?

In Computer Science, general-purpose programming languages (GPLs) can be classified according to the paradigm(s) they support. For example, Eiffel is object-oriented and supports the contract-based-design paradigm, Prolog is declarative, and Lisp is functional. The paradigm characterises the underlying syntactic and semantic structures and principles that govern these GPLs. In particular, object orientation is imperative in nature and imposes viewing the world in terms of classes and communicating objects, whereas the declarative style relies on term substitution and rewriting. As a consequence, a statement in Eiffel has very little in common with a Prolog sentence due to the very different view supported by each language. A programming paradigm directly translates into different concepts encoded in the GPL syntax definition (known as a metamodel in the Model-Driven Engineering world). Very naturally, the idea of combining several paradigms at the level of GPLs led to more expressive, powerful programming languages such as Java (which is imperative, object-oriented, concurrent, and real-time and, recently, functional) and Maude (which is declarative, object-oriented and also concurrent and real-time).

What is a *paradigm* then? The science philosopher Kuhn [175], while investigating how science evolves through paradigm shifts, defines it as an open-ended contribution that frames the thinking of an object of study with concepts, results and procedures that structure future achievements. Though seemingly far from the concerns in the discipline of Computer Science, this definition does highlight the emergence of a *structure* (a formalism) that captures the object of discourse, and the notion of *procedures* (the processes) that guide achievements.

1.4.2 The dimensions of Multi-Paradigm Modelling

The application of MPM requires *modeling everything explicitly*, using the *most appropriate formalism(s)*, at the *most appropriate level(s) of abstraction* [266]. This suggests that a *paradigm* can be understood as an arrangement of the properties in each of the dimensions described above: the *formalisms* and the *levels of abstraction* in the modelling activities.

Oftentimes, formalisms are general-purpose, and hard to be used by modellers (domain users, or domain experts) who need to start by picking the most adequate formalism based on its well-known semantics, e.g., Petri Nets for workflows and concurrency, or Statecharts for describing event-based systems, among others. However, having to master mathematical notation poses a steep learning curve. To alleviate this problem, specialised languages, called *Domain-Specific Modelling Languages*, are created to simplify the act of expressing the modeller's specification intent. The constructs in these languages are designed to be closer to the way domain experts are used to conceptualize problems. The systematic approach of building new modelling languages, is called *Modelling Language Engineering* (MLE) and must, itself, follow an engineering process [?].

To tackle complexity during the course of system development, three basic abstraction approaches are commonly combined: *Abstraction/Refinement*, *Architectural decomposition*, and *View decomposition*.

- **Model abstraction (and its dual, refinement)** is used when focusing on a particular set of *properties* of interest. While abstraction implies removing unnecessary detail, in opposition to refinement, the same set of chosen properties should hold both on the abstract and detailed models. Verifying the property on the abstract model is, expectably, cheaper (or simpler) than in the detailed model. Yet, note that the more detailed model does have some advantages as it will allow the correct assessment of a larger set of properties which can not be covered otherwise.
- **Architectural decomposition (and its dual, component composition)** is used when the problem can be broken into parts, each with an appropriate *interface*. Such an encapsulation reduces a problem to *(i)* a number of sub-problems, each requiring the satisfaction of its own properties, and each leading to the design of a component and *(ii)* the design of an appropriate architecture connecting the components in a way that the composition satisfies the original required properties. This a breakdown often comes naturally at some levels of abstraction, using appropriate formalisms (which support hierarchy), for example, thanks to locality or continuity in the problem/solution domain. Note that the above describes a top-down workflow where decomposition of the requirements leads to the design of components followed by the architectural composition of these components. A bottom-up workflow is also possible, where existing components are combined to satisfy full-system requirements.

- **View decomposition (and its dual, view merge)** is used to enable the collaboration between multiple stakeholders, each with different concerns. Each viewpoint allows the evaluation of a stakeholder-specific set of properties. When concrete views are merged, the conjunction of all the views' properties must hold. In the software realm, IEEE Standard 1471 defines the relationships between viewpoints and their realisations, views. Note that the views may be described in different formalisms.

One particular combination of the former approaches leads to Contract-Based System Design [84]. Indeed, modelling activities are combined into *processes* (or workflows) that relate the various MPM activities. Processes may be *descriptive*, charting the sequence of activities carried out as well as the artefacts involved, *proscriptive* by declaratively specifying constraints on the allowed activities and their combinations, and *prescriptive* allowing enactment. Processes are often supported by toolchains whereby different tools support different activities. It can be said that a MPM framework aims to support (meta-)tool builders who assist practitioners to reason about CPS and figure out which formalisms, abstractions, workflows and supporting methods, techniques and tools are *most appropriate* to carry out their task(s).

Ultimately, by selecting, organising and managing the three dimensions above (formalisms, abstractions, and processes), MPM facilitates the communication between experts to help them better grasp the essence of how their CPS are built. Moreover, it also facilitates a rigorous comparison of distinct approaches to MPM based on their core MPM components. The implications and challenges that MPM brings to formalisms and to abstraction mechanisms need to be discussed further.

1.5 A foundation for MPM4CPS

This book introduces a representative set of modelling formalisms, each with a characteristic collection of features. The set is by no means complete but rather intended to showcase the wide variety of features available in well-established formalisms, often supported by scale-able tools. These may be used to choose a most appropriate formalism for a particular task at hand, as a starting point for looking into more formalisms, with other desirable features (such as the inclusion of spatial distribution as found in Cellular Automata or Partial Differential Equations), or as a basis for the design of Domain-Specific Modelling Languages (DSMLs) to maximally constrain a modeller to a specific application domain.

The formalisms introduced in this book may also be combined, leading to "hybrid" languages, when a particular combination of features is required that is not available in a single formalism. Note, however, that some of the formalisms introduced in this book are already hybrid in the above sense. Bond Graphs, for example, unify modelling of systems in various physical domains by focusing on power flow, and Modelica combines features of Object-Orientation with those of computationally a-causal (equation-based) modelling. The Architecture Analysis and Design Language (AADL), which focuses on embedded systems, with architecture at its core, brings together different viewpoints, making it suitable for documentation, analysis and code synthesis.

The material is presented in a bottom-up fashion. Starts by presenting the formalisms to model physical components. Then mechanisms encapsulate and re-use description of CPS components are presented as a means to tame the complexity of large descriptions. We then present formalisms analyse the behaviour of CPS. Finally, these formalisms are put together through the use of architectural descriptions and processes.

1.5.1 Modelling physical components

When modelling a physical system, the first decision to make is whether the properties of interest of the system depend on the spatial dimension. The heating of a metal object due to an electrical current flowing through it, for example, is determined by the interaction between the electrical and thermal physical domains. It depends on the geometry of that object as well as on the object's material properties such as density, electrical conductivity, relative permittivity, heat capacity, and thermal conductivity, and their distribution across the entire object. The object's dynamics can then be described using the mathematical expression of the relationships between the physical quantities of interest. Due to the dependence on spatial coordinates, this requires the use of "distributed parameter" models. These are typically expressed using the Partial Differential Equation (PDE) formalism.

When the parameters of an object are sufficiently homogeneous over its geometry, the properties of interest may not depend on the spatial dimension. In that case, the parameters may be aggregated over an entire object, and it may be reduced to its dimensionless essence. A rigid body in the mechanical domain with a constant density over its geometry may, for example, be reduced to a simple "point mass". Its dynamics can be described using Newton's Laws or a Hamiltonian or Lagrangian formulation. Such "lumped parameter" models are typically expressed using the Ordinary Differential Equation (ODE) or Differential Algebraic Equation (DAE) formalisms.

Often, the physical components of a Cyber-Physical System span distinct physical domains (electrical, mechanical, thermodynamic, hydraulic, etc.). The Bond Graphs formalism described in Chapter 2 unifies the different domains at a "lumped parameter" level of detail. It recognises the analogy between physical processes in different physical domains, such as energy storage and dissipation. A system is modelled as a Bond Graph connecting nodes representing physical elements. These nodes encode how physical quantities such as voltage and current are related in, for example, a resistor. The Bond Graph's edges—called Power Bonds—denote the power flow between the nodes. Special nodes—junctions— encode conservation laws, generalisations of Kirchoff's current and voltage laws in the electrical domain. The chapter introduces a systematic procedure for modelling multi-domain physical systems. It starts from Idealised Physical Models and converts these into Bond Graph models. These Bond Graph models are computationally a-causal and can be translated to a set of Differential-Algebraic Equations (DAEs). Computational a-causal models consist of equations relating signals (variables, functions of continuous-time), without specifying which variables are known (inputs) and which are unknown (outputs), nor how these equations need to be solved (i.e., how the unknowns are computed from the knowns). Such DAEs can be represented in (mathematical) Equation-Based modelling languages such as Modelica. Modelica is described in Chapter 3 The Bond Graph chapter then shows how computational causality can be assigned, effectively converting to a Continuous-Time Causal Block Diagram (CT-CBD). Causal Block Diagrams are described in Chapter 4. Causality assignment on a Bond Graph model may give insight based on physics, into flaws in the model. This aid in "model debugging" is thanks to the (physical) domain-specificity of the Bond Graph formalism.

1.5.2 Joining the 'Physical' with the 'Cyber'

Cyber-Physical Systems are composed of networked physical and computational components. To allow for a modular and hierarchical design of such systems, maximising model re-use and enabling the construction of model libraries, the Modelica language, described in Chapter 3 combines features of Object-Orientation such as encapsulation and inheritance with those of computationally a-causal (equation-based) modelling. Computationally a-causal models allow the modeller to express the fundamental laws of physics using mathematical equations. The semantics of Modelica is given by expanding object-oriented constructs such as inheritance, by instantiating classes, and by flattening the hierarchy. This results in a set of (hybrid) Differential-Algebraic Equations. For each particular simulation experiment context, computational causality can be assigned by a Modelica compiler. This effectively generates a model in the Continuous-Time Causal Block Diagram formalism. Most Modelica compilers will further (time-)discretise these models, either symbolically through "inline integration" or by calling upon external numerical solvers. This ultimately leads to an executable simulation code. Note that it is possible to create a Modelica library with Bond Graph components. In this case, the Bond Graph causality assignment procedure will not be used. Rather, Modelica's causality assignment will be applied to the entire model, including non-Bond Graph parts. Through the code-based specification of functions, Modelica also allows traditional object-oriented code to be represented. It is this combination of code, equations, and hybrid constructs such as "when" (which allows the introduction of discrete events, so-called "state events", based on conditions over continuous behaviour such a crossing a threshold value) that makes Modelica suited to build models of Cyber-Physical Systems, spanning their physical, network and computational parts.

As mentioned earlier, declarative, computationally a-causal models need to ultimately be transformed into a causal form, which allows for their computational solution. In Chapter 4, a family of Causal Block Diagram (CBD) formalisms is introduced. Causal Block Diagrams consist of a network of computational blocks. Each block specifies the computationally causal relationship between its input and output signals. The block diagram network specifies how outputs of one block are connected to inputs of other blocks. Such a connection denotes that the values at connected output and input ports must at all times be equal. The three CBD variants are built up gradually. The Algebraic Causal Block Diagram (ALG-CBD) formalism has no notion of time: values are propagated through a CBD according to a computation "schedule" (i.e., the order in which block computations are invoked) derived from the dependency structure encoded in the block diagram network. Special care needs to be taken to detect and properly solve dependency cycles known as "algebraic loops". A discrete (Natural Number) notion of time is then added, as well as a delay/memory block, resulting in Discrete-Time Causal Block Diagrams (DT-CBDs). These are equivalent to Synchronous Data Flow (SDF) models. Finally, the Real Numbers are introduced as a time base to give Continuous-Time Causal Block Diagrams (CT-CBDs). These have the same expressiveness as mathematical equations and need to be discretised to allow for their computational solution. Numerical discretisation techniques are used to turn a CT-CBD into a DT CBD.

Very often, it is reasonable to abstract away many of the details of the behaviour of a system and to only focus on pertinent "events". Such Discrete-Event abstractions see a system as changing its internal state, either reacting to input events of autonomously changing its state after a certain time (due to an internal "time event") and possibly producing output events at certain times. As only the events are what changes the state of the system, and in between event instances, nothing pertinent is assumed to happen, the evolution of the state over time is piecewise constant. Unlike in Discrete-Time (DT) formalisms, in Discrete-Event (DE) formalisms, time advances in leaps and bounds, from pertinent event to pertinent event. One advantage of DE abstraction is performance: a simulator will directly step to the next time at which an event occurs whereas a DT simulator would have to step through time in fixed increments even if nothing noteworthy happens (i.e., the state remains unchanged). The DE abstraction is commonly used to study the competition of different processes for shared resources. If resources are constrained, this inevitably leads to queueing. The abstraction is hence useful for simulation-based performance analysis. Utilization of resources, time spent to complete an activity, the distribution of queue length and queueing time are all examples of the typical performance measures that are obtained from discrete-event simulations. Note that DE formalisms are often deterministic. Through the inclusion of distributions for parameter values such as Inter Arrival Time rather than unique values, and using Monte-Carlo simulation, distributions of performance measures are obtained. Thus, repeatable (as pseudo-random number generators – which are deterministic– are used to sample from distributions) stochastic simulations are obtained. Many DE formalisms were developed over the years. The Discrete EVent Specification (DEVS) formalism described in Chapter 5 is a DE formalism that is primitive and expressive enough to act as a DE "assembly language": models in all DE simulation formalisms can be mapped onto a DEVS equivalent. As such, it can be used to architecturally connect models in different formalisms by first mapping all components onto DEVS. The resulting architecture only contains DEVS components. As DEVS is modular and supports hierarchical architectural composition, the resulting model has a precise meaning. DEVS's support for hierarchy makes it suitable to build model libraries and to subsequently build up highly complex models.

A different way of modularly combining state automata is found in the Statecharts formalism described in Chapter 6. The Statecharts formalism consists of hierarchies of state automata, of parallel composition of these automata, of a notion of time, and an event broadcast mechanism. The popularity of Statecharts is partly due to its intuitive visual notation. The main purpose of Statecharts is to not only simulate models, but also to synthesise from them, autonomous, timed and reactive software and/or hardware.

When a parallel composition is made of state automata, many interleavings are possible. One option is to choose a unique interleaving, leading to a unique, deterministic behaviour trace. This is what is done in the DEVS and Statecharts formalisms. To model true concurrency, this artificial sequentialisation is not always appropriate. Rather, a non-deterministic choice should be allowed. This leads, not to a single behaviour trace, but to a collection of possible behaviour traces. This collection of traces may be summarised in a compact representation in the form of a state reachability graph. The satisfaction of interesting properties may then be checked over the collection of traces. An example is the reachability of a certain undesirable state. Such properties are also expressed in an appropriate property language. Non-deterministic languages, as described above, usually have a weak notion of time: not the Natural or Real numbers are used as a time basis, but only a (partial) ordering of event instants. The focus is on concurrency and synchronisation. Rather than simulation or synthesis, such formalisms are mostly used for the analysis of properties, across all possible behaviours of a system. This makes them suited for, for instance, safety analysis. One such formalism is Petri Nets, as described in Chapter 7. Petri Nets encode state as a "marking", an n-dimensional vector of Natural numbers. Each element of the vector corresponds to the number of "tokens" in a Petri Net Place. The evolution of the state is encoded in a Petri Net graph which, apart from Places, contains transitions and Arcs. Thanks to the use of Natural numbers, the number of possible states can be infinite (but countable). Petri Nets are a simple formalism that, like DEVS, is often used as a common semantic domain onto which to map diverse other, often domain-specific, formalisms. The chapter also demonstrates how Petri Nets can be combined with other formalisms. In particular, the co-simulation of Petri Nets with Functional Mockup Units (encoding discretised continuous models) is introduced. This effectively leads to non-deterministic hybrid models.

1.5.3 Tooling support for MPM4CPS

Complex engineered systems consist of heterogenous components arranged in an architecture. Furthermore, multiple viewpoints on the same system may be of interest and ultimately (part of) a system model needs to be "deployed" on an often embedded software/hardware architecture. The Architecture Analysis and Design Language (AADL) described in Chapter 8.1 is foremost an Architecture Description Language. It allows one to provide a description of the overall system and the environment into which it will operate. From such a description, other models in other formalisms such as those described in this book can be generated. These can be further augmented to study various aspects of the system, which is essential for its optimisation, verification and validation. After a brief introduction to ADLs and their role in MPM4CPS, the AADL is presented and its use illustrated through the modelling, analysis and code generation for a simple Lego Mindstorm robot for carrying objects in a warehouse. A simple top-down architecture-centric design process is followed, starting from the capture of stakeholder goals and system requirements, followed by system design, design analysis and verification and finally automated code synthesis.

It becomes apparent from the above that complex systems modelling involves not only different abstractions, architectures and views, modelled using varying formalisms but also complex development workflows. Chapter 9 looks into the topic of process (workflow) modelling. Often complex, concurrent development processes, modelled in the form of a Process Model (PM) are built up of primitive activities which take as input, modelling artefacts and produce modified or new modelling artefacts as output. The activities may require human or computer resources, possibly leading to delays as described in the chapter on DEVS. As the artefacts manipulated by activities are models in various formalisms, it makes sense to "type" them with the appropriate formalisms. To chart the many formalisms used, and to show how they are related, a Formalism Transformation Graph (FTG) is introduced. The FTG+PM, combining FTG with PM, allows one to characterise the essence of Multi-Paradigm Modelling solution patterns to CPS development problems. One advantage of the explicit representation of the FTG+PM is that is can be used as a basis for the synthesis of MPM tools.

1.6 Summary

The field of CPS is affected by the complexity of different approaches, processes, and modelling languages. Indeed, this field is well-known to have an inherently multi-displinary nature and, since there is no single well accepted modelling approach, multi-paradigm modelling has been advanced as a solution. Yet, to date, literature still lacks a solid reference that introduces distinct CPS modelling and analysis techniques towards a multi-paradigm approach.

This first chapter motivates Multi-Paradigm Modelling for CPS and the need for its clear foundations. Staring from an introduction to the concept of System, it introduces distinct classes of systems, discusses their characteristics, and then derives a definition of CPS. As CPS are complex to build, we dedicate part of the chapter to discussing and drilling down on their common sources of complexity. Overall, designing a system means that one has to make use of several kinds of abstractions to describe different properties and system's concerns, using languages (with existing, modelling formalisms and paradigms) and processes to be able to avoid unnecessary complexity. The chapter then defines what a model is, and what does it mean to use multiple formalisms when engaging in multi-paradigm modelling. Finally, a structure of the book is described, further explaining how the formalisms and techniques presented fit together.

Chapter 2
Bond Graphs: A Unifying Framework for Modelling of Physical Systems

Jan F. Broenink

Abstract This chapter introduces a formalism to model the dynamic behaviour of physical systems known as bond graphs. A important property of this formalism is that systems from different domains (cf. electrical, mechanical, hydraulical, acoustical, thermodynamical, material) are described in the same way an integrated under the unifying concept of energy exchange. Bond graph models are directed graphs where parts are interconnected by bonds, along which exchange of energy occurs. We present a method to systematically build a bond graph starting from an ideal physical model and present methods to perform the causal analysis of bond graphs and procedures to generate equations to enable simulation.

Learning Objectives

After reading this chapter, we expect you to be able to:

- Use bond graphs as an abstraction to model bi-directional energy exchange between components in a domain-neutral fashion
- Be able to translate domain dependent diagrams into ideal physical models and subsequently into bond-graph models
- Translate bond graph models into systems of differential equations for simulation and analysis

2.1 Introduction

Bond graphs are a domain-independent graphical description of dynamic behaviour of physical systems. This means that systems from different domains (cf. electrical, mechanical, hydraulical, acoustical, thermodynamical, material) are described in the same way. The basis is that bond graphs are based on energy and energy exchange. Analogies between domains are more than just equations being analogous: the used *physical concepts* are analogous.

Bond-graph modelling is a powerful tool for modelling engineering systems, especially when different physical domains are involved. Furthermore, bond-graph submodels can be re-used *elegantly*, because bond-graph models are *non-causal*. The submodels can be seen as *objects*; bond-graph modelling is a form of *object-oriented* physical systems modelling.

Bond graphs are labelled and directed graphs, in which the vertices represent submodels and the edges represent an ideal energy connection between power ports. The *vertices* are idealised descriptions of physical phenomena: it are *concepts*, denoting the relevant (i.e. dominant and interesting) aspects of the dynamic behaviour of the system. It can be bond graphs itself, thus allowing hierarchical models, or it can be a set of

Jan F. Broenink
University of Twente, Netherlands
e-mail: j.f.broenink@utwente.nl

P. Carreira et al. (eds.), *Foundations of Multi-Paradigm Modelling for Cyber-Physical Systems*,
https://doi.org/10.1007/978-3-030-43946-0_2

equations in the variables of the ports (two at each port). The *edges* are called *bonds*. They denote point-to-point connections between submodel ports.

When preparing for simulation, the bonds are embodied as two-signal connections with opposite directions. Furthermore, a bond has a power direction and a computational causality direction. Proper assigning the power direction resolves the sign-placing problem when connecting submodels structures. The internals of the submodels give preferences to the computational direction of the bonds to be connected. The eventually assigned computational causality dictates which port variable will be computed as a result (output) and consequently, the other port variable will be the cause (input). Therefore, it is necessary to rewrite equations if another computational form is specified then is needed. Since bond graphs can be mixed with block-diagram parts, bond-graph submodels can have power ports, signal inputs and signal outputs as their interfacing elements. Furthermore, aspects like the physical domain of a bond (energy flow) can be used to support the modelling process.

The concept of bond graphs was originated by [228]. The idea was further developed by Karnopp and Rosenberg in their textbooks ([161, 162, 160]), such that it could be used in practice [268, 86]. By means of the formulation by Breedveld [43, 44] of a framework based on thermodynamics, bond-graph model descriptions evolved to a systems theory.

In the next section, we will introduce the bond graph method by some examples, where we start from a given network composed of ideal physical models. Transformation to a bond graph leads to a domain independent model. In Section 3, we will introduce the foundations of bond graphs, and present the basic bond graph elements in Section 4. We will discuss a systematic method for deriving bond graphs from engineering systems in Section 5. How to enhance bond-graph models to generate the model equations and for analysis is presented in Section 6, and is called Causal Analysis. The equations generation and block diagram expansion of causal bond graphs is treated in Sections 7 and 8. Section 9 discusses simulation issues. In Section 10 we review this chapter, and also include some hints for further reading.

2.2 Bond-Graph Examples

To introduce bond graphs, we will discuss examples of two different physical domains, namely an RLC circuit (electrical domain) and a damped mass-spring system (mechanical domain, translation). The RLC circuit is given in Figure 2.1.

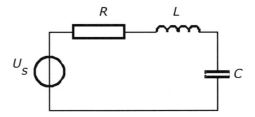

Fig. 2.1: The RLC cicuit

In electrical networks, the port variables of the bond-graph elements are the electrical voltage over the element port and electrical current through the element port. Note that a port is an interface of an element to other elements; it is the connection point of the bonds. The power being exchanged by a port with the rest of the system is the product of voltage and current: $P = ui$. The equations of a resistor, capacitor and inductor are:

$$u_R = iR$$

$$u_C = \frac{1}{C} \int i \, dt$$

$$i_L = \frac{1}{I} \int u \, dt$$

In order to facilitate the conversion to bond graphs, we draw the different elements of the electric domain in such a way that their ports become visible. For brevity, we only show this for the Capacitor Figure 2.2. To this port, we connect a power bond or bond for short. This bond denotes the energy exchange between the elements. A bond is drawn as an edge with half an arrow. The direction of this half arrow denotes the positive direction of the energy flow. In principle, the voltage source delivers power and the other elements absorb power.

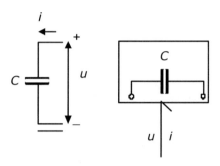

Fig. 2.2: Electric Capacitor: Circuit (left) and with bond (right)

Considering the circuit of Figure 2.1, we see that the voltage over the elements are different and through all elements flows the same current. We indicate this current with i and connect the bonds of all elements with this current i (2.3). Changing the electric symbols into corresponding bond graph mnemonics, result in the bond graph of the electrical circuit. The common i is changed to a '1', a so-called *1-junction*. Writing the specific variables along the bonds makes the bond graph an electric bond graph. The voltage is mapped onto the domain-independent *effort* variable and the current maps onto the domain-independent *flow* variable (the current always on the side of the arrow). The 1-junction means that the current (flow) through all connected bonds is the same, and that the voltages (efforts) sum to zero, considering the sign. This sign is related to the power direction (i.e. direction of the half arrow) of the bond. This summing equation is the Kirchhoff voltage law.

Parallel connections, in which the voltage over all connected elements is the same, are denoted by a u in the port-symbol network. The bond–graph mnemonic is a 0, the so-called *0-junction*. A 0-junction means that the voltage (effort) over all connected bonds is the same, and that the currents (flows) sum to zero, considering the sign. This summing equation is the Kirchhoff current law.

The second example is the damped mass-spring system, a mechanical system shown in 2.4. In mechanical diagrams, the port variables of the bond graph elements are the *force* on the element port and *velocity* of the element port. For the rotational mechanical domain, the port variables are the *torque* and *angular velocity*. Again, two variables are involved. The *power* being exchanged by a port with the rest of the system is the product of force and velocity: $P = Fv$ ($P = T\omega$ for the rotational case). The equations of a damper, spring and mass are (we use damping coefficient a, spring coefficient K_s, mass m and applied force F_a):

$$F_d = \alpha v$$

$$F_s = K_s \int v \, dt = \frac{1}{C_s} \int v \, dt$$

$$F_m = m\frac{dv}{dt} \quad \text{or} \quad v = \frac{1}{m} \int F_m \, dt$$

$$F_a = force$$

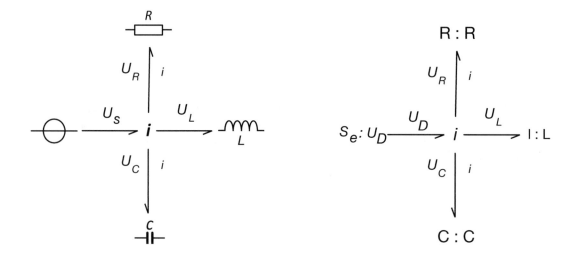

Fig. 2.3: Bond graph with electrical symbols (left) and with standard symbols (right). The standard bond-graph symbols are defined in 2.4

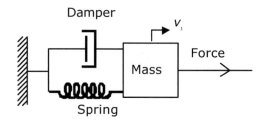

Fig. 2.4: The damped mass-spring system

In the same way as with the electrical circuit, we can redraw the element such that their ports become visible (2.5). The loose ends of the example all have the same velocity, which is indicated by a v. This junction element also implies that the forces sum up to zero, considering the sign (related to the power direction). The force is mapped onto an effort and the velocity onto a flow. For the rotational mechanical domain , the torque is mapped onto an effort and the angular velocity onto a flow. This implies that force is related to electric voltage and that velocity is related to electric current.

We see the following analogies between the mechanical and electrical elements:

- The damper is analogous to the resistor.
- The spring is analogous to the capacitor; the mechanical compliance corresponds with the electrical capacity.
- The mass is analogous to the inductor.
- The force source is analogous to the voltage source.
- The common velocity is analogous to the loop current.

Besides points with common velocity, also points with common force exist in mechanical systems. Then forces are all equal and velocities sum up to zero, considering the sign (related to the power direction). These common force points are denoted as *0-junctions* in a bond graph (an example is a concatenation of a mass, a spring and a damper: the three elements are connected in 'series'). A further elaboration on analogies can be found in the next section, where the foundations of bond graphs are discussed.

Through these two examples, we have introduced most bond graph symbols and indicated how in two physical domains the elements are transformed into bond graph mnemonics. One group of bond graph elements was not yet introduced: namely the *transducers*. Examples are the electric transformer, an electric motor and toothed wheels. In the next section, we will discuss the foundations of bond graphs.

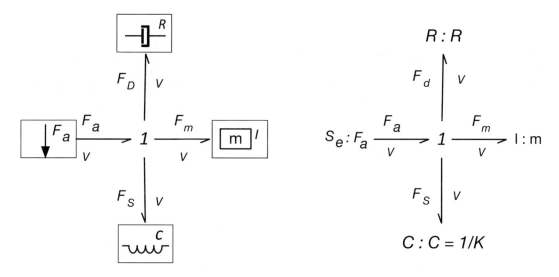

Fig. 2.5: Bond graph with mechanical symbols (left) and with standard symbols (right)

2.3 Foundation of Bond Graphs

Analogies between different systems were shown in the previous section: Different systems can be represented by the same set of differential equations. These analogies have a *physical foundation*: the underlying physical concepts are analogous, and consequently, the resulting differential equations are analogous. The physical concepts are based on energy and energy exchange. Behaviour with respect to energy is domain independent. It is the same in all engineering disciplines, as can be concluded when comparing the RLC circuit with the damped mass spring system. This leads to identical bond graphs.

2.3.1 Starting Points

Before discussing the specific properties of bond graphs and the elementary physical concepts, we first recall the assumptions general for network like descriptions of physical systems, like electrical networks, mechanical or hydraulic diagrams:

- The conservation law of energy is applicable.
- It is possible to use a *lumped approach*.

This implies that it is possible to separate system properties from each other and to denote them distinctly, while the connections between these submodels are ideal. *Separate system properties* mean physical concepts and the ideal connections represent the energy flow, i.e. the bonds between the submodels. This *idealness* property of the connections means that in these connections no energy can be generated or dissipated. This is called *power continuity*. This structure of connections is a conceptual structure, which does not necessary have a size. This concept is called reticulation [228] or tearing [173]. See also [304].

The system's submodels are concepts, idealised descriptions of physical phenomena, which are recognised as the dominating behaviour in components (i.e. real-life, tangible system parts). This implies that a model of a concrete part is not necessary only one concept, but can consist of a set of interconnected concepts.

2.3.2 Bonds and Ports

The contact point of a submodel where an ideal connection will be connected to is called a power port or port for short. The connection between two submodels is called a power bond or bond; it is drawn as a single line (2.6). This bond denotes an ideal energy flow between the two connected submodels. The energy entering the bond on one side immediately leaves the bond at the other side (power continuity).

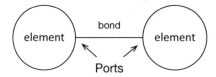

Fig. 2.6: The energy flow between two submodels represented by a bond.

The *energy flow* along a bond has the physical dimension of power, being the product of two variables. In each physical domain, there is such a combination of variables, for which a physical interpretation is useful. In electrical networks, the two variables are voltage and current. In mechanical systems, the variable pairs are force and velocity for translation and torque and angular velocity for rotation. In hydraulics, it is pressure and volume flow. For thermodynamic systems, temperature and entropy flow are used. These pairs of variables are called (power-) conjugated variables.

In order to understand the connection as established by a bond, this bond can be interpreted in two different ways, namely:

1. As an interaction of energy.
    The connected subsystems form a load to each other by their energy exchange. A power bond embodies a connection where a physical quantity is exchanged.
2. As a bilateral signal flow.
    The connection is interpreted as two signals, an effort and flow, flowing in opposite direction, thus determining the computational direction of the bond variables. With respect to one of the connected submodels, the effort is the input and the flow the output, while for the other submodel input and output are of course established by the flow and effort respectively.

These two ways of conceiving a bond is essential in bond graph modelling. Modelling is started by indicating the physical structure of the system. The bonds are first interpreted as interactions of energy, and then the bonds are endowed with the computational direction, interpreting the bonds as bilateral signal flows. During modelling, it need *not* be decided yet what the computational direction of the bond variables is. Note that, determining the computational direction during modelling restricts submodel reuse. It is however necessary to derive the mathematical model (set of differential equations) from the graph. The process of determining the computational direction of the bond variables is called *causal analysis*. The result is indicated in the graph by the so-called *causal stroke*, indicating the direction of the effort, and is called the causality of the bond (2.7).

In equation form, 2.7 can be written as:

$$element1.e := element2.e \qquad\qquad element2.e := element1.e$$
$$element2.f := element1.f \qquad\qquad element1.f := element2.f$$

2.4 Bond-Graph Elements

The constitutive equations of the bond graph elements are introduced via examples from the electrical and mechanical domains. The nature of the constitutive equations lay demands on the causality of the connected bonds. Bond graph elements are drawn as letter combinations (mnemonic codes) indicating the type of element. The bond graph elements are the following:

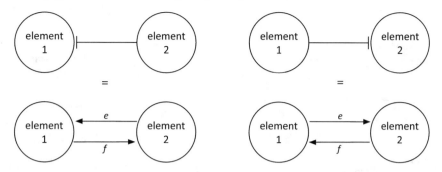

Fig. 2.7: Determine the signal direction of the effort and flow (we do not use the power direction at the bonds, so it is not shown here)

C storage element for a q-type variable, e.g. capacitor (stores charge), spring (stores displacement).
 I storage element for a p-type variable, e.g. inductor (stores flux linkage), mass (stores momentum).
 R Resistor dissipating free energy, e.g. electric resistor, mechanical friction.
Se, Sf sources, e.g. electric mains (voltage source), gravity (force source), pump (flow source).
TF transformer, e.g. an electric transformer, toothed wheels, lever.
GY gyrator, e.g. electromotor, centrifugal pump.
0, 1 0- and 1-junctions, for ideal connecting two or more submodels.

2.4.1 Storage Elements

Storage elements store all kinds of free energy. As indicated above, there are two types of storage elements: C-elements and I-elements. The *q*-type and *p*-type variables are *conserved quantities* and are the result of an accumulation (or integration) process. They are the *(continuous) state variables* of the system.

In C-elements, like a capacitor or spring, the conserved quantity, q, is stored by accumulating the net flow, f, to the storage element. This results in the differential equation:

$$\dot{q} = f$$

which is called a *balance equation*, and forms a part of the constitutive equations of the storage element. In the other part of the constitutive equations, the state variable, q, is related to the effort, e:

$$e = e(q)$$

This relation depends on the specific shape of the particular storage element.

In 2.8, examples of C-elements are given together with the equivalent block diagram. The equations for a linear capacitor and linear spring are:

$$\dot{q} = i, \qquad u = \frac{1}{C}q$$

$$\dot{x} = v, \qquad F = Kx = \frac{1}{C}x$$

For a capacitor, C [F] is the capacitance and for a spring, K [N/m] is the stiffness and C [m/N] the compliance. For all other domains, a C-element can be defined.

The effort variable is equal when two C-storage elements connected in parallel with a resistor in between are in equilibrium. Therefore, the domain-independent property of an effort is *determination of equilibrium*.

In I-elements, like a inductor or mass, the conserved quantity, p, is stored by accumulating the net flow, e, to the storage element. This results in the differential equation:

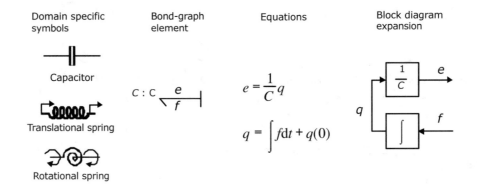

Fig. 2.8: Examples of C-elements

$$\dot{p} = e$$

which is called a *balance equation*, and forms a part of the constitutive equations of the storage element. In the other part of the constitutive equations, the state variable, q, is related to the effort, e:

$$f = f(q)$$

This relation depends on the specific shape of the particular storage element.

In 2.9, examples of I-elements are given together with the equivalent block diagram. The equations for a linear inductor and linear mass are:

$$\dot{\lambda} = u, \ i = \tfrac{1}{L}\lambda \tag{2.1}$$

$$\dot{p} = F, \ v = \tfrac{1}{m}p \tag{2.2}$$

For an inductor, L [H] is the inductance and for a mass, m [kg] is the mass. For all other domains, an I-element can be defined.

The flow variable is equal when two I-storage elements connected in parallel with a resistor in between, are in equilibrium. Therefore, at I-elements, the domain-independent property of the flow is *determination of equilibrium*. F or example, when two bodies, moving freely in space each having a different momentum, are being coupled (collide and stick together), the momentum will divide among the masses such that the velocity of both masses is the same (this is the conservation law of momentum).

Note that when at the two types of storage elements, the role of effort and flow are exchanged: the C- element and the I-element are each other's *dual form*.

The block diagrams in 2.8 and 2.9, and also in the next Figures 10 to 16, show the computational direction of the signals involved. They are indeed the expansion of the corresponding causal bond graph. The equations are given in computational form, consistent with the causal bond graph and the block diagram.

2.4.2 Resistors

Resistors, R-elements, dissipate free energy. Examples are dampers, frictions and electric resistors (2.10). In real-life mechanical components, friction is always present. Energy from an arbitrary domain flows irreversibly to the thermal domain (and heat is produced). This means that the energy flow towards the resistor is always

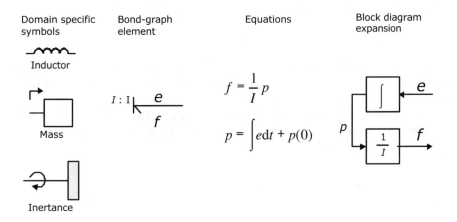

Fig. 2.9: Examples of I-elements

positive. The constitutive equation is an algebraic relation between the effort and flow, and lies principally in the first or third quadrant.

$$e = f(f)$$

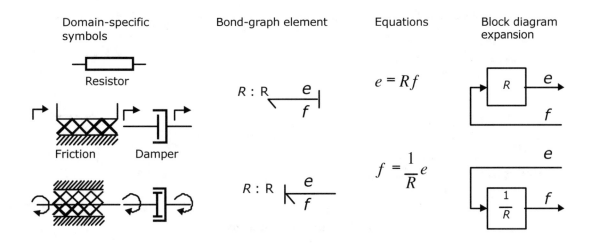

Fig. 2.10: Examples of resistors

An electrical resistor is mostly linear (at constant temperature), namely Ohm's law. The electrical resistance value is in [Ω].

$$u = Ri$$

Mechanical friction mostly is non-linear. The resistance function is a combination of dry friction and viscous friction. Dry friction is a constant friction force and viscous friction is the linear term. Sometimes, also stiction is involved, a tearing-loose force only applicable when starting a movement. All these forms of friction can be modelled with the R-element. The viscous friction has as formula (R in [Ns/m]:

$$F = Rv$$

If the resistance value can be controlled by an external signal, the resistor is a modulated resistor, with mnemonic MR. An example is a hydraulic tap: the position of the tap is controlled from the outside, and it determines the value of the resistance parameter.

If the thermal domain is modelled explicitly, the production of thermal energy should explicitly be indicated. Since the dissipator irreversibly produces thermal energy, the thermal port is drawn as a kind of source of thermal energy. The R becomes an RS.

2.4.3 Sources

Sources represent the interaction of a system with its environment. Examples are external forces, voltage and current sources, ideal motors, etc. (2.11). Depending on the type of the imposed variable, these elements are drawn as Se or Sf.

Besides as a 'real' source, source elements are used to give a variable a fixed value, for example, in case of a point in a mechanical system with a fixed position, a Sf with value 0 is used (fixed position means velocity zero). When a system part needs to be excited, often a known signal form is needed, which can be modelled by a modulated source driven by some signal form. An example is shown in Figure 2.12.

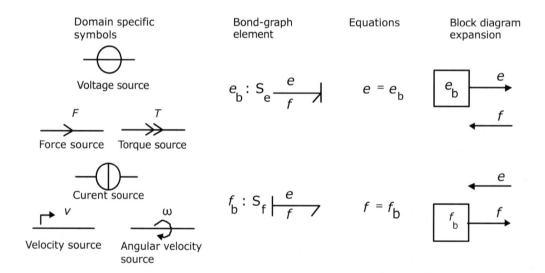

Fig. 2.11: Examples of sources

Fig. 2.12: Example of a modulated voltage source

2.4.4 Transformers and Gyrators

An ideal *transformer* is represented by TF and is power continuous (i.e. no power is stored or dissipated). The transformation can within the same domain (toothed wheel, lever) or between different domains (electromotor, winch), see Figure 13. The equations are:

$$e_1 = ne_2 \qquad (2.3)$$
$$f_2 = nf_1 \qquad (2.4)$$

Efforts are transduced to efforts and flows to flows. The parameter n is the *transformer ratio*. Due to the power continuity, only *one* dimensionless parameter, n, is needed to describe both the effort transduction and the flow transduction. The parameter n is unambiguously defined as follows: e_1 and f_1 belong to the bond pointing towards the TF. This way of defining the transformation ratio is standard in leading publications [160],[44],[269],[67]. If n is not constant, the transformer is a *modulated transformer*, a MTF. The transformer ratio now becomes an input signal to the MTF.

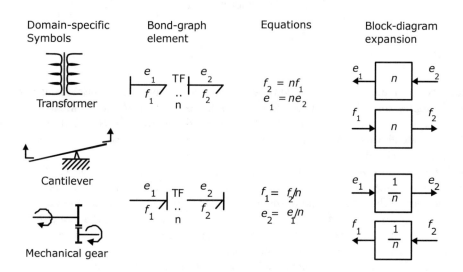

Fig. 2.13: Examples of transformers

An ideal *gyrator* is represented by GY, and is also power continuous (i.e. no power is stored or is dissipated. Examples are an electromotor, a pump and a turbine. Real-life realisations of gyrators are mostly transducers representing a domain-transformation (Figure 14). The equations are:

$$e_1 = rf_2 \qquad (2.5)$$
$$e_2 = rf_1 \qquad (2.6)$$

The parameter r is the *gyrator ratio*, and due to the power continuity, only one parameter to describe both equations. No further definition is needed since the equations are symmetric (it does not matter which bond points inwards, only that one bond points towards and the other points form the gyrator). r has a physical dimension, since r is a relation between effort and flow (it has the same dimension as the parameter of the R element). If r is not constant, the gyrator is a *modulated gyrator*, a MGY.

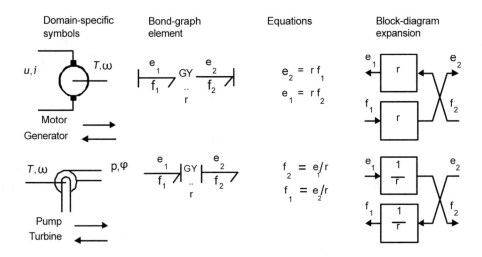

Fig. 2.14: Examples of gyrators

2.4.5 Junctions

Junctions couple two or more elements in a power continuous way: there is *no* energy storage or dissipation in a junction. Examples are a series connection or a parallel connection in an electrical network, a fixed coupling between parts of a mechanical system. Junctions are *port-symmetric*: the ports can be exchanged in the constitutive equations. Following these properties, it can be proven that there exist only two pairs of junctions: the 1-junction and the 0-junction.

The *0-junction* represents a node at which all efforts of the connecting bonds are equal (2.15). An example is a parallel connection in an electrical circuit. Due to the power continuity, the sum of the flows of the connecting bonds is zero, considering the sign. The power direction (i.e. direction of the half arrow) determines the sign of the flows: all inward pointing bonds get a plus and all outward pointing bonds get a minus. (Figure X). This summation is the Kirchhoff current law in electrical networks: all currents connecting to one node sum to zero, considering their signs: all inward currents are positive and all outward currents are negative.

We can depict the 0-junction as the representation of an effort variable, and often the 0-junction will be interpreted as such. The 0-junction is more than the (generalised) Kirchhoff current law, namely also the equality of the efforts (like electrical voltages being equal at a parallel connection).

The *1-junction* (2.16) is the dual form of the 0-junction (roles of effort and flow are exchanged). The 1-junction represents a node at which all flows of the connecting bonds are equal. An example is a series connection in an electrical circuit. The efforts sum to zero, as a consequence of the power continuity. Again, the power direction (i.e. direction of the half arrow) determines the sign of the efforts: all inward pointing bonds get a plus and all outward pointing bonds get a minus. This summation is the Kirchhoff voltage law in electrical networks: the sum of all voltage differences along one closed loop (a mesh) is zero. In the mechanical domain, the 1-junction represents a force balance (also called the principle of d'Alembert), and is a generalisation of Newton's third law, action = - reaction).

Just as with the 0-junction, the 1-junction is more than these summations, namely the equality of the flows. Therefore, we can depict the 1-junction as the representation of a flow variable, and often the 1-junction will be interpreted as such.

2.4.6 Positive Orientation

By definition, the power is positive in the direction of the power bond (i.e. direction of the half arrow). A port that has an incoming bond connected to, consumes power if this power is positive (i.e. both effort and flow are

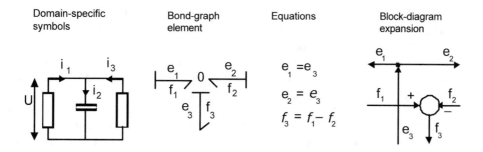

| Domain-specific symbols | Bond-graph element | Equations | Block-diagram expansion |

Fig. 2.15: Examples of gyrators

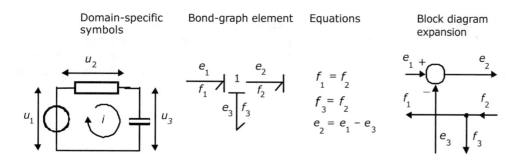

| Domain-specific symbols | Bond-graph element | Equations | Block diagram expansion |

Fig. 2.16: Examples of gyrators

either positive or negative, as the product of effort and flow is the power). In other words: the power flows in the direction of the half arrow if it is positive and the other way if it is negative.

R-, C- and I-elements have an incoming bond (half arrow towards the element) as standard, which results in positive parameters when modelling real-life components. For source elements, the standard is outgoing, as sources mostly deliver power to the rest of the system. A real-life source then has a positive parameter. For TF- and GY-elements (transformers and gyrators), the standard is to have one bond incoming and one bond outgoing, to show the 'natural' flow of energy. Furthermore, using the standard definition of the parameter at the transformer (incoming bond is connected to port 1 and the ratio n is e_1/e_2) positive parameters will be the result. Note that a gyrator does *not* need such a definition, since its equations are symmetric.

It is possible, however, that negative parameters occur. Namely, at transformers and sources in the mechanical domain when there is a reverse of velocity or the source acts in the negative direction.

Using the definitions discussed in this section, the bond-graph definition is unambiguous, implying that in principle there is no need for confusion. Furthermore, this systematic way will help resolving possible sign-placing problems often encountered in modelling, especially in mechanical systems.

2.4.7 Duality and Dual Domains

As indicated in 2.4.1, the two storage elements are each other's dual form. The role of effort and flow in a C-element and I-element are exchanged. Leaving one of the storage elements (and also one of the sources) out of the list of bond graph elements, to make this list as small as possible, can be useful from a mathematical viewpoint, but does not enhance the insight in physics.

Decomposing an I-element into a GY and a C, though, gives more insight. The only storage element now is the C-element. The flow is only a time derivative of a conserved quantity, and the effort determines the equilibrium. This implies that the physical domains are actually pairs of two dual domains: in mechanics, we have *potential* and *kinetic* domains for both rotation and translation), in electrical networks, we have the *electrical* and *magnetic* domains. However, in the thermodynamic domain, no such dual form exists (Breedveld, 1982). This is consistent with the fact that *no* thermal I-type storage exists (as a consequence of the second law of thermodynamics: in a thermally isolated system, the entropy never decreases).

2.5 Systematic Procedure to Derive a Bond-Graph Model

In the previsous section, we have discussed the basic bond-graph elements and the bonds, so we can transform a domain-dependent *ideal-physical model*, written in domain-dependent symbols, into a bond graph. For this transformation, there is a systematic procedure, which will be presented in the next section.

To generate a bond-graph model starting from an ideal-physical model, a *systematic method* exist, which we will present here as a procedure. This procedure consists roughly of the identification of the domains and basic elements, the generation of the connection structure (called the *junction structure*), the placement of the elements, and possibly simplifying the graph. The procedure is different for the mechanical domain compared to the other domains. These differences are indicated between parenthesis. The reason is that elements need to be connected to *difference variables* or *across variables*. The efforts in the non-mechanical domains and the velocities (flows) in the mechanical domains are the across variables we need.

2.5.1 The Eight Steps of the Systematic Procedure

Steps 1 and 2 concern the identification of the domains and elements.

1. Determine which physical domains exist in the system and identify all basic elements like C, I, R, Se, Sf, TF and GY. Give every element a unique name to distinguish them from each other.
2. Indicate in the ideal-physical model per domain a reference effort (reference velocity with positive direction for the mechanical domains).
   Note that only the references in the mechanical domains have a direction.

   Steps 3 through 6 describe the generation of the connection structure (called the *junction structure*).

3. Identify all *other* efforts (mechanical domains: velocities) and give them unique names.
4. Draw these efforts (mechanical: velocities), and *not* the references, graphically by 0-junctions (mechanical: 1-junctions). Keep if possible, the same layout as the IPM.
5. Identify all effort differences (mechanical: velocity (= flow) differences) needed to connect the ports of all elements enumerated in step 1 to the junction structure. Give these differences a unique name, preferably showing the difference nature. The difference between e_1 and e_2 can be indicated by $e_1 2$.
6. Construct the effort differences using a 1-junction (mechanical: flow differences with a 0-junction) according to Figure 2.17, and draw them as such in the graph.

   The junction structure is now ready and the elements can be connected.

7. Connect the port of all elements found at step 1 with the 0-junctions of the corresponding efforts or effort differences (mechanical: 1-junctions of the corresponding flows or flow differences).
8. Simplify the resulting graph by applying the following simplification rules (2.18):

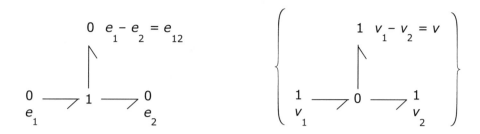

Fig. 2.17: Construction of effort differences (flow differences)

- A junction between two bonds can be left out, if the bonds have a 'through' power direction (one bond incoming, the other outgoing).
- A bond between two the same junctions can be left out, and the junctions can join into one junction.
- Two separately constructed identical effort or flow differences can join into one effort or flow difference.

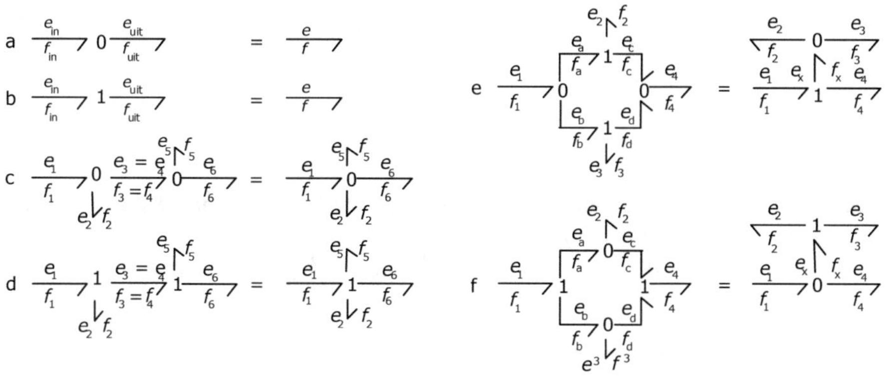

Fig. 2.18: Simplification rules for the junction structure: (a, b): elimination of a junction between bonds; (c, d): contraction of two the same junctions; (e,f): two separately constructed identical differences fuse to one difference.

2.5.2 Illustration of the Systematic Procedure

We will illustrate these steps with a concrete example consisting of an electromotor fed by electric mains, a cable drum and a load (2.19).

A possible ideal-physical model (IPM) is given in Figure 2.20. The mains is modelled as an ideal voltage source. At the electromotor, the inductance, electric resistance of the coils, bearing friction and rotary inertia are taken into account. The cable drum is the transformation from rotation to translation, which we consider as ideal. The load consists of a mass and the gravity force. Starting from the IPM of Figure 2.20, we will construct a bond graph using the 8 steps mentioned above.

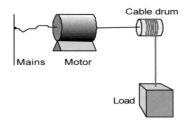

Fig. 2.19: Sketch of the hoisting device

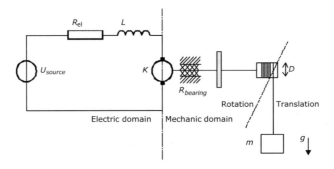

Fig. 2.20: Possible ideal-physical model augmented with the domain information of step 1

Step 1

This system contains:

- An electric domain part with a voltage source (Se), a resistor (R), an inductor (I) and the electric port of the electromotor (GY port).
- A rotation mechanic domain part with the rotation port of the electromotor (GY port), bearing friction (R), inertia (I), and the axis of the cable drum (TF port).
- A translation mechanic domain part with the cable of the cable drum (TF port), the mass of the load (I) and the gravity force acting on the mass (Se).

In Figure 2.20, the domains are indicated and all elements have a unique name.

Step 2

The references are indicated in the ideal physical model: the voltage u_0, the rotational velocity ω_0 and the linear velocity v_0. The two velocities also get a positive orientation (i.e. a direction in which the velocity is positive). This result is shown in Figure 2.21.

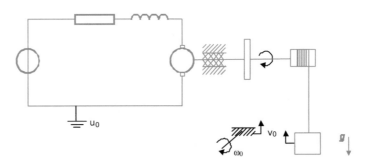

Fig. 2.21: References added to the IPM.

Step 3

The other voltages, angular velocities and linear velocities are sought for and are indicated in the IPM (2.22). These variables are respectively $u_1, u_2, u_3, \omega_1, v_1$.

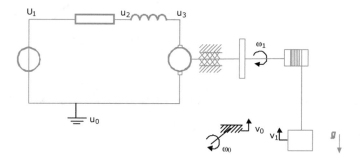

Fig. 2.22: The IPM augmented with relevant voltages, velocities, and angular velocities.

Step 4

The variables found in step 3 are depicted with 0- respectively 1-junctions in Figure 2.23, in a layout compatible to the IPM. The references are not drawn, because they are so to speak eliminated (references have the value 0 and do not contribute to the dynamic behaviour).

Fig. 2.23: First skeleton of the bond graph: Voltages are shown as 0-junctions and velocities as 1-junctions.

Step 5

When checking all ports of the elements found in step 1 for voltage differences, angular velocity differences and linear velocity differences, only $u_1 2$ and $u_2 3$ are identified. No velocity differences are needed.

Step 6

The difference variables are drawn in the bond graph, see Figure 2.24. After this step, the junction structure is generated and the elements can be connected.

Step 7

All elements are connected to the appropriate junctions, as shown in Figure 2.25. Note that non-mechanical domain elements are always connected to 0-junctions (efforts or effort differences) and that mechanical domain elements are always connected to 1-junctions.

Step 8

As last action, the bond graph needs to be simplified, to eliminate superfluous junctions (according to the rules given in Figure 2.18). The resulting bond graph is the outcome of the systematic method, see Figure 2.26.

Obviously, this systematic method is not the *only* method for deriving bond graphs from ideal physical models (IPMs). Another method is the so-called *inspection method*, where parts of the IPM are recognised that can be

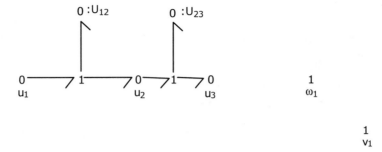

Fig. 2.24: Junction Structure ready: Difference variables (u_{12} and u_{23}) shown in the bond graph.

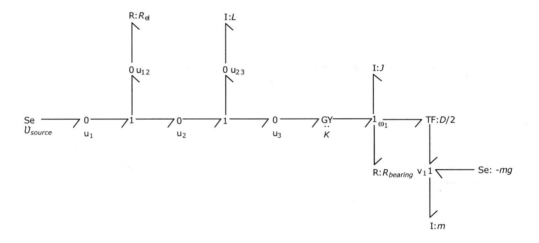

Fig. 2.25: The IPM augmented with relevant voltages, velocities, and angular velocities.

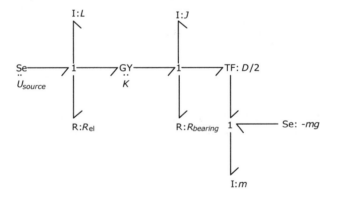

Fig. 2.26: The simplified bond graph, the result of the systematic method.

represented by one junction. An example is a series connection in an electrical network, which is drawn as one 1-junction. This is the case in the example above: The voltage source, inductor, electric resistor and electric port of the motor are directly connected to one 1-junction. Although the inspection method is shorter than the systematic method, it is rather error prone.

2.6 Causal Analysis

Causal analysis is the determination of the signal direction of the bonds. The energetic connection (bond) is now interpreted as a bi-directional signal flow. The result is a causal bond graph, which can be seen as a compact block diagram. Causal analysis is in general completely covered by modelling and simulation software packages that support bond graphs like Enport [242], MS1 [191], CAMP [129] and 20-sim [46, 48, 49, 50]. Therefore, in practice, causal analysis need not be done by hand. Besides derivation of equations, causal analysis can give insight in the correctness and competency of the model. This last reason especially motivates the discussion of causal analysis in this chapter.

2.6.1 Causal Constraints

Dependent on the kind of equations of the elements, the element ports can impose constraints on the connected bonds. There are four different constraints, which will be treated before a systematic procedure for causal analysis of bond graphs is discussed.

2.6.1.1 Fixed causality

Fixed causality is the case, when the equations only allow one of the two port variables to be the outgoing variable. This occurs at sources: an effort source (Se) has by definition always its effort variable as signal output, and has the causal stroke outwards. This causality is called *effort-out causality* or *effort causality*. A flow source (Sf) clearly has a *flow-out causality* or *flow causality*.

Another situation where fixed causality occurs is at nonlinear elements, where the equations for that port cannot be inverted (for example, division by zero). This is possible at R, GY, TF, C and I elements. Thus, there are two reasons to impose a fixed causality:

1. There is no relation between the port variables.
2. The equations are not invertible ('singular').

2.6.1.2 Constrained Causality

At TF, GY, 0- and 1-junction, relations exist between the causalities of the different ports of the element. These relations are *causal constraints*, since the causality of a particular port imposes the causality of the other ports. At a TF, one of the ports has effort-out causality and the other has flow-out causality. At a GY, both ports have either effort-out causality or flow-out causality.

At a 0-junction, where all efforts are the same, *exactly one* bond must bring in the effort. This implies that 0-junctions always have exactly one causal stroke at the side of the junction. The causal condition at a 1-junction is the dual form of the 0-junction. All flows are equal, thus *exactly one* bond will bring in the flow, implying that *exactly one* bond has the causal stroke away from the 1-junction.

2.6.1.3 Preferred Causality

At the storage elements, the causality determines whether an integration or differentiation with respect to time will be the case. Integration has preference above a differentiation. At the integrating form, an initial condition must be specified. Besides, integration with respect to time is a process, which can be realised physically. Numerical differentiation is not physically realisable, since information at future time points is needed. Another drawback of differentiation occurs when the input contains a step function: the output will then become infinite. Therefore, integrating causality is seen as the *preferred causality*. This implies that a C-element has effort-out causality and an I-element has flow-out causality at its preference. These preferences are also illustrated in Figure 2.8 and Figure 2.9, when looking at the block-diagram expansion.

Effort-out vs. flow-out causality

When a voltage u is imposed on an electrical capacitor (a C-element), the current i is the result of the constitutive equation of the capacitor:

$$i = C \frac{\mathrm{d}u}{\mathrm{d}t}$$

A differentiation is thus happening. We have a problem when the voltage instantly steps to another value, since the current will be infinite (the derivative of a step is infinite). This is not the case when the current is imposed on a capacitor. Now, an integral is used:

$$u = u_0 + \int i\,\mathrm{d}t$$

The first case is flow-out causality (effort imposed, flow the result), and the second case is effort-out causality, which is the preferred causality. Furthermore, an effort-out causality also results in a state variable with initial condition u_0. At an inductor, the dual form of the C-element is the case: flow-out causality will result in an integral causality, being the preference.

2.6.1.4 Indifferent Causality

Indifferent causality is used, when there are no causal constraints! At a linear R, it does not matter which of the port variables is the output. Consider an electrical resistor. Imposing a current (flow) yields:

$$u = Ri$$

It is also possible to impose a voltage (effort) on the linear resistor:

$$i = \frac{u}{R}$$

There is no difference choosing the current as incoming variable and the voltage as outgoing variable, or the other way around.

2.6.2 Causal Analysis Procedure

In terms of causal constraints, we can say that the Se and Sf have a fixed causality, the C and I have a preferred causality, the TF, GY, 0 and 1 have constrained causality, and the R has an indifferent causality (provided that the equations of these basic elements all are invertible). These causal forms have been shown in 2.4. When the equations are not invertible, a fixed causality must be used.

The procedure for assigning causality on a bond graph starts with those elements that have the strongest causality constraint namely fixed causality (deviation of the causality condition cannot be granted by rewriting the equations, since rewriting is not possible). Via the bonds (i.e. connections) in the graph, one causality assignment can cause other causalities to be assigned. This effect is called *causality propagation*: after one assignment, the causality propagates through the bond graph due to the causal constraints.

The causality assignment algorithm is as follows:

1a Chose a fixed causality of a source element, assign its causality, and propagate this assignment through the graph using the causal constraints. Go on until all sources have their causalities assigned.

1b Chose a not yet causal port with fixed causality (non-invertible equations), assign its causality, and propagate this assignment through the graph using the causal constraints. Go on until all ports with fixed causality have their causalities assigned.

2 Chose a not yet causal port with preferred causality (storage elements), assign its causality, and propagate this assignment through the graph using the causal constraints. Go on until all ports with preferred causality have their causalities assigned.

3 Chose a not yet causal port with indifferent causality, assign its causality, and propagate this assignment through the graph using the causal constraints. Go on until all ports with indifferent causality have their causalities assigned.

Often, the bond graph is completely causal after step 2, without any causal conflict (all causal conditions are satisfied). If this is *not* the case, then the moment in the procedure where a conflict occurs or where the graph becomes completely causal, can give insight in the correctness and competence of the model. Before discussing these issues, first an example will be treated.

Causality algorithm

To exemplify the causality algorithm, the same example as in 2.5 is used. In Figure 2.27, the completely causal bond graph of the hoisting device is shown. Numbers at the causal strokes indicate the order in which the bonds were made causal.

At step 1a, we assign causality 1 and 2. It does not matter with which source we start. No causality can be propagated. Step 1b is not applicable, since there are no ports of that category. At step 2, we started with the inductor. Propagation is from stroke 3 until stroke 6. The next storage element (preferred causality) is inertia J. Propagation of this causality (number 7), completes the causality of the graph, implying that we do not need step 3. However, the mass of the load does not get his preference. What the consequences are, is subject of the next section.

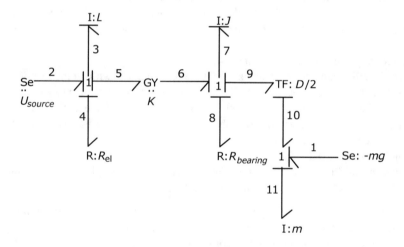

Fig. 2.27: The causal bond graph of the hoisting device.

2.6.3 Model Insight via Causal Analysis

We discuss here those situations whereby conflicts occur in the causal analysis procedure or when step 3 of the algorithm appears to be necessary. The place in procedure where a conflict appears or the bond graph becomes completely causally augmented, can give insight in the correctness of the model.

Often, the bond graph is completely causal after *step 2*, without any causal conflict (*all* causal conditions are satisfied). Each storage element represents a state variable, and the set of equations is an explicit set of ordinary differential equations (not necessarily linear or time invariant).

When the bond graph is completely causal after *step 1a*, the model does not have any dynamics. The behaviour of all variables now is determined by the fixed causalities of the sources. Arises a causal conflict at step 1a or at step 1b, then the problem is ill posed. The model must be changed, by adding some elements. An example of a causal conflict at step 1a is two effort sources connected to one 0-junction. Both sources 'want' to determine the one effort variable.

At a conflict at *step 1b*, a possible adjustment is changing the equations of the fixed-causality element such that these equations become invertible, and thus the fixedness of the constraint disappears. An example is a diode or a valve having zero current resp. flow while blocking. Allowing a small resistance during blocking, the equations become invertible.

When a conflict arises at *step 2*, a storage element receives a non-preferred causality. This means that this storage element does *not* represent a *state variable*. The initial value of this storage element cannot be chosen freely. Such a storage element often is called a *dependent storage element*. This indicates that a storage element was *not* taken into account during modelling, which should be there from physical systems viewpoint. It can be deliberately omitted, or it might be forgotten. At the hoisting device example, the load of the hoist (I-element) is such a dependent storage element. Elasticity in the cable was not modelled. If it had been modelled, a C-storage element connected to a 0-junction between the cable drum and load would appear, and would take away the causal conflict.

When *step 3* of the causality algorithm is necessary, a so-called *algebraic loop* is present in the graph. This loop causes the resulting set differential equations to be *implicit*. Often this is an indication that a storage element was not modelled, which should be there from a physical systems viewpoint.

In general, different ways to handle the causal conflicts arising at step 2 or step 3 are possible:

1. Add elements.
   For example, you can withdraw the decision to neglect certain elements. The added elements can be parasitic, for example, to add elasticity (C-element) in a mechanical connection, which was modelled as rigid. Additionally adding a damping element (R) reduces the simulation time considerably, which is being advised.
2. Change the bond graph such that the conflict disappears.
   For a step-2 conflict, the dependent storage element is taken together with an independent storage element, having integral causality. For a step-3 conflict, sometimes resistive elements can be taken together to eliminate the conflict. This can be performed via transformations in the graph. The complexity of this operation depends on the size and kind of submodels along the route between the storage elements or resistors under concern.
3. The bond graph is *not* changed.
   For simulation a special (implicit) integration routine is needed. The implicit equations are computed using an iteration scheme, mostly as part of the numerical integration method.

Algebraic loops and loops between a dependent and an independent storage element are called *zero-order causal paths* (ZCPs). Besides these two kinds, there are three other kinds, having an increasing complexity and resulting in more complex equations. These occur for instance in rigid-body mechanical systems (van Dijk and Breedveld, 1991).

By interpreting the result of causal analysis, several properties of the model can be recognised, which could otherwise only be done after deriving equations.

2.6.4 Order of the Set State Equations

The causal analysis also gives information on the order of the set equations. The number of initial conditions equals the number of storage elements with integral causality, which was also the preference during causality assignment. This number is called the order of the system. In the example (2.28), the order of the system is 2.

The order of the set state equations is smaller than or equal to the order of the system, because storage elements can depend on each other. These kind of dependent storage elements each have their own initial value, but they together represent one state variable. Their input signals are equal, or have a factor in between (2.28).

A recipe exists to check whether this kind of dependent storage elements show up: Perform causal analysis again, but chose differential causality as preference. For the example, this is done in Figure 2.29.

Those storage elements that get both at differential preference and at integral preference their preferred causality are the *real* storage elements and contribute to the state of the system. The order of the set of state equations, is by definition the amount of storage elements that get in both cases their preferred causality. At

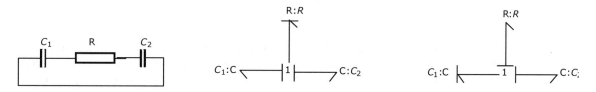

Fig. 2.28: System with order of state equations smaller than order of the system. a) IPM; b) causality using integral preference; c) causality using differential preference.

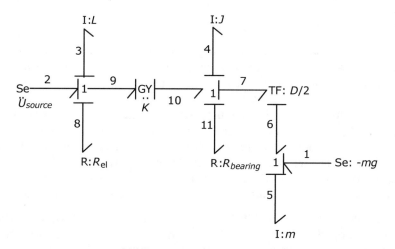

Fig. 2.29: Differential causality as preference applied to the hoisting device.

the hoisting device, this is 2 (namely L and J). The storage elements that get in both cases *not* their preferred causality, are the dependent storage elements. Those storage elements that get *only* at the integral preferred causality case their preferred causality are called *semi-dependent storage elements*, to distinguish them from the dependent storage elements. This indicates that a storage element was not taken into account during modelling, which should be there from physical systems viewpoint.

2.6.5 Matrix Form of Linear Systems

If the system is linear, we can write the resulting set of state equations in the standard form, namely,

$$\dot{\mathbf{x}} = \mathbf{A}\mathbf{x} + \mathbf{B}\mathbf{u}$$

where $\mathbf{A}$ is the system matrix, and $\mathbf{A}$ is the input matrix. The order of the system is the dimension of the square matrix $\mathbf{A}$ and the order of the set of state equations is the rank of $\mathbf{A}$.

When dependent states or algebraic loops are present, the matrix description is as follows:

$$\mathbf{E}\dot{\mathbf{x}} = \mathbf{A}\mathbf{x} + \mathbf{B}\mathbf{u}$$

where $\mathbf{E}$ is a square matrix of size equal to the amount of storage elements plus the amount of algebraic loops. For each differential storage element and for each algebraic loop, $\mathbf{E}$ contains one row of zeros. The state vector $\mathbf{x}$ is extended with the algebraic loop variables, to enable implicit integration methods to solve these kinds of systems. The hoisting device gives such a set of equations.

2.7 Generation of Equations

A causal bond graph contains all information to derive the set of state equations. It is either a set of ordinary first-order differential equations, ODEs, when the model is explicit (no causal conflicts), or a set of differential and algebraic equations, DAEs, when the model is implicit (a causal conflict in step 2 of the procedure or step 3 is necessary, cf. 2.6.2).

The procedure to derive the equations is covered by bond-graph software like Enport [242], MS1 [191], CAMP [129] and 20-sim [46, 48, 49, 50]. Therefore, in practice, generation of equations need not be done by hand. However, we do discuss the generation of equations on the one hand to be complete and on the other hand to indicate what exactly has to be done.

We use the following procedure to generate equations:

1. We first write the set of mixed differential and algebraic equations. These are the constitutive relations of all elements in computational form, or causal form. This comprises of $2n$ equations of a bond graph having n bonds. n equations compute an effort and n equations compute a flow, or derivatives of them.
2. We then eliminate the algebraic equations. We can organise this elimination process by first eliminate the identities coming from the sources and junctions. Thereafter, we substitute the multiplications with a parameter, stemming from resistors and transducers (TF, GY). At last, we substitute the summation equations of the junctions into the differential equations of the storage elements. Within this process, it is efficient to *first* mark the state variables. In principle, the state variables are the contents of the storage elements (p or q type variables). However, if we write the constitutive relations of storage elements as one differential equation, we can also use the *efforts* at C-elements and *flows* at I-elements.

If we are going to generate the equations by hand, we can take the first elimination step into account while formulating the equations by, at the sources, directly use the signal function at the bond. Furthermore, we can write the variable determining the junction along all bonds connected to that junction. The *variable determining the junction* is that variable, which gets assigned to bond variables of all the other bonds connected to that junction via the identities of the junction equations. At a 0-junction, this is the effort of the only bond with its causal stroke towards the 0-junction. At a 1-junction, this is the flow of the only bond with its causal stroke away from the 1-junction.

In case of dependent storage elements, we have to take care that the accompanying state variable gets not eliminated. These are the so-called *semi state variables*. When we mark the state variables, including the semi state variables in this situation, on beforehand, we can prevent the wrong variable from being eliminated. In case of algebraic loops, implicit equations will be encountered. We choose one of the variables in these loops as *algebraic loop breaker* and that variable becomes a semi state variable. See also 2.6.5. The equation consisting the semi state variable of a storage element gets eliminated at the second elimination step: it is a multiplication. The semi state variable itself must *not* be eliminated.

Deriving the diferential equations of the hoisting device

The set of *mixed differential and algebraic equations* of the hoisting device is shown below. The efforts and flows are numbered in the same order as their causality was assigned. We have 22 equations, of which 11 compute a flow and 11 compute an effort. Since we want to generate the set of equations as differential equations, we write at the storage elements in integral form the equations as differential equations and not as integral equations. This is called *deferred integration*, see Figure 2.30.

As state variables, we have f_3 and f_7, whereas f_{11} is the semi state variable belonging to the mass having derivative causality (it is an I element). Note that the equation of the mass is written in causal form: e_{11} is the output. After eliminating the identities at the junctions and sources, we have 12 equations. Note that the flows of the two effort sources (f_1 and f_2) are not used elsewhere, so we leave them out.

After substitution of the multiplication with resistors and transducers, our set of equations reduces to the following 6 equations: 2 differential equations, 1 constraint equation, computing the semi state variable f_{11} and 4 junction equations. Now, only state variables, junction variables and input variables are used.

After substitution of the summations at the junctions into the differential equations and the constraint equations, the result is shown in Figure 2.31

This system is a linear system, so we can write the equations in matrix form (according to the second form of 2.6.5), shown in Figure 2.32.

$$e_2 = u_{source}$$

$$\frac{df_3}{dt} = \frac{1}{L}e_3$$

$$e_4 = R_{el}f_4$$

$$f_2 = f_3$$

$$f_4 = f_3$$

$$f_5 = f_3$$

$$e_3 = e_2 - e_4 - e_5$$

$$e_5 = Kf_6$$

$$e6 = Kf_5$$

$$\frac{df_7}{dt} = \frac{1}{J}e_7$$

$$e_8 = R_{bearing}f_8$$

$$f_6 = f_7$$

$$f_8 = f_7$$

$$f_9 = f_7$$

$$e_7 = e_6 - e_8 - e_9$$

$$e_9 = -\frac{D}{2}e_{10}$$

$$f_{10} = -\frac{D}{2}f_9$$

$$f_1 = f_{10}$$

$$f_{11} = f_{10}$$

$$e_{10} = e_{11} - e_1$$

$$e_1 = -mg$$

$$e_{11} = m\frac{df_{11}}{dt}$$

Fig. 2.30: Equations of the hoisting device.

$$\frac{df_3}{dt} = \frac{1}{L}u_{source} - \frac{R_{el}}{L}f_3 - \frac{K}{L}f_7$$

$$\frac{df_7}{dt} = \frac{K}{J}f_3 - \frac{R_{bearing}}{J}f_7 + \frac{D}{2}m\frac{df_{11}}{dt} + \frac{D}{2}mg$$

$$f_{11} = -\frac{D}{2}f_7$$

Fig. 2.31: Differential and constraint equations of the hoisting device.

$$\begin{pmatrix} 1 & 0 & 0 \\ 0 & 1 & -\dfrac{D}{2} \\ 0 & 0 & 0 \end{pmatrix} \frac{d}{dt}\begin{pmatrix} f_3 \\ f_7 \\ f_{11} \end{pmatrix} = \begin{pmatrix} -\dfrac{R_{el}}{L} & -\dfrac{K}{L} & 0 \\ \dfrac{K}{J} & -\dfrac{R_{bearing}}{J} & 0 \\ 0 & \dfrac{D}{2} & 1 \end{pmatrix}\begin{pmatrix} f_3 \\ f_7 \\ f_{11} \end{pmatrix} + \begin{pmatrix} \dfrac{1}{L} & 0 \\ 0 & \dfrac{D}{2} \\ 0 & 0 \end{pmatrix}\begin{pmatrix} u_{source} \\ mg \end{pmatrix}$$

Fig. 2.32: DAE of hoisting device in matrix form

When the model first was made explicit by adding elements, according to alternative 1 of 2.6.3, the causal bond graph and the equations are given below. We add the elasticity of the rope: A C-element connected to a 0-junction is inserted on the bond between the TF of the cable drum and the 1-junction of the payload (see Figure 2.33). The efforts and flows are numbered in the same order as their causality was assigned. We have 26 equations, of which 13 compute a flow and 13 compute an effort.

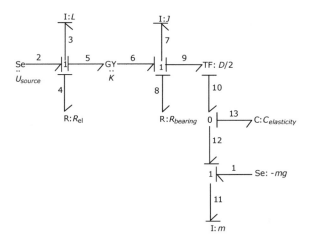

Fig. 2.33: Causal bond graph of hoisting device with elasticity added

As state variables, we have f_3, f_7, e_{11} and f_{13}. So *four* independent state variables. Applying the procedure of deriving equations leads to the set of state equations written in matrix form shown in Figure 2.34.

$$\frac{d}{dt}\begin{pmatrix} f_3 \\ f_7 \\ e_{11} \\ f_{13} \end{pmatrix} = \begin{pmatrix} -\dfrac{\mathrm{Re}\,l}{L} & -\dfrac{K}{L} & 0 & 0 \\ \dfrac{K}{J} & -\dfrac{R_{bearing}}{J} & -\dfrac{D}{2} & 0 \\ -\dfrac{D}{2C_{elasticity}} & -\dfrac{1}{C_{elasticity}} & 0 & 0 \\ 0 & 0 & \dfrac{1}{m} & 0 \end{pmatrix}\begin{pmatrix} f_3 \\ f_7 \\ e_{11} \\ f_{13} \end{pmatrix} + \begin{pmatrix} \dfrac{1}{L} & 0 \\ 0 & 0 \\ 0 & 0 \\ 0 & -\dfrac{1}{m} \end{pmatrix}\begin{pmatrix} u_{source} \\ mg \end{pmatrix}$$

Fig. 2.34: ODE of hoisting device in matrix form

2.8 Expansion to Block Diagrams

To show that a causal bond graph is a compact block diagram, we treat in this section the expansion of a causal bond graph to a block diagram. Furthermore, a block diagram representation of a system might be more familiar than a bond graph representation. Thus this work might help understanding bond graphs.

The expansion of a causal bond graph into a block diagram consists of three steps:

1. Expand all bonds to *bilateral signal flows* (two signals with opposite directions). The *causal stroke* determines in which direction the effort flows. The bond graph elements can be encircled to connect the signals to.
2. Replace the bond-graph elements by their block-diagram representations (see 2.4). Deduce the signs of the summations of the junctions from the directions of the bond arrows (half arrows, see 2.4.5). Often, it is efficient to determine those signs after all bond-graph elements are written in block diagram form. The block diagram is ready in principle.
3. Redraw the block diagram in *standard form*: all integrators in an ongoing stream (form left to right) and all other operations as feedback loops. Of course, this is not always possible. Blocks might be taken together.

Since block diagrams represent mathematical operations, for which *commutative* and *associative* properties apply, these properties can be used to manipulate the block diagram such that the result looks appealing enough.

As an example, we show the block diagram of the hoisting device, using the three steps to construct the block diagram. In Figure 2.35 the result after step 2 is shown, and in Figure 2.36 the block diagram in standard form is presented. The block diagram of the model with elasticity is shown in Figure 2.37.

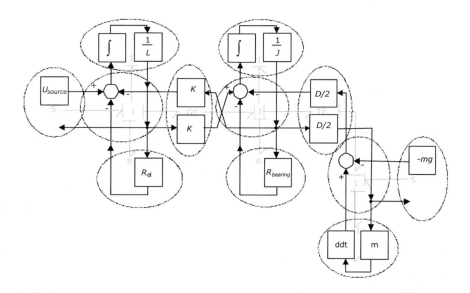

Fig. 2.35: Bond graph expanded to a block diagram in the layout of the bond graph (the bond graph is shown in grey).

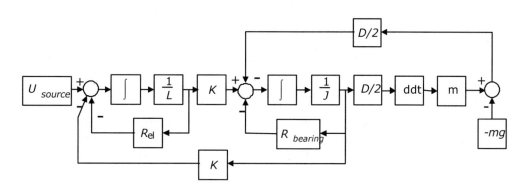

Fig. 2.36: The block diagram redrawn in standard form.

2.9 Simulation

The resulting set of equations coming from a bond-graph model is called the simulation model. It consists of first-order ordinary differential equations (ODEs), possibly extended with algebraic constraint equations (DAEs). Hence, it can be simulated using standard numerical integration methods. However, because numerical integration is an approximation of the actual integration process, it is useful to check the simulation model on

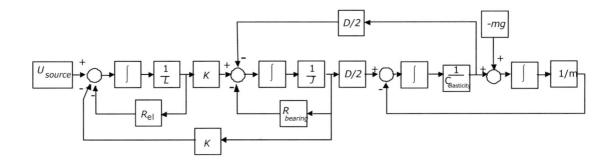

Fig. 2.37: The block diagram of the Hoist with elasticity

aspects significant for simulation. As a result, an appropriate integration method can be chosen: the computational work is minimal and the results stay within a specified error margin.

Since at causal analysis, one can decide whether or not to change the bond-graph model to obtain an explicit simulation model, it is useful to know about the consequences for simulation of relevant characteristics of the simulation model. The following 4 aspects of simulation models are relevant for choosing an numerical integration method:

1. Presence of implicit equations.
   Implicit models (DAEs) can only be simulated with implicit integration methods. The iteration procedure of the implicit integration method is also used to calculate the implicit model. Explicit models (ODEs) can be simulated with both explicit as implicit integration methods. Sometimes, implicit integration methods need more computation time than explicit integration methods. During causal analysis, one can see whether a simulation model will be explicit or implicit, see 2.6.3.
2. Presence of discontinuities.
   Integration methods with special provisions for events will perform best. If that is not available, variable step methods can be used. Multistep methods become less accurate, since they need information from the past, which is useless after a discontinuity. While constructing the model, presence of discontinuities can be marked.
3. Numerical stiffness.
   $S(t)$, the stiffness ratio, is a measure for the distance between real parts of eigenvalues, λ, namely

$$S(t) = \frac{max(|Re\lambda(t)|)}{min(|Re\lambda(t)|)}$$

   Stiff models (large S) need stiff integration methods. The time step is now determined by the stability instead of the accuracy (namely, eigenvalues are now used to determine the step size). When the high frequency parts are faded out, they do not influence the step size anymore. Hence, the step size can grow to limits determined by the lower frequencies.
4. Oscillatory parts.
   When a model has *no* damping, it should not be simulated with a stiff method. Stiff methods perform badly for eigenvalues on the imaginary axis (i.e. no damping) of the complex eigenvalue plane.

Eigenvalues can be localised in a causal bond graph, especially when all elements are linear. There is a bond graph version of Mason's Loop rule to determine the transfer function from a bond graph [53]. As a side effect, the eigenvalues can be calculated. We will not discuss the procedure to obtain eigenvalues from a causal bond graph by hand.

2.10 Summary

In this chapter, we have introduced bond graphs to model physical systems in a *domain independent* way. Only *macroscopic* systems are treated, thus *quantum effects* do not play a significant role. Domain independence has its basics in the fact that physical concepts are analogous for the different physical domains. Six different elementary concepts exist: storage of energy, dissipation, transduction to other domains, distribution, transport, input or output of energy.

Another starting point is that it is possible to write models as directed graphs: parts are interconnected by bonds, along which exchange of energy occurs. A bond represents the energy flow between the two connected submodels. This energy flow can be described as the product of 2 variables (effort and flow), letting a bond be conceived as a *bilateral signal* connection. During modelling, the first interpretation is used, while during analysis and equations generation the second interpretation is used.

Furthermore, we presented a method to systematically build a bond graph starting from an ideal physical model. Causal analysis gives, besides the computational direction of the signals at the bonds, also information about the correctness of the model. We presented methods to derive the causality of a bond graph. In addition, procedures to generate equations and block diagrams out of a causal bond graph are presented.

Due to the introductory nature of this text, some procedures presented, without a deep motivation and possible alternatives. It was also *not* the incentive to elaborate on physical systems modelling. We did not discuss multiple connections (arrays of bonds written as one multibond) and multiport elements (to describe transducers), neither different causal analysis algorithms. Those different causality algorithms give slightly different sets of DAEs especially when applied to certain classes of models (for instance multibody systems with kinematic loops).

Finally, bond-graph modelling is in fact a form of *object-oriented physical-systems modelling*, a term which is often used. This can be seen as follows: bond-graph models are declarative, they can be hierarchically structured, and fully support encapsulation (due to the non-causal way of specifying equations, and the notion of ports). Moreover, due to allowing hierarchy, the notion of definition and use of models are distinguished (i.e. the class concept and instantiation). Since bond graphs came into existence before the term object oriented was used in the field of physical systems modelling, bond graphs can be seen as an object-oriented physical-systems modelling paradigm avant-la-lettre.

A bond-graph library was written in Modelica, a contemporary object-oriented modelling language [47]. The basic bond-graph elements and block-diagram elements have been specified in Modelica, using the essential object-orientation features inheritance and encapsulation. Equations have been specified in an a-causal format. Thus, it can be said that the Modelica modelling concepts are consistent with bond-graph concepts. Furthermore, automatic Modelica code generation from bond graphs appeared to be rather straightforward [46].

2.11 Literature and Further Reading

For a more thorough analysis of bond graphs, see Paynter [228] and Breedveld [43, 44], while an extensive discussion on textbook level is given by Karnopp, Margolis and Rosenberg [160]. Cellier [67] wrote a textbook on continuous system modelling in which besides bond graphs also other modelling methods are used. Current research on bond graphs is reported at the International Conference on Bond Graph modelling, every two years (Granda and Cellier [129]). Journals regularly publishing bond graph papers are the Journal of the Franklin Institute, which also had special issues on bond graphs and the Journal of Dynamic Systems, Measurement and Control.

2.12 Self-Assessment

1. What is a bond? Why is starting with bonds as means of connecting subsystems beneficial and crucial?
2. What are conserved quantities? Give examples of each type of conserved quantity.
3. What is a causal constraint? Why is the order of using causal constraints in the causal analysis procedure as is?
4. What insight gives conducting the causal analysis procedure?

5. What properties of a causal bond-graph model (or any model written as differential equations) influence the choice of the numerical integration algorithm needed for simulation of that model?

Acknowledgements

Most of the inspiration for this chapter came from the Dutch course material of Breedveld and Van Amerongen (1994). I sincerely acknowledge Peter Breedveld and Job van Amerongen for their valuable suggestions and discussions.

Chapter 3
Modelica: Equation-Based, Object-Oriented Modelling of Physical Systems

Peter Fritzson

Abstract The field of equation-based object-oriented modelling languages and tools continues its success and expanding usage all over the world primarily in engineering and natural sciences but also in some cases social science and economics. The main properties of such languages, of which Modelica is a prime example, are: acausal modelling with equations, multi-domain modelling capability covering several application domains, object-orientation supporting reuse of components and evolution of models, and architectural features facilitating modelling of system architectures including creation and connection of components. This enables ease of use, visual design of models with combination of lego-like predefined model building blocks, ability to define model libraries with reusable components enables. This chapter gives an introduction and overview of Modelica as the prime example of an equation-based object-oriented language.

Learning Objectives

After reading this chapter, we expect you to be able to:

- Create Modelica models that represent the dynamic behaviour of physical components at a *lumped parameter* abstraction level
- Employ Object-Oriented constructs (such as class specialisation, nesting and packaging) known from software languages for the reuse and management of complexity
- Assemble complex physical models employing the component and connector abstractions
- Understand how models are translated into differential algebraic equations for execution

3.1 Introduction

Modelica is primarily a modelling language that allows specification of mathematical models of complex natural or man-made systems, e.g., for the purpose of computer simulation of dynamic systems where behaviour evolves as a function of time. Modelica is also an object-oriented equation-based programming language, oriented toward computational applications with high complexity requiring high performance. The four most important features of Modelica are:

- Modelica is primarily based on equations instead of assignment statements. This permits acausal modelling that gives better reuse of classes since equations do not specify a certain data flow direction. Thus a Modelica class can adapt to more than one data flow context.

Peter Fritzson

Linköping University, Department of Computer and Information Science, SE-58183 Linköping, Sweden
e-mail: peter.fritzson@liu.se

© The Author(s) 2020

P. Carreira et al. (eds.), *Foundations of Multi-Paradigm Modelling for Cyber-Physical Systems*,
https://doi.org/10.1007/978-3-030-43946-0_3

- Modelica has multi-domain modelling capability, meaning that model components corresponding to physical objects from several different domains such as, e.g., electrical, mechanical, thermodynamic, hydraulic, biological, and control applications can be described and connected.
- Modelica is an object-oriented language with a general class concept that unifies classes, generics–known as templates in C++ –and general subtyping into a single language construct. This facilitates reuse of components and evolution of models.
- Modelica has a strong software component model, with constructs for creating and connecting components. Thus the language is ideally suited as an architectural description language for complex physical systems, and to some extent for software systems.

These are the main properties that make Modelica both powerful and easy to use, especially for modelling and simulation. We will start with a gentle introduction to Modelica from the very beginning.

3.2 Getting Started with Modelica

Modelica programs are built from classes, also called models. From a class definition, it is possible to create any number of objects that are known as instances of that class. Think of a class as a collection of blueprints and instructions used by a factory to create objects. In this case the Modelica compiler and run-time system is the factory.

A Modelica class contains elements which for example can be variable declarations and equation sections containing equations. Variables contain data belonging to instances of the class; they make up the data storage of the instance. The equations of a class specify the behaviour of instances of that class.

There is a long tradition that the first sample program in any computer language is a trivial program printing the string "Hello World". Since Modelica is an equation-based language, printing a string does not make much sense. Instead, our HelloWorld Modelica program solves a trivial *differential equation*:

$$\dot{x} = -a.x \tag{3.1}$$

The variable x in this equation is a dynamic variable (here also a state variable) that can change value over time. The time derivative $\dot{x}$ is the time derivative of x, represented as der(x) in Modelica. Since all Modelica programs, usually called *models*, consist of class declarations, our HelloWorld program is declared as a class:

```
class HelloWorld
   Real x(start = 1);
   parameter Real a = 1;
   equation
   der(x) = -a * x;
end HelloWorld;
```

Use your favorite text editor or Modelica programming environment to type in this Modelica code[1], or open the DrModelica electronic document (part of OpenModelica) containing most examples and exercises in this book. Then invoke the simulation command in your Modelica environment. This will compile the Modelica code to some intermediate code, usually C code, which in turn will be compiled to machine code and executed together with a numerical ordinary differential equation (ODE) solver or differential algebraic equation (DAE) solver to produce a solution for x as a function of time. The following command in the OpenModelica environment produces a solution between time 0 seconds and time 2 seconds:

simulate(HelloWorld,stopTime=2)[2]

Since the solution for x is a function of time, it can be plotted by a plot command:

---

[1] There is a downloadable open source Modelica environment OpenModelica from OSMC at www.openmodelica.org, Dymola is a product from Dassault Systems www.dymola.com, see www.modelica.org for more products.

[2] Before simulating an existing model, you need to open it or load it, e.g. loadModel(HelloWorld) in OpenModelica. The Open-Modelica command for simulation is simulate. The corresponding using Dymola is simulateModel("HelloWorld", stopTime=2).

plot(x)[3]

(or the longer form plot(x,xrange={0,2}) that specifies the x-axis), giving the curve in Figure 3.31:

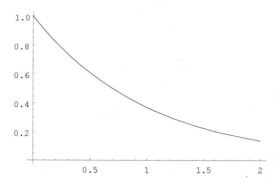

Fig. 3.1: Plot of a simulation of the simple HelloWorld model.

Now we have a small Modelica model that does something, but what does it actually mean? The program contains a declaration of a class called HelloWorld with two variables and a single equation. The first attribute of the class is the variable x, which is initialised to a start value of 1 at the time when the simulation starts. All variables in Modelica have a start attribute with a default value which is normally set to 0. Having a different start value is accomplished by providing a so-called modifier within parentheses after the variable name, i.e., a modification equation setting the start attribute to 1 and replacing the original default equation for the attribute.

The second attribute is the variable a, which is a constant that is initialised to 1 at the beginning of the simulation. Such a constant is prefixed by the keyword parameter in order to indicate that it is constant during simulation but is a model parameter that can be changed between simulations, e.g., through a command in the simulation environment. For example, we could rerun the simulation for a different value of a, without re-compilation if a is changed through the environment's parameter update facility.

Also note that each variable has a type that precedes its name when the variable is declared. In this case both the variable x and the "variable" a have the type Real.

The single equation in this HelloWorld example specifies that the time derivative of x is equal to the constant $-a$ times x. In Modelica the equal sign = always means equality, i.e., establishes an equation, and not an assignment as in many other languages. Time derivative of a variable is indicated by the pseudo-function der( ).

Our second example is only slightly more complicated, containing five rather simple equations:

$$m\dot{v}_x = -\frac{x}{L}F \tag{3.2}$$

$$m\dot{v}_y = -\frac{y}{L}F - mg \tag{3.3}$$

$$\dot{x} = v_x \tag{3.4}$$

$$\dot{y} = v_y \tag{3.5}$$

$$x^2 + y^2 = L^2 \tag{3.6}$$

This example is actually a mathematical model of a physical system, a planar pendulum, as depicted in Figure 3.2.

---

[3] plot is the OpenModelica command for plotting simulation results. The corresponding Dymola command would be plot({"x"}) respectively.

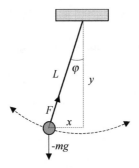

Fig. 3.2: A planar pendulum.

$$v_x \tag{3.7}$$

$$x^2 + y^2 = L^2 \tag{3.8}$$

$$v_y \tag{3.9}$$

The equations are Newton's equations of motion for the pendulum mass under the influence of gravity, together with a geometric constraint, the 5th equation $x^2 + y^2 = L^2$, that specifies that its position (x,y) must be on a circle with radius L. The variables v_x and v_y are its velocities in the x and y directions respectively. A Modelica model of the pendulum appears below:

```
class Pendulum "Planar Pendulum"
 constant Real PI=3.141592653589793;
 parameter Real m=1, g=9.81, L=0.5;
 Real F;
 output    Real x(start=0.5),y(start=0);
 output    Real vx,vy;
equation
 m*der(vx)=-(x/L)*F;
 m*der(vy)=-(y/L)*F-m*g;
 der(x)=vx;
 der(y)=vy;
 x^2+y^2=L^2;
end Pendulum;
```

The interesting property of this model, however, is the fact that the 5th equation is of a different kind: a so-called *algebraic equation* only involving algebraic formulas of variables but no derivatives. The first four equations of this model are differential equations as in the HelloWorld example. Equation systems that contain both differential and algebraic equations are called *differential algebraic equation systems* (DAEs).

We simulate the Pendulum model and plot the x-coordinate, shown in Figure 3.3:

simulate(Pendulum, stopTime=4)

plot(x);

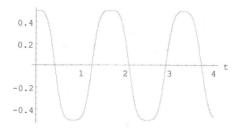

Fig. 3.3: Plot of a simulation of the Pendulum DAE model.

You can also write down DAE equation systems without physical significance, with equations containing formulas selected more or less at random, as in the class DAEexample below:

```
class DAEexample
  Real x(start=0.9);
  Real y;
equation
  der(y) + (1+0.5*sin(y)) * der(x) = sin(time);
  x-y = exp(-0.9*x) * cos(y);
end DAEexample;
```

This class contains one differential and one algebraic equation. Try to simulate and plot it yourself, to see if any reasonable curve appears. Finally, an important observation regarding Modelica models: The number of *variables* must be equal to the number of *equations*!

This statement is true for the three models we have seen so far, and holds for all solvable Modelica models. By variables we mean something that can vary, i.e., not named constants and parameters to be described in a section further below.

3.2.1 Variables and Predefined Types

This example shows a slightly more complicated model, which describes a Van der Pol[4] oscillator. Notice that here the keyword model is used instead of class with almost the same meaning.

```
model VanDerPol "Van der Pol oscillator model"
  Real x(start = 1) "Descriptive string for x"; // x starts at 1
  Real y(start = 1) "Descriptive string for y"; // y starts at 1
parameter
  Real lambda = 0.3;
equation
  der(x) = y; // This is the first equation
  der(y) = -x + lambda*(1 - x*x)*y; // The 2nd differential equation
end VanDerPol;
```

This example contains declarations of two dynamic variables (here also state variables) x and y, both of type Real and having the start value 1 at the beginning of the simulation, which normally is at time 0. Then follows a declaration of the parameter constant lambda, which is a so-called model parameter.

A *parameter* is a special kind of constant which is implemented as a static variable that is initialised once and never changes its value during a specific execution. A parameter is a constant variable that makes it simple for a user to modify the behaviour of a model, e.g., changing the parameter lambda which strongly influences the behaviour of the Van der Pol oscillator. By contrast, a fixed Modelica constant declared with the prefix constant never changes and can be substituted by its value wherever it occurs.

---

[4] Balthazar van der Pol was a Dutch electrical engineer who initiated modern experimental dynamics in the laboratory during the 1920's and 1930's. Van der Pol investigated electrical circuits employing vacuum tubes and found that they have stable oscillations, now called limit cycles. The van der Pol oscillator is a model developed by him to describe the behaviour of nonlinear vacuum tube circuits

Finally, we present declarations of three dummy variables just to show variables of data types different from Real: the boolean variable bb, which has a default start value of false if nothing else is specified, the string variable dummy which is always equal to "dummy string", and the integer variable fooint always equal to 0.

```
Boolean bb;
String dummy = "dummy string";
Integer fooint = 0;
```

Modelica has built-in "primitive" predefined data types to support floating-point, integer, boolean, and string values. These predefined primitive types contain data that Modelica understands directly, as opposed to class types defined by programmers. There is also the Complex type for complex numbers computations, which is predefined in a library. The type of each variable must be declared explicitly. The predefined basic data types of Modelica are:

Boolean	either true or false
Integer	32-bit two's complement, usually corresponding to the C int data type
Real	64-bit floating-point usually corresponding to the C double data type
String	string of text characters
enumeration(...)	enumeration type of enumeration literals
Clock	clock type for clocked synchronous models
Complex	for complex number computations, a basic type predefined in a library

Finally, there is an equation section starting with the keyword equation, containing two mutually dependent equations that define the dynamics of the model. To illustrate the behaviour of the model, we give a command to simulate the Van der Pol oscillator during 25 seconds starting at time 0:

simulate(VanDerPol, stopTime=25)

A phase plane plot of the state variables for the Van der Pol oscillator model (Figure 3.4):

plotParametric(x,y, stopTime=25)

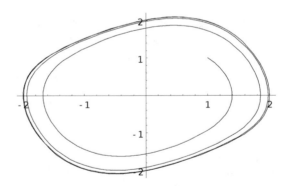

Fig. 3.4: Parametric plot of a simulation of the Van der Pol oscillator model.

The names of variables, functions, classes, etc. are known as *identifiers*. There are two forms in Modelica. The most common form starts with a letter, followed by letters or digits, e.g. x2. The second form starts with a single-quote, followed by any characters, and terminated by a single-quote, e.g. '2nd * 3'.

3.2.2 Comments

Arbitrary descriptive text, e.g., in English, inserted throughout a computer program, are *comments* to that code. Modelica has three styles of comments, all illustrated in the previous VanDerPol example.

Comments make it possible to write descriptive text together with the code, which makes a model easier to use for the user, or easier to understand for programmers who may read your code in the future. That programmer may very well be yourself, months or years later. You save yourself future effort by commenting your own code. Also, it is often the case that you find errors in your code when you write comments since when explaining your code you are forced to think about it once more.

The first kind of comment is a string within string quotes, e.g., "a comment" , optionally appearing after variable declarations, or at the beginning of class declarations. Those are "definition comments" that are processed to be used by the Modelica programming environment, e.g., to appear in menus or as help texts for the user. From a syntactic point of view they are not really comments since they are part of the language syntax. In the previous example such definition comments appear for the VanDerPol class and for the x and y variables.

The other two types of comments are ignored by the Modelica compiler, and are just present for the benefit of Modelica programmers. Text following // up to the end of the line is skipped by the compiler, as is text between /* and the next */. Hence the last type of comment can be used for large sections of text that occupies several lines.

Finally, we should mention a construct called annotation, a kind of structured "comment" that can store information together with the code, described in the section about annotations at the end of this chapter.

3.2.3 Constants

Constant literals in Modelica can be integer values such as 4, 75, 3078; floating-point values like 3.14159, 0.5,2.735E-10, 8.6835e+5; string values such as "hello world", "red"; and enumeration values such as Colors.red,Sizes.xlarge.

Named constants are preferred by programmers for two reasons. One reason is that the name of the constant is a kind of documentation that can be used to describe what the particular value is used for. The other, perhaps even more important reason, is that a named constant is defined at a single place in the program. When the constant needs to be changed or corrected, it can be changed in only one place, simplifying program maintenance.

Named constants in Modelica are created by using one of the prefixes constant or parameter in declarations, and providing a declaration equation as part of the declaration. For example:

```
constant Real PI = 3.141592653589793;
constant String redcolor = "red";
constant Integer one = 1;
parameter Real mass = 22.5;
```

Parameter constants (often called *parameters*) can usually be accessed and changed from Modelica tool graphical user interfaces. Most Modelica tools support re-simulating a model without re-compilation after changing a parameter, which usually makes the time waiting for results much shorter.

Parameters can be declared without a declaration equation (but a default start value is recommended) since their value can be defined, e.g., interactively or by reading from a file, before simulation starts. This is not possible for constants with prefix constant. For example:

```
parameter Real mass, gravity, length;
```

A parameter with a default start value as shown below means that the parameter has a value, but that this value is expected to be replaced by a more realistic value, e.g., by the user, at a real simulation:

```
parameter Real mass(start=0), gravity(start=0), length(start=0);
```

Constants and parameters are very similar, but also have some differences. Parameters are typically much more common in application models. When in doubt whether to use constant or parameter, use parameter, except when declaring a constant in a package/library where constant is allowed but not parameter.

3.2.4 Variability

We have seen that some variables can change value at any point in time whereas named constants are more or less constant. In fact, there is a general concept of four levels of variability of variables and expressions in Modelica:

- Expressions or variables with *continuous-time variability* can change at any point in time.
- *Discrete-time variability* means that value changes can occur only at so-called events. There are two kinds: *unclocked discrete-time variability* and *clocked discrete-time variability*. In the latter case the value can only change at clock tick events.
- *Parameter variability* means that the value can be changed at initialisation before simulation, but is fixed during simulation.
- *Constant variability* means the value is always fixed. However, a named constant definition equation can be replaced by a so-called redeclaration or modification.

3.2.5 Default Start Values

If a numeric variable lacks a specified definition value or start value in its declaration, it is usually initialised to zero at the start of the simulation. Boolean variables have start value false, and string variables the start value empty string "" if nothing else is specified.

Exceptions to this rule are function *results* and *local* variables in functions, where the default initial value at function call is *undefined*.

3.3 Object-Oriented Mathematical Modelling

Traditional object-oriented programming languages like Simula, C++, Java, and Smalltalk, as well as procedural languages such as Fortran or C, support programming with operations on stored data. The stored data of the program include variable values and object data. The number of objects often changes dynamically. The Smalltalk view of object-orientation emphasises sending messages between (dynamically) created objects.

The Modelica view on object-orientation is different since the Modelica language emphasises *structured* mathematical modelling. Object-orientation is viewed as a structuring concept that is used to handle the complexity of large system descriptions. A Modelica model is primarily a declarative mathematical description, which simplifies further analysis. Dynamic system properties are expressed in a declarative way through equations.

The concept of *declarative* programming is inspired by mathematics, where it is common to state or declare what *holds*, rather than giving a detailed stepwise *algorithm* on *how* to achieve the desired goal as is required when using procedural languages. This relieves the programmer from the burden of keeping track of such details. Furthermore, the code becomes more concise and easier to change without introducing errors.

Thus, the declarative Modelica view of object-orientation, from the point of view of object-oriented mathematical modelling, can be summarised as follows:

- Object-orientation is primarily used as a *structuring* concept, emphasising the declarative structure and reuse of mathematical models. Our three ways of structuring are hierarchies, component-connections, and inheritance.
- Dynamic model properties are expressed in a declarative way through *equations*[5].
- An object is a collection of *instance* variables and equations that share a set of data.

However:

- Object-orientation in mathematical modelling is *not* viewed as dynamic message passing.

---

[5] Algorithms are also allowed, but in a way that makes it possible to regard an algorithm section as a system of equations.

The declarative object-oriented way of describing systems and their behaviour offered by Modelica is at a higher level of abstraction than the usual object-oriented programming since some implementation details can be omitted. For example, we do not need to write code to explicitly transport data between objects through assignment statements or message passing code. Such code is generated automatically by the Modelica compiler based on the given equations.

Just as in ordinary object-oriented languages, classes are blueprints for creating objects. Both variables and equations can be inherited between classes. Function definitions can also be inherited. However, specifying behaviour is primarily done through equations instead of via methods. There are also facilities for stating algorithmic code including functions in Modelica, but this is an exception rather than the rule.

3.3.1 Classes and Instances

Modelica, like any object-oriented computer language, provides the notions of classes and objects, also called instances, as a tool for solving modelling and programming problems. Every object in Modelica has a class that defines its data and behaviour. A class has three kinds of members:

- Data variables associated with a class and its instances. Variables represent results of computations caused by solving the equations of a class together with equations from other classes. During numeric solution of time-dependent problems, the variables stores results of the solution process at the current time instant.
- Equations specify the behaviour of a class. The way in which the equations interact with equations from other classes determines the solution process, i.e., program execution.
- Classes can be members of other classes. Here is the declaration of a simple class that might represent a point in a three-dimensional space:

```
class Point "Point in a three-dimensional space"
public
  Real x;
  Real y, z;
end Point;
```

The Point class has three variables representing the x, y, and z coordinates of a point and has no equations. A class declaration like this one is like a blueprint that defines how instances created from that class look like, as well as instructions in the form of equations that define the behaviour of those objects. Members of a class may be accessed using dot (.) notation. For example, regarding an instance myPoint of the Point class, we can access the x variable by writingmyPoint.x.

Members of a class can have two levels of visibility. The public declaration of x, y, and z, which is default if nothing else is specified, means that any code with access to a Point instance can refer to those values. The other possible level of visibility, specified by the keyword protected, means that only code inside the class as well as code in classes that inherit this class, are allowed access.

Note that an occurrence of one of the keywords public or protected means that all member declarations following that keyword assume the corresponding visibility until another occurrence of one of those keywords, or the end of the class containing the member declarations has been reached.

3.3.1.1 Creating Instances

In Modelica, objects are created implicitly just by declaring instances of classes. This is in contrast to object-oriented languages like Java or C++, where object creation is specified using the new keyword when allocating on the heap. For example, to create three instances of our Point class we just declare three variables of type Point in a class, here Triangle:

```
class Triangle
  Point point1;
  Point point2;
  Point point3;
```

```
end Triangle;
```

There is one remaining problem, however. In what context should Triangle be instantiated, and when should it just be interpreted as a library class not to be instantiated until actually used?

This problem is solved by regarding the class at the *top* of the instantiation hierarchy in the Modelica program to be executed as a kind of "main" class that is always implicitly instantiated, implying that its variables are instantiated, and that the variables of those variables are instantiated, etc. Therefore, to instantiate Triangle, either make the class Triangle the "top" class or declare an instance of Triangle in the "main" class. In the following example, both the classTriangle and the class Foo1 are instantiated.

```
class Foo1
  ...
end Foo1;

class Foo2
  ...
end Foo2; ...

class Triangle
  Point point1;
  Point point2;
  Point point3;
end Triangle;

class Main
  Triangle pts;
  Foo1 f1;
end Main;
```

The variables of Modelica classes are instantiated per object. This means that a variable in one object is distinct from the variable with the same name in every other object instantiated from that class. Many object-oriented languages allow class variables. Such variables are specific to a class as opposed to instances of the class, and are shared among all objects of that class. The notion of class variables is not yet available in Modelica.

3.3.1.2 Initialization

Another problem is initialisation of variables. As mentioned previously in the section about variability, if nothing else is specified, the default start value of all numerical variables is zero, apart from function results and local variables where the initial value at call time is unspecified. Other start values can be specified by setting the start attribute of instance variables. Note that the start value only gives a suggestion for initial value the solver may choose a different value unless the fixed attribute is true for that variable. Below a start value is specified in the example class Triangle:

```
class Triangle
  Point point1(start=Point(1,2,3));
  Point point2;
  Point point3;
end Triangle;
```

Alternatively, the start value of point1 can be specified when instantiating Triangle as below:

```
class Main
  Triangle pts(point1.start=Point(1,2,3));
  Foo1 f1;
end Main;
```

A more general way of initialising a set of variables according to some constraints is to specify an equation system to be solved in order to obtain the initial values of these variables. This method is supported in Modelica through the initial equation construct.

An example of a continuous-time controller initialised in steady-state, i.e., when derivatives should be zero, is given below:

```
model Controller
parameter
  Real a=1, b=2;
  Real y;
equation
  der(y) = a*y + b*u;
initial
  equation
    der(y)=0;
end Controller;
```

This has the following solution at initialization:

```
der(y) = 0;
y = -(b/a)*u;
```

3.3.1.3 Specialised Classes

The class concept is fundamental to Modelica, and is used for a number of different purposes. Almost anything in Modelica is a class. However, in order to make Modelica code easier to read and maintain, special keywords have been introduced for specific uses of the class concept. The keywords model, connector, record, block, type, package, and function can be used to denote a class under appropriate conditions, called restrictions. Some of the specialised classes also have additional capabilities, called enhancements. For example, a function class has the enhancement that it can be called, whereas a record is a class used to declare a record data structure and has the restrictions that it may not contain equations and may only contain public declarations.

```
record Person
  Real age;
  String name;
end Person;
```

A model is almost the same as a class, i.e., those keywords are completely interchangeable. A block is a class with fixed causality, which means that for each connector variable of the class it is specified whether it has input or output causality. Thus, each connector variable in a block class interface must be declared with a causality prefix keyword of either input or output.

A connector class is used to declare the structure of "ports" or interface points of a component, may not contain equations, and has only public sections, but has the additional property to allow connect(...) to instances of connector classes. A type is a class that can be an alias or an extension to a predefined type, enumeration, or array of a type. For example:

```
type vector3D = Real[3];
```

The idea of specialised classes has some advantages since they re-use most properties of the general *class concept*. The notion of specialised classes also gives the user a chance to express more precisely what a class is intended for, and requires the Modelica compiler to check that these usage constraints are actually fulfilled.

Fortunately, the notion of specialised class is quite uniform since all basic properties of a class, such as the syntax and semantics of definition, instantiation, inheritance, and generic properties, are identical for all kinds of specialised classes. Furthermore, the construction of Modelica translators is simplified since only the syntax and semantics of the general class concept has to be implemented along with some additional checks on restrictions and enhancements of the classes.

The package and function specialised classes in Modelica have much in common with the class concept but also have many additional properties, so called enhancements. Especially functions have quite a lot of enhancements, e.g., they can be called with an argument list, instantiated at run-time, etc. An operator class is similar to a package which is like a container for other definitions, but may only contain declarations of functions and is intended for user-defined overloaded operators.

3.3.1.4 Reuse of Classes by Modifications

The class concept is the key to reuse of modelling knowledge in Modelica. Provisions for expressing adaptations or modifications of classes through so-called modifiers in Modelica make reuse easier. For example, assume that we would like to connect two filter models with different time constants in series.

Instead of creating two separate filter classes, it is better to define a common filter class and create two appropriately modified instances of this class, which are connected. An example of connecting two modified low-pass filters is shown after the example low-pass filter class below:

```
model LowPassFilter
  parameter
    Real T=1 "Time constant of filter";
    Real u, y(start=1);
  equation
    T*der(y) + y = u;
end LowPassFilter;
```

The model class can be used to create two instances of the filter with different time constants and "connecting" them together by the equation $F2.u = F1.y$ as follows:

```
model FiltersInSeries
  LowPassFilter F1(T=2), F2(T=3);
equation
  F1.u = sin(time);
  F2.u = F1.y;
end FiltersInSeries;
```

Here we have used modifiers, i.e., attribute equations such as T=2 and T=3, to modify the time constant of the low-pass filter when creating the instances F1 and F2. The independent time variable is a built-in predefined variable denoted time. If the FiltersInSeries model is used to declare variables at a higher hierarchical level, e.g., F12, the time constants can still be adapted by using hierarchical modification, as for F1 and F2 below:

```
model ModifiedFiltersInSeries
  FiltersInSeries F12(F1(T=6), F2.T=11);
end ModifiedFiltersInSeries;
```

3.3.1.5 Built-in Classes

The built-in predefined type classes of Modelica correspond to the types Real, Integer, Boolean, String, enumeration(...), and Clock, and have most of the properties of a class, e.g., can be inherited, modified, etc. Only the value attribute can be changed at run-time, and is accessed through the variable name itself, and not through dot notation, i.e., use x and not x.value to access the value. Other attributes are accessed through dot notation.

For example, a Real variable has a set of default attributes such as unit of measure, initial value, minimum and maximum value. These default attributes can be changed when declaring a new class, for example:

```
class Voltage = Real(unit= "V", min=-220.0, max=220.0);
```

3.3.2 Inheritance

One of the major benefits of object-orientation is the ability to extend the behaviour and properties of an existing class. The original class, known as the *superclass* or *base class*, is extended to create a more specialised version of that class, known as the *subclass* or *derived class*. In this process, the behaviour and properties of the original class in the form of variable declarations, equations, and other contents are reused, or inherited, by the subclass.

Let us regard an example of extending a simple Modelica class, e.g., the class Point introduced previously. First we introduce two classes named ColorData and Color, where Color inherits the data variables to represent the color from class ColorData and adds an equation as a constraint. The new class ColoredPoint inherits from multiple classes, i.e., uses multiple inheritance, to get the position variables from class Point, and the color variables together with the equation from class Color.

```
record ColorData
  Real red;
  Real blue;
  Real green;
end ColorData;

class Color
  extends ColorData;
equation
  red + blue + green = 1;
end Color;

class Point
  public
    Real x;
    Real y, z;
end Point;

class ColoredPoint
  extends Point;
  extends Color;
end ColoredPoint;
```

3.3.3 Generic Classes

In many situations it is advantageous to be able to express generic patterns for models or programs. Instead of writing many similar pieces of code with essentially the same structure, a substantial amount of coding and software maintenance can be avoided by directly expressing the general structure of the problem and providing the special cases as *parameter* values.

Such generic constructs are available in several programming languages, e.g., templates in C++, generics in Ada, and type parameters in functional languages such as Haskell or Standard ML. In Modelica the class construct is sufficiently general to handle generic modelling and programming in addition to the usual class functionality.

There are essentially two cases of generic class parameterisation in Modelica: *class parameters* can either be *instance parameters*, i.e., have instance declarations (components) as values, or be *type parameters*, i.e., have types as values. Note that by class parameters in this context we do not usually mean model parameters prefixed by the keyword parameter, even though such "variables" are also a kind of class parameter. Instead we mean *formal parameters to the class*. Such formal parameters are prefixed by the keyword replaceable. The special case of replaceable local functions is roughly equivalent to virtual methods in some object-oriented programming languages.

3.3.3.1 Class Parameters Being Instances

First we present the case when class parameters are variables, i.e., declarations of instances, often called components. The class C in the example below has three class parameters *marked* by the keyword replaceable. These class parameters, which are components (variables) of classC, are declared as having the (default) types GreenClass, YellowClass, and GreenClass respectively. There is also a red object declaration which is not replaceable and therefore not a class parameter (Figure 3.5).

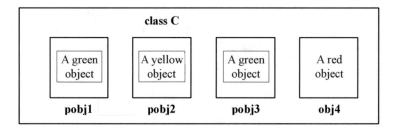

Fig. 3.5: Three class parameters pobj1, pobj2, and pobj3 that are instances (variables) of class C. These are essentially slots that can contain objects of different colors.

Here is the class C with its three class parameters pobj1, pobj2, and pobj3 and a variable obj4 that is not a class parameter:

```
class C
  replaceable GreenClass pobj1(p1=5);
  replaceable YellowClass pobj2;
  replaceable GreenClass pobj3;
  RedClass obj4;
equation
  ...
end C;
```

Now a class C2 is defined by providing two declarations of pobj1 and pobj2 as actual arguments to class C, being red and green respectively, instead of the defaults green and yellow. The keyword redeclare must precede an actual argument to a class formal parameter to allow changing its type. The requirement to use a keyword for a redeclaration in Modelica has been introduced in order to avoid accidentally changing the type of an object through a standard modifier.

In general, the type of a class component cannot be changed if it is not declared as replaceable and a redeclaration is provided. A variable in a redeclaration can replace the original variable if it has a type that is a subtype of the original type or its type constraint. It is also possible to declare type constraints (not shown here) on the substituted classes.

```
class C2 = C(redeclare RedClass pobj1, redeclare GreenClass pobj2);
```

Such a class C2 obtained through redeclaration of pobj1 and pobj2 is of course equivalent to directly defining C2 without reusing class C, as below.

```
class C2
  RedClass pobj1(p1=5);
  GreenClass pobj2;
  GreenClass pobj3;
  RedClass obj4;
equation
  ...
end C2;
```

3.3.3.2 Class Parameters being Types

A class parameter can also be a type, which is useful for changing the type of many objects. For example, by providing a type parameter ColoredClass in class C below, it is easy to change the color of all objects of type ColoredClass.

```
class C
  replaceable class ColoredClass = GreenClass;
  ColoredClass obj1(p1=5);
```

```
  replaceable YellowClass obj2;
  ColoredClass obj3;
  RedClass obj4;
equation
  ...
end C;
```

Figure 3.6 depicts how the type value of the ColoredClass class parameter is propagated to the member object declarations obj1 and obj3.

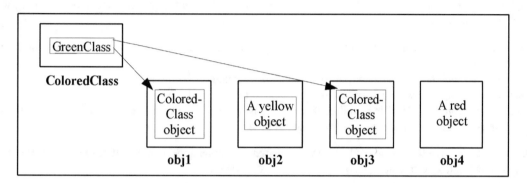

Fig. 3.6: The class parameter ColoredClass is a type parameter that is propagated to the two member instance declarations of obj1 and obj3.

We create a class C2 by giving the type parameter ColoredClass of class C the value BlueClass.

```
class C2 = C(redeclare class ColoredClass = BlueClass);
```

This is equivalent to the following definition of C2:

```
class C2
  BlueClass obj1(p1=5);
  YellowClass obj2;
  BlueClass obj3;
  RedClass obj4;
equation
  ...
end C2;
```

3.4 Equations

As we already stated, Modelica is primarily an equation-based language in contrast to ordinary programming languages, where assignment statements proliferate. Equations are more flexible than assignments since they do not prescribe a certain data flow direction or execution order. This is the key to the physical modelling capabilities and increased reuse potential of Modelica classes.

Thinking in equations is a bit unusual for most programmers. In Modelica the following holds:

1. Assignment statements in conventional languages are usually represented as equations in Modelica.
2. Attribute assignments are represented as equations.
3. Connections between objects generate equations.

Equations are more powerful than assignment statements. For example, consider a resistor equation where the resistance R multiplied by the current i is equal to the voltage v:

$$R \times i = v \tag{3.10}$$

This equation can be used in three ways corresponding to three possible assignment statements: computing the current from the voltage and the resistance, computing the voltage from the resistance and the current, or computing the resistance from the voltage and the current. This is expressed in the following three assignment statements:

```
i := v/R;
v := R * i;
R := v/i;
```

Equations in Modelica can be informally classified into four different groups depending on the syntactic context in which they occur:

1. *Normal equations* occurring in equation sections, including the connect-equation, which is a special form of equation.
2. *Declaration equations*, which are part of variable or constant declarations.
3. *Modification equations*, which are commonly used to modify attributes.
4. *Initial equations*, specified in initial equation sections or as start attribute equations. These equations are used to solve the initialization problem at startup time.

As we already have seen in several examples, normal equations appear in equation sections started by the keyword equation and terminated by some other allowed keyword:

```
equation
... <equations> ...
<some other allowed keyword>
```

The above resistor equation is an example of a normal equation that can be placed in an equation section. Declaration equations are usually given as part of declarations of fixed or parameter constants, for example:

```
constant Integer one = 1;
parameter Real mass = 22.5;
```

An equation always holds, which means that the mass in the above example never changes value during simulation. It is also possible to specify a declaration equation for a normal variable, e.g.:

```
Real speed = 72.4;
```

However, this does not make much sense since it will constrain the variable to have the same value throughout the computation, effectively behaving as a constant. Therefore a declaration equation is quite different from a variable initialiser in other languages since it always holds and can be solved together with other equations in the total system of equations.

Concerning attribute assignments, these are typically specified using modification equations. For example, if we need to specify an initial value for a variable, meaning its value at the start of the computation, then we give an attribute equation for the start attribute of the variable, e.g.:

```
Real speed(start = 72.4);
```

3.4.1 Repetitive Equation Structures

Before reading this section you might want to take a look at the section further down about arrays, and the section further down about statements and algorithmic for-loops.

Sometimes there is a need to conveniently express sets of equations that have a regular, i.e., repetitive structure. Often this can be expressed as array equations, including references to array elements denoted using square bracket notation. However, for the more general case of repetitive equation structures Modelica provides a loop construct. Note that this is not a loop in the algorithmic sense of the word it is rather a shorthand notation for expressing a set of equations.

For example, consider an equation for a polynomial expression:

$$y = a[1] + a[2] \times x + a[3] \times x^2 + ... + a[n+1] \times x^n \qquad (3.11)$$

The polynomial equation Eq. 3.11 above can be expressed as a set of equations with regular structure in Modelica, with y equal to the scalar product of the vectors a and xpowers, both of length $n + 1$:

```
xpowers[2] = xpowers[1]*x;
xpowers[3] = xpowers[2]*x;
...
xpowers[n+1] = xpowers[n]*x;
y = a * xpowers;
```

The regular set of equations involving xpowers can be expressed more conveniently using the for-loop notation:

```
for i in 1:n loop
   xpowers[i+1] = xpowers[i]*x;
end for;
```

In this particular case a vector equation provides an even more compact notation:

```
powers[2:n+1] = xpowers[1:n]*x;
```

Here the vectors x and xpowers have length $n + 1$. The colon notation 2:n+1 means extracting a vector of length n, starting from element 2 up to and including element $n + 1$.

3.4.2 Partial Differential Equations

Partial differential equations (abbreviated PDEs) contain derivatives with respect to other variables than time, for example of spatial Cartesian coordinates such as x and y. Models of phenomena such as heat flow or fluid flow typically involve PDEs. At the time of this writing PDE functionality is not yet part of the official Modelica language.

3.5 Acausal Physical modelling

Acausal modelling is a declarative modelling style, meaning modelling based on equations instead of assignment statements. Equations do not specify which variables are inputs and which are outputs, whereas in assignment statements variables on the left-hand side are always outputs (results) and variables on the right-hand side are always inputs. Thus, the causality of equation-based models is unspecified and becomes fixed only when the corresponding equation systems are solved. This is called *acausal modelling*. The term *physical modelling* reflects the fact that *acausal modelling* is very well suited for representing the *physical structure* of modelled systems.

The main advantage with acausal modelling is that the solution direction of equations will adapt to the data flow context in which the solution is computed. The data flow context is defined by stating which variables are needed as *outputs,* and which are external *inputs* to the simulated system.

The acausality of Modelica library classes makes these more reusable than traditional classes containing assignment statements where the input-output causality is fixed.

To illustrate the idea of acausal physical modelling we give an example of a simple electrical circuit (Figure 3.7). The connection diagram[6] of the electrical circuit shows how the components are connected. It may be drawn with component placements to roughly correspond to the physical layout of the electrical circuit on a printed circuit board. The physical connections in the real circuit correspond to the logical connections in the diagram. Therefore the term *physical modelling* is quite appropriate.

---

[6] A connection diagram emphasises the connections between components of a model, whereas a composition diagram specifies which components a model is composed of, their subcomponents, etc. A class diagram usually depicts inheritance and composition relations.

The Modelica SimpleCircuit model below directly corresponds to the circuit depicted in the connection diagram of Figure 3.7. Each graphic object in the diagram corresponds to a declared instance in the simple circuit model. The model is acausal since no signal flow, i.e., cause-and-effect flow, is specified. Connections between objects are specified using the connect-equation construct, which is a special syntactic form of equation that we will examine later. The classes Resistor, Capacitor, Inductor, VsourceAC, and Ground will be presented in more detail later in this chapter.

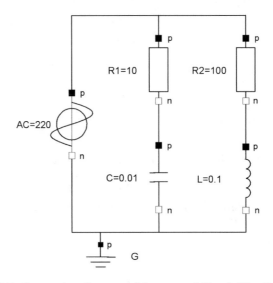

Fig. 3.7: Connection diagram of the acausal SimpleCircuit model.

```
model SimpleCircuit
  Resistor R1(R=10);
  Capacitor C(C=0.01);
  Resistor R2(R=100);
  Inductor L(L=0.1);
  VsourceAC AC;
  Ground G;
equation
  connect(AC.p, R1.p);
  connect(R1.n, C.p);
  connect(C.n, AC.n);
  connect(R1.p, R2.p);
  connect(R2.n, L.p);
  connect(L.n, C.n);
  connect(AC.n, G.p);
end SimpleCircuit;
```

As a comparison, we show the same circuit modelled using causal block-oriented modelling depicted as a diagram in Figure 3.8. Here the physical topology is lost–the structure of the diagram has no simple correspondence to the structure of the physical circuit board.

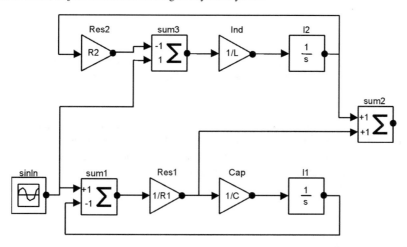

Fig. 3.8: The simple circuit model using causal block-oriented modelling with explicit signal flow.

This model is causal since the signal flow has been deduced and is clearly shown in the diagram. Even for this simple example the analysis to convert the intuitive physical model to a causal block-oriented model is nontrivial. Another disadvantage is that the resistor representations are context dependent. For example, the resistors R1 and R2 have different definitions, which makes reuse of model library components hard. Furthermore, such system models are usually hard to maintain since even small changes in the physical structure may result in large changes to the corresponding block-oriented system model.

3.6 The Modelica Software Component Model

For a long time, software developers have looked with envy on hardware system builders, regarding the apparent ease with which reusable hardware components are used to construct complicated systems. With software there seems too often to be a need or tendency to develop from scratch instead of reusing components. Early attempts at software components include procedure libraries, which unfortunately have too limited applicability and low flexibility. The advent of object-oriented programming has stimulated the development of software component frameworks such as CORBA, the Microsoft COM/DCOM component object model, and JavaBeans. These component models have considerable success in certain application areas, but there is still a long way to go to reach the level of reuse and component standardisation found in hardware industry.

The reader might wonder what all this has to do with Modelica. In fact, Modelica offers quite a powerful software component model that is on par with hardware component systems in flexibility and potential for reuse. The key to this increased flexibility is the fact that Modelica classes are based on equations. What is a software component model? It should include the following three items:

1. Components
2. A connection mechanism
3. A component framework

Components are connected via the connection mechanism, which can be visualised in connection diagrams. The component framework realizes components and connections, and ensures that communication works and constraints are maintained over the connections. For systems composed of acausal components the direction of data flow, i.e., the causality is automatically deduced by the compiler at composition time.

3.6.1 Components

Components are simply instances of Modelica classes. Those classes should have well-defined interfaces, sometimes called ports, in Modelica called connectors, for communication and coupling between a component and the outside world.

A component is modelled independently of the environment where it is used, which is essential for its reusability. This means that in the definition of the component including its equations, only local variables and connector variables can be used. No means of communication between a component and the rest of the system, apart from going via a connector, should be allowed. However, in Modelica access of component data via dot notation is also possible. A component may internally consist of other connected components, i.e., hierarchical modelling.

3.6.2 Connection Diagrams

Complex systems usually consist of large numbers of connected components, of which many components can be hierarchically decomposed into other components through several levels. To grasp this complexity, a pictorial representation of components and connections is quite important. Such graphic representation is available as connection diagrams, of which a schematic example is shown in Figure 3.9. We have earlier presented a connection diagram of a simple circuit in Figure 3.7.

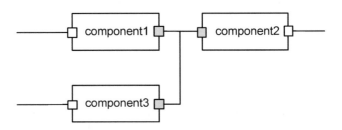

Fig. 3.9: Schematic picture of a connection diagram for components.

Each rectangle in the diagram example represents a physical component, e.g., a resistor, a capacitor, a transistor, a mechanical gear, a valve, etc. The connections represented by lines in the diagram correspond to real, physical connections. For example, connections can be realised by electrical wires, by the mechanical connections, by pipes for fluids, by heat exchange between components, etc. The connectors, i.e., interface points, are shown as small square dots on the rectangle in the diagram. Variables at such interface points define the interaction between the component represented by the rectangle and other components.

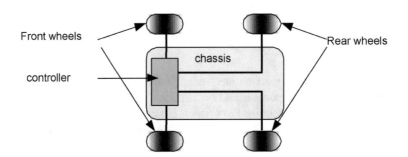

Fig. 3.10: A connection diagram for a simple car model.

A simple car example of a connection diagram for an application in the mechanical domain is shown in Figure 3.10.

The simple car model below includes variables for subcomponents such as wheels, chassis, and control unit. A "comment" string after the class name briefly describes the class. The wheels are connected to both the chassis and the controller. Connect-equations are present, but are not shown in this partial example.

```
class Car "A car class to combine car components"
  Wheel w1,w2,w3,w4 "Wheel one to four";
  Chassis chassis "Chassis";
  CarController controller "Car controller";
  ...
end Car;
```

3.6.3 Connectors and Connector Classes

Modelica connectors are instances of connector classes, which define the variables that are part of the communication interface that is specified by a connector. Thus, connectors specify external interfaces for interaction.

For example, Pin is a connector class that can be used to specify the external interfaces for electrical components (Figure 3.11) that have pins. The types Voltage and Current used within Pin are the same as Real, but with different associated units. From the Modelica language point of view, the types Voltage and Current are similar to Real, and are regarded as having equivalent types. Checking unit compatibility within equations is optional.

```
  type Voltage = Real(unit="V");
  type Current = Real(unit="A");
```

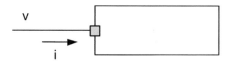

Fig. 3.11: A component with one electrical Pin connector.

The Pin connector class below contains two variables. The flow prefix on the second variable indicates that this variable represents a flow quantity, which has special significance for connections as explained in the next section.

```
  connector Pin
    Voltage v;
    flow Current i;
  end Pin;
```

3.6.3.1 Connections

Connections between components can be established between connectors of equivalent type. Modelica supports equation-based acausal connections, which means that connections are realised as equations. For acausal connections, the direction of data flow in the connection need not be known. Additionally, causal connections can be established by connecting a connector with an output attribute to a connector declared as input.

Two types of coupling can be established by connections depending on whether the variables in the connected connectors are potential (default), or declared using the flow prefix:

1. Equality coupling, for potential variables, according to Kirchhoff's first law.
2. Sum-to-zero coupling, for flow variables, according to Kirchhoff's current law.

For example, the keyword flow for the variable i of type Current in the Pin connector class indicates that all currents in connected pins are summed to zero, according to Kirchhoff's current law.

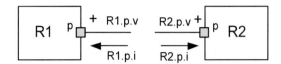

Fig. 3.12: Connecting two components that have electrical pins.

Connection equations are used to connect instances of connection classes. A connection equation connect(R1.p,R2.p), with R1.p and R2.p of connector class Pin, connects the two pins (Figure 3.12) so that they form one node. This produces two equations, namely:

```
R1.p.v = R2.p.v;
R1.p.i + R2.p.i = 0;
```

The first equation says that the voltages of the connected wire ends are the same. The second equation corresponds to Kirchhoff's second law, saying that the currents sum to zero at a node (assuming positive value while flowing into the component). The sum-to-zero equations are generated when the prefix flow is used. Similar laws apply to flows in piping networks and to forces and torques in mechanical systems.

We should also mention the concept of *implicit connections*, e.g. useful to model force fields, which is represented by the Modelica inner/outer construct.

3.6.3.2 Implicit Connections with inner/outer

So far we have focused on explicit connections between connectors where each connection is explicitly represented by a connect-equation and a corresponding line in a connection diagram. However, when modelling certain kinds of large models with many interacting components this approach becomes rather clumsy because of the large number of potential connections–a connection might be needed between each pair of components. This is especially true for system models involving *force fields*, which lead to a maximum of $n \times n$ connections between the n components influenced by the force field or $1 \times n$ connections between a central object and n components if inter-component interaction is neglected.

For the case of $1 \times n$ connections, instead of using a large number of explicit connections, Modelica provides a convenient mechanism for *implicit connections* between an object and n of its components through the inner and outer declaration prefixes.

A rather common kind of implicit interaction is where a *shared attribute* of a single environment object is *accessed* by a number of components within that environment. For example, we might have an environment incl uding house components, each accessing a shared environment temperature, or a circuit board environment with electronic components accessing the board temperature.

A Modelica environment-component example model along these lines is shown below, where a shared environment temperature variable T0 is declared as a *definition declaration* marked by the keyword inner. This declaration is implicitly accessed by the *reference declarations* of T0 marked by the prefix outer in the components comp1 and comp2.

```
model Environment
  inner
    Real T0; //Definition of actual environment temperature T0
    Component comp1, comp2; // Lookup match comp1.T0 = comp2.T0 = T0
  parameter
    Real k=1;
  equation
    T0 = sin(k*time);
end Environment;
```

```
model Component
  outer
    Real T0; // Reference to temperature T0 defined in the environments
      Real T;
  equation
    T = T0;
end Component;
```

3.6.3.3 Expandable Connectors for Information Buses

It is common in engineering to have so-called information buses with the purpose to transport information between various system components, e.g. sensors, actuators, control units. Some buses are even standardised (e.g., IEEE), but usually rather generic to allow many kinds of different components.

This is the key idea behind the expandable connector construct in Modelica. An expandable connector acts like an information bus since it is intended for connection to many kinds of components. To make this possible it automatically expands the expandable connector type to accommodate all the components connected to it with their different interfaces. If an element with a certain name and type is not present, it is added.

All fields in an expandable connector are seen as connector instances even if they are not declared as such, i.e., it is possible to connect to e.g. a Real variable.

Moreover, when two expandable connectors are connected, each is augmented with the variables that are only declared in the other expandable connector. This is repeated until all connected expandable connector instances have matching variables, i.e., each of the connector instances is expanded to be the union of all connector variables. If a variable appears as an input in one expandable connector, it should appear as a non-input in at least one other expandable connector instance in the connected set. The following is a small example:

```
expandable connector EngineBus
end EngineBus;

block Sensor
  RealOutput speed;
end Sensor;

block Actuator
  RealInput speed;
end Actuator;

model Engine
  EngineBus bus;
  Sensor sensor;
  Actuator actuator;
equation
  connect(bus.speed, sensor.speed); // provides the non-input
  connect(bus.speed, actuator.speed);
end Engine;
```

3.6.3.4 Stream Connectors

In thermodynamics with fluid applications where there can be bi-directional flows of matter with associated quantities, it turns out that the two basic variable types in a connector–potential variables and flow variables–are not sufficient to describe models that result in a numerically sound solution approach. Such applications typically have bi-directional flow of matter with convective transport of specific quantities, such as specific enthalpy and chemical composition

If we would use conventional connectors with flow and potential variables, the corresponding models would include nonlinear systems of equations with Boolean unknowns for the flow directions and singularities around zero flow. Such equation systems cannot be solved reliably in general. The model formulations can be simplified when formulating two different balance equations for the two possible flow directions. This is however not possible only using flow and potential variables.

This fundamental problem is addressed in Modelica by introducing a third type of connector variable, called *stream variable*, declared with the prefix stream. A stream variable describes a quantity that is carried by a flow variable, i.e., a purely convective transport phenomenon.

If at least one variable in a connector has the stream prefix, the connector is called a *stream connector* and the corresponding variable is called a *stream variable*. For example:

```
connector FluidPort ...
  flow Real m_flow "Flow of matter; m_flow>0 if flow into component";
  stream Real h_outflow "Specific variable in component if m_flow < 0"
end FluidPort

model FluidSystem ...
  FluidComponent m1, m2, ..., mN;
  FluidPort c1, c2, ..., cM;
equation connect(m1.c, m2.c); ...
  connect(m1.c, cM); ...
end FluidSystem;
```

3.6.4 Partial Classes

A common property of many electrical components is that they have two pins. This means that it is useful to define a "blueprint" model class, e.g., called TwoPin, that captures this common property. This is a *partial class* since it does not contain enough equations to completely specify its physical behaviour, and is therefore prefixed by the keyword partial. Partial classes are usually known as *abstract classes* in other object-oriented languages. Since a partial class is incomplete it cannot be used as a class to instantiate a data object.

```
partial class TwoPin8 "Superclass of elements with two electrical pins"
  Pin p, n;
  Voltage v;
  Current i;
equation
  v = p.v - n.v;
  0 = p.i + n.i;
  i = p.i;
end TwoPin;
```

This TwoPin class is referred to by the name Modelica.Electrical.Analog.Interfaces.OnePort in the Modelica standard library since this is the name used by electrical modelling experts. Here we use the more intuitive name TwoPin since the class is used for components with two physical ports and not one. The OnePort naming is more understandable if it is viewed as denoting composite ports containing two subports.

The TwoPin class has two pins, p and n, a quantity v that defines the voltage drop across the component, and a quantity i that defines the current into pin p, through the component, and out from pin n (Figure 3.13). It is useful to label the pins differently, e.g., p and n, and using graphics, e.g. filled and unfilled squares respectively, to obtain a well-defined sign for v and i although there is no physical difference between these pins in reality.

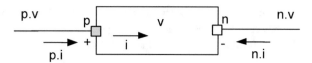

Fig. 3.13: Generic TwoPin class that describes the general structure of simple electrical components with two pins.

The equations define generic relations between quantities of simple electrical components. In order to be useful, a constitutive equation must be added that describes the specific physical characteristics of the component.

3.6.5 Reuse of Partial Classes

Given the generic partial class TwoPin, it is now straightforward to create the more specialised Resistor class by adding a constitutive equation:

```
R * i = v;
```

This equation describes the specific physical characteristics of the relation between voltage and current for a resistor (Figure 3.14).

Fig. 3.14: A resistor component.


```
class Resistor "Ideal electrical resistor"
  extends TwoPin;
  parameter Real R(unit="Ohm") "Resistance";
equation
    R*i = v;
end class
```

A class for electrical capacitors can also reuse TwoPin in a similar way, adding the constitutive equation for a capacitor (Figure 3.15).

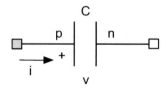

Fig. 3.15: A capacitor component.


```
class Capacitor "Ideal electrical capacitor"
  extends TwoPin;
  parameter Real C(Unit="F") "Capacitance";
equation
  C*der(v) = i;
end Capacitor;
```

During system simulation the variables i and v specified in the above components evolve as functions of time. The solver of differential equations computes the values of $v(t)$ and $i(t)$ (where t is time) such that $C.\dot{v}(t) = i(t)$ for all values of t, fulfilling the constitutive equation for the capacitor.

3.7 Component Library Design and Use

In a similar way as we previously created the resistor and capacitor components, additional electrical component classes can be created, forming a simple electrical component library that can be used for application models such as the SimpleCircuit model. Component libraries of reusable components are actually the key to effective modelling of complex systems.

Below, we show an example of designing a small library of electrical components needed for the simple circuit example, as well as the equations that can be extracted from these components.

Resistor

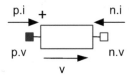

Fig. 3.16: Resistor component.

Four equations can be extracted from the resistor model depicted in Figure 3.14 and Figure 3.16. The first three originate from the inherited TwoPin class, whereas the last is the constitutive equation of the resistor.

```
0 = p.i + n.i
v = p.v - n.v
i = p.i
v = R*i
```

Capacitor

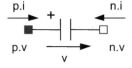

Fig. 3.17: Capacitor component.

The following four equations originate from the capacitor model depicted in Figure 3.15 and Figure 3.17, where the last equation is the constitutive equation for the capacitor.

```
0 = p.i + n.i;
v = p.v - n.v;
i = p.i;
i = C * der(v);
```

Inductor

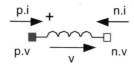

Fig. 3.18: Inductor component

The inductor class depicted in Figure 3.18 and shown below gives a model for ideal electrical inductors.

```
class Inductor "Ideal electrical inductor"
   extensd TwoPin;
   parameter Real L(unit="H") "Inductance";
equation
   v = L*der(i);
end Inductor;
```

These equations can be extracted from the inductor class, where the first three come from TwoPin as usual and the last is the constitutive equation for the inductor.

```
0 = p.i + n.i;
v = p.v - n.v;
i = p.i;
v = L * der(i);
```

Voltage Source

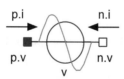

Fig. 3.19: Voltage source component VsourceAC, where $v(t) = VA \times sin(2 \times PI \times f \times time)$.

A classVsourceAC for the sin-wave voltage source to be used in our circuit example is depicted in Figure 3.19 and can be defined as below. This model as well as other Modelica models specify behaviour that evolves as a function of time. Note that the built-in predefined variable time is used. In order to keep the example simple the constant PI is explicitly declared even though it is usually imported from the Modelica standard library.

```
class VsourceAC "Sin-wave voltage source"
   extends TwoPin;
   parameter Voltage VA = 220 "Amplitude";
   parameter Real f(unit="Hz") = 50 "Frequency";
   constant Real PI = 3.141592653589793;
equation
   v = VA*sin(2*PI*f*time);
end VsourceAC;
```

In this TwoPin-based model, four equations can be extracted from the model, of which the first three are inherited from TwoPin:

```
0 = p.i + n.i;
v = p.v - n.v;
i = p.i;
v = VA*sin(2*PI*f*time);
```

Ground

Fig. 3.20: Ground component.

Finally, we define a class for ground points that can be instantiated as a reference value for the voltage levels in electrical circuits. This class has only one pin (Figure 3.20).

```
class Ground "Ground"
  Pin p;
equation
  p.v = 0;
end Ground;
```

A single equation can be extracted from the Ground class.

```
p.v = 0
```

3.7.1 The Simple Circuit Model

Having collected a small library of simple electrical components, we can now put together the simple electrical circuit shown previously and in Figure 3.21.

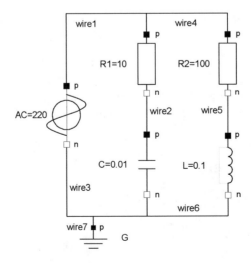

Fig. 3.21: The SimpleCircuit model.

The two resistor instances R1 and R2 are declared with modification equations for their respective resistance parameter values. Similarly, an instance C of the capacitor and an instance L of the inductor are declared with modifiers for capacitance and inductance respectively. The voltage source AC and the ground instance G have no modifiers. Connect-equations are provided to connect the components in the circuit.

```
class SimpleCircuit
  Resistor R1(R=10);
  Capacitor C(C=0.01);
  Resistor R2(R=100);
  Inductor L(L=0.1);
  VsourceAC AC;
  Ground G;
equation
  connect(AC.p, R1.p);   // Wire1, Capacitor circuit
  connect(R1.n, C.p);    //     Wire 2
  connect(C.n, AC.n);    //     Wire 3
  connect(R1.p, R2.p);   // Wire4, Inductor circuit
  connect(R2.n, L.p);    //     Wire 5
  connect(L.n, C.n);     //     Wire 6
  connect(AC.n, G.p);    // Wire7, Ground
end SimpleCircuit;
```

3.7.2 Arrays

An array is a collection of variables all of the same type. Elements of an array are accessed through simple Integer indexes ranging from a lower bound of 1 to an upper bound being the size of the respective dimension, or by Boolean indexes from false to true, or by enumeration indexes. An array variable can be declared by appending dimensions within square brackets after a class name, as in Java, or after a variable name, as in the C language. For example:

```
Real[3]     positionvector = {1,2,3};
Real[3,3]   identitymatrix = {{1,0,0}, {0,1,0}, {0,0,1}};
Real[3,3,3] arr3d;
```

This declares a three-dimensional position vector, a transformation matrix, and a three-dimensional array. Using the alternative syntax of attaching dimensions after the variable name, the same declarations can be expressed as:

```
Real   positionvector[3] = {1,2,3};
Real   identitymatrix[3,3] = {{1,0,0}, {0,1,0}, {0,0,1}};
Real   arr3d[3,3,3];
```

In the first two array declarations, declaration equations have been given, where the array constructor { } is used to construct array values for defining positionvector and identitymatrix. Indexing of an array A is written A[i,j,...], where 1 is the lower bound and size(A,k) is the upper bound of the index for the kth dimension. Submatrices can be formed by utilising the : notation for index ranges, for example, A[i1:i2, j1:j2], where a range i1:i2 means all indexed elements starting with i1 up to and including i2.

Array expressions can be formed using the arithmetic operators +, -, *, and /, since these can operate on either scalars, vectors, matrices, or (when applicable) multidimensional arrays with elements of type Real or Integer. The multiplication operator * denotes scalar product when used between vectors, matrix multiplication when used between matrices or between a matrix and a vector, and element-wise multiplication when used between an array and a scalar. As an example, multiplying positionvector by the scalar 2 is expressed by:

```
positionvector * 2
```

which gives the result:

```
{2, 4, 6}
```

In contrast to Java, arrays of dimensionality greater than 1 in Modelica are always rectangular as in Matlab or Fortran.

A number of built-in array functions are available, of which a few are shown in the table below.

transpose(A)	Permutes the first two dimensions of array A.
zeros(n1,n2,n3,...)	Returns an $n_1 \times n_2 \times n_3 \times \ldots$ zero-filled integer array.
ones(n1,n2,n3,...)	Returns an $n_1 \times n_2 \times n_3 \times \ldots$ one-filled integer array.
fill(s,n1,n2,n3, ...)	Returns the $n_1 \times n_2 \times n_3 \times \ldots$ array with all elements filled with the value of the scalar expression s.
min(A)	Returns the smallest element of array expression A.
max(A)	Returns the largest element of array expression A.
sum(A)	Returns the sum of all the elements of array expression A.

A scalar Modelica function of a scalar argument is automatically generalised to be applicable also to arrays element-wise. For example, if A is a vector of real numbers, then cos(A) is a vector where each element is the result of applying the function cos to the corresponding element in A. For example:

```
cos({ 1, 2, 3 }) = {cos(1), cos(2), cos(3)}
```

General array concatenation can be done through the array concatenation operator cat(k,A,B,C,...) that concatenates the arrays A,B,C,... along the kth dimension. For example, cat(1, {2,3}, {5,8,4}) gives the result {2,3,5,8,4}.

$$m \times n$$

The common special cases of concatenation along the first and second dimensions are supported through the special syntax forms [A;B;C;...] and [A,B,C,...] respectively. Both of these forms can be mixed. In order to achieve compatibility with Matlab array syntax, being a *de facto* standard, scalar and vector arguments to these special operators are promoted to become matrices before performing the concatenation. This gives the effect that a matrix can be constructed from scalar expressions by separating rows by semicolons and columns by commas. The example below creates an matrix:

[expr$_{11}$, expr$_{12}$, ... expr$_{1n}$;

expr$_{21}$, expr$_{22}$, ... expr$_{2n}$;

...

expr$_{m1}$, expr$_{m2}$, ... expr$_{mn}$]

It is instructive to follow the process of creating a matrix from scalar expressions using these operators. For example:

[1,2;

3,4]

First, each scalar argument is promoted to become a matrix, giving:

[{{1}}, {{2}};

{{3}}, {{4}}]

Since [... , ...] for concatenation along the second dimension has higher priority than [... ; ...], which concatenates along the first dimension, the first concatenation step gives:

[{{1, 2}};

{{3, 4}}]

$$2 \times 2$$

Finally, the row matrices are concatenated giving the desired matrix:

{{1, 2},

{3, 4}}

$$1 \times 1$$

The special case of just one scalar argument can be used to create a matrix. For example:

[1]

gives the matrix:

{{1}}

3.8 Algorithmic Constructs

Even though equations are eminently suitable for modelling physical systems and for a number of other tasks, there are situations where non-declarative algorithmic constructs are needed. This is typically the case for algorithms, i.e., procedural descriptions of how to carry out specific computations, usually consisting of a number of statements that should be executed in the specified order.

3.8.1 Algorithm Sections and Assignment Statements

In Modelica, algorithmic statements can occur only within algorithm sections, starting with the keyword algorithm. Algorithm sections may also be called algorithm equations, since an algorithm section can be viewed as a group of equations involving one or more variables, and can appear among equation sections. Algorithm sections are terminated by the appearance of one of the keywords equation, public, protected,algorithm, end, or annotation.

```
algorithm
   ...
   <statements>
   ...
   <some other keyword>
```

An algorithm section embedded among equation sections can appear as below, where the example algorithm section contains three assignment statements.

```
equation
   x = y*2;
   z = w;
algorithm
   x1 := z+x;
   x2 := y-5;
   x1 := x2+y;
equation
   u = x1+x2;
   ...
```

Note that the code in the algorithm section, sometimes denoted algorithm equation, uses the values of certain variables from outside the algorithm. These variables are so called *input variables* to the algorithm–in this example x, y, and z. Analogously, variables assigned values by the algorithm define the *outputs of the algorithm*–in this example x1 and x2. This makes the semantics of an algorithm section quite similar to a function with the algorithm section as its body, and with input and output formal parameters corresponding to inputs and outputs as described above.

3.8.2 Statements

In addition to assignment statements, which were used in the previous example, a few other kinds of "algorithmic" statements are available in Modelica: if-then-else statements, for-loops, while-loops, return statements, etc. The summation below uses both a while-loop and an if-statement, where size(a,1) returns the size of the first dimension of array a. The elseif- and else-parts of if-statements are optional.

```
sum := 0;
n := size(a,1);
while n>0 loop
  if a[n]>0 then
    sum := sum + a[n];
  elseif a[n] > -1 then
    sum := sum - a[n] -1;
  else
    sum := sum - a[n];
  end if;
  n := n-1;
end while;
```

Both for-loops and while-loops can be immediately terminated by executing a break-statement inside the loop. Such a statement just consists of the keyword break followed by a semicolon.

Consider once more the computation of the polynomial previously presented in the section on repetitive equation structures:

$$y \quad := \quad a[1] + a[2] * x + a[3] * x^{\wedge}1 + \ldots + a[n + 1] * x^{\wedge}n;$$

When using equations to model the computation of the polynomial it was necessary to introduce an auxiliary vector xpowers for storing the different powers of x. Alternatively, the same computation can be expressed as an algorithm including a for-loop as below. This can be done without the need for an extra vector–it is enough to use a scalar variable xpower for the most recently computed power of x.

```
algorithm
  y := 0;
  xpower := 1;
  for i in 1:n+1 loop
    y := y + a[i]*xpower;
    xpower := xpower*x;
  end for;
```

3.8.3 Functions

Functions are a natural part of any mathematical model. A number of mathematical functions like abs, sqrt, mod,sin, cos, exp, etc. are both predefined in the Modelica language and available in the Modelica standard math library Modelica.Math. The arithmetic operators +, -, *, / can be regarded as functions that are used through a convenient operator syntax. Thus it is natural to have user-defined mathematical functions in the Modelica language. The body of a Modelica function is an algorithm section that contains procedural algorithmic code to be executed when the function is called. Formal parameters are specified using the input keyword, whereas results are denoted using the output keyword. This makes the syntax of function definitions quite close to Modelica block class definitions.

Modelica functions are *mathematical functions*, i.e., without global side-effects and with no memory (except impure Modelica functions marked with the keyword impure). A Modelica function always returns the same results given the same arguments. Below we show the algorithmic code for polynomial evaluation in a function named polynomialEvaluator.

```
function polynomialEvaluator
  input Real a[:];      // Array, size defined at function call time
  input Real x := 1.0;  // Default value 1.0 for x
  output Real y;
protected
  Real xpower;
algorithm
  y := 0;
  xpower := 1;
  for i in 1:size(a,1) loop
    y := y + a[i]*xpower;
    xpower := xpower*x;
  end for;
end polynomialEvaluator;
```

Functions are usually called with positional association of actual arguments to formal parameters. For example, in the call below the actual argument {1,2,3,4} becomes the value of the coefficient vector a, and 21 becomes the value of the formal parameter x. Modelica function formal parameters are read-only, i.e., they may not be assigned values within the code of the function. When a function is called using positional argument association, the number of actual arguments and formal parameters must be the same. The types of the actual argument expressions must be compatible with the declared types of the corresponding formal parameters. This allows passing array arguments of arbitrary length to functions with array formal parameters with unspecified length, as in the case of the input formal parameter a in the polynomialEvaluator function.

```
p = polynomialEvaluator({1, 2, 3, 4}, 21);
```

The same call to the function polynomialEvaluator can instead be made using named association of actual arguments to formal parameters, as in the next example. This has the advantage that the code becomes more self-documenting as well as more flexible with respect to code updates.

For example, if all calls to the function polynomialEvaluator are made using named parameter association, the order between the formal parameters a and x can be changed, and new formal parameters with default values can be introduced in the function definitions without causing any compilation errors at the call sites. Formal parameters with default values need not be specified as actual arguments unless those parameters should be assigned values different from the defaults.

```
p = polynomialEvaluator(a={1, 2, 3, 4} , x=21);
```

Functions can have multiple results. For example, the function f below has three result parameters declared as three formal output parameters r1, r2, and r3.

```
function f
  input Real x;
  input Real y;
  output Real r1;
  output Real r2;
  output Real r3;
  ...
end f;
```

Within algorithmic code multiresult functions may be called only in special assignment statements, as the one below, where the variables on the left-hand side are assigned the corresponding function results.

```
(a, b, c) := f(1.0, 2.0);
```

In equations a similar syntax is used:

```
(a, b, c) = f(1.0, 2.0);
```

A function is returned from by reaching the end of the function or by executing a return-statement inside the function body.

3.8.4 Operator Overloading and Complex Numbers

Function and operator overloading allow several definitions of the same function or operator, but with a different set of input formal parameter types for each definition. This allows, e.g., to define operators such as addition, multiplication, etc., of complex numbers, using the ordinary + and * operators but with new definitions, or provide several definitions of a solve function for linear matrix equation solution for different matrix representations such as standard dense matrices, sparse matrices, symmetric matrices, etc.

In fact, overloading already exists predefined to a limited extent for certain operators in the Modelica language. For example, the plus (+) operator for addition has several different definitions depending on the data type:

1. 1+2 — means integer addition of two Integer constants giving an integer result, here 3.
2. 1.0+2.0 — means floating point number addition of two Real constants giving a floating-point number result, here 3.0.
3. "ab"+"2" — means string concatenation of two string constants giving a string result, here "ab2".
4. {1,2} +{3,4} — means integer vector addition of two integer constant vectors giving a vector result, here {4,6}.

Overloaded operators for user-defined data types can be defined using operator record and operator function declarations. Here we show part of a complex numbers data type example:

```
operator record Complex "Record defining a Complex number"

  Real re "Real part of complex number";
  Real im "Imaginary part of complex number";

  encapsulated operator 'constructor'
    import Complex;
    function fromReal
      input Real re;
      output Complex result = Complex(re=re, im=0.0);
        annotation(Inline=true);
    end fromReal;
  end 'constructor';

  encapsulated operator function '+'
    import Complex;
    input Complex c1;
    input Complex c2;
    output Complex result "Same as: c1 + c2";
      annotation(Inline=true);
    algorithm
      result := Complex(c1.re + c2.re, c1.im + c2.im);
    end '+';
end Complex;
```

In the above example, we start as usual with the real and imaginary part declarations of the re and im fields of the Complex operator record definition. Then comes a *constructor definition* fromReal with only one input argument instead of the two inputs of the default Complex constructor implicitly defined by the Complex record definition, followed by overloaded operator definition for '+'.

How can these definitions be used? Take a look at the following small example:

```
Real    a;
Complex b;
Complex c = a + b;   // Addition of Real number a and Complex number b
```

The interesting part is in the third line, which contains an addition a+b of a Real number a and a Complex number b. There is no built-in addition operator for complex numbers, but we have the above overloaded operator definition of '+' for two complex numbers. An addition of two complex numbers would match this definition right away in the lookup process.

However, in this case we have an addition of a real number and a complex number. Fortunately, the lookup process for overloaded binary operators can also handle this case if there is a constructor function in the Complex record definition that can convert a real number to a complex number. Here we have such a constructor called fromReal.

Note that the Complex is already predefined in a Modelica record type called Complex. This is distributed as part of the Modelica standard library (MSL) but defined at the top-level outside of MSL. It is referred to as Complex or .Complex. and is automatically preloaded by most tools, enabling direct usage without the need for any import statements (except in encapsulated classes). There is a library ComplexMath with operations on complex numbers.

The following are some examples of using the predefined type Complex:

```
Real a = 2;
Complex j = Modelica.ComplexMath.j;
Complex b = 2 + 3*j;
Complex c = (2*b + a)/b;
Complex d = Modelica.ComplexMath.sin(c);
Complex v[3] = {b/2, c, 2*d};
```

3.8.5 External Functions

It is possible to call functions defined outside of the Modelica language, implemented in C or Fortran. If no external language is specified the implementation language is assumed to be C. The body of an external function is marked with the keyword external in the Modelica external function declaration.

```
function log input
  Real x;
  output Real y;
  external
end log;
```

The external function interface supports a number of advanced features such as in–out parameters, local work arrays, external function argument order, explicit specification of row-major versus column-major array memory layout, etc. For example, the formal parameter Ares corresponds to an in–out parameter in the external function leastSquares below, which has the value A as input default and a different value as the result. It is possible to control the ordering and usage of parameters to the function external to Modelica. This is used below to explicitly pass sizes of array dimensions to the Fortran routine called dgels. Some old-style Fortran routines like dgels need work arrays, which is conveniently handled by local variable declarations after the keyword protected.

```
function leastSquares "Solves a linear least squares problem" input
    Real A[:,:];
  input Real B[:,:];
  output Real Ares[size(A,1),size(A,2)] := A;
  // Factorization is returned in Ares for later use
  output Real x[size(A,2),size(B,2)];
protected
  Integer lwork = min(size(A,1),size(A,2)) +
                  max(max(size(A,1),size(A,2)),size(B,2))*32;
  Real work[lwork];
  Integer info;
  String transposed="NNNN"; // Workaround for passing CHARACTER data to
                            // Fortran routine
  external "FORTRAN 77"
```

```
        dgels(transposed, 100, size(A,1), size(A,2), size(B,2), Ares,
              size(A,1), B, size(B,1), work, lwork, info);
  end leastSquares;
```

3.8.6 Algorithms Viewed as Functions

The function concept is a basic building block when defining the semantics or meaning of programming language constructs. Some programming languages are completely defined in terms of mathematical functions. This makes it useful to try to understand and define the semantics of algorithm sections in Modelica in terms of functions. For example, consider the algorithm section below, which occurs in an equation context:

```
algorithm
  y := x;
  z := 2*y;
  y := z+y;
  ...
```

This algorithm can be transformed into an equation and a function as below, without changing its meaning. The equation equates the output variables of the previous algorithm section with the results of the function f. The function f has the inputs to the algorithm section as its input formal parameters and the outputs as its result parameters. The algorithmic code of the algorithm section has become the body of the function f.

```
  (y,z) = f(x);
  ...
function f
  input Real x;
  output Real y,z;
algorithm
  y := x;
  z := 2*y;
  y := z+y;
end f;
```

3.9 Discrete Event and Hybrid modelling

Macroscopic physical systems in general evolve continuously as a function of time, obeying the laws of physics. This includes the movements of parts in mechanical systems, current and voltage levels in electrical systems, chemical reactions, etc. Such systems are said to have continuous dynamics.

On the other hand, it is sometimes beneficial to make the approximation that certain system components display discrete behaviour, i.e., changes of values of system variables may occur instantaneously and discontinuously at specific points in time.

In the real physical system the change can be very fast, but not instantaneous. Examples are collisions in mechanical systems, e.g., a bouncing ball that almost instantaneously changes direction, switches in electrical circuits with quickly changing voltage levels, valves and pumps in chemical plants, etc. We talk about system components with discrete-time dynamics. The reason to make the discrete approximation is to simplify the mathematical model of the system, making the model more tractable and usually speeding up the simulation of the model several orders of magnitude.

For this reason, it is possible to have variables in Modelica models of *discrete-time variability*, i.e., the variables change value only at specific points in time, so-called *events*. There are two kinds of discrete-time variables: *unclocked discrete-time variables* which keep their values constant between events and *clocked discrete-time variables* which are undefined between associated clock-tick events, see Figure 3.22. Examples of discrete-time variables are Real variables declared with the prefix discrete, clock variables and clocked variables, or Integer, Boolean, and enumeration variables which are discrete-time by default and cannot be continuous-time.

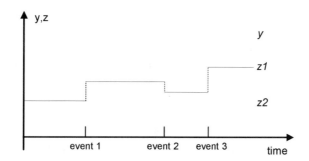

Fig. 3.22: Discrete-time variables $z1$ and $z2$ change value only at event instants, whereas continuous-time variables like y may change value both between and at events. An unclocked discrete-time variable $z1$ is defined also between events, whereas a clocked discrete-time variable $z2$ is defined only at associated clock-tick events.

Since the discrete-time approximation can only be applied to certain subsystems, we often arrive at system models consisting of interacting continuous and discrete components. Such a system is called a *hybrid system* and the associated modelling techniques *hybrid modelling*. The introduction of hybrid mathematical models creates new difficulties for their solution, but the disadvantages are far outweighed by the advantages.

Modelica provides four kinds of constructs for expressing hybrid models: conditional expressions or equations to describe discontinuous and conditional models, when-equations to express equations that are valid only at discontinuities (e.g., when certain conditions become true), clocked synchronous constructs, and clocked state machines. For example, if-then-else conditional expressions allow modelling of phenomena with different expressions in different operating regions, as for the equation describing a limiter below.

```
y = if v > limit then limit else v;
```

A more complete example of a conditional model is the model of an ideal diode. The characteristic of a real physical diode is depicted in Figure 3.23, and the ideal diode characteristic in parameterized form is shown in Figure 3.24.

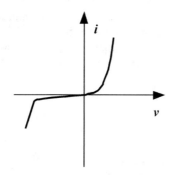

Fig. 3.23: Real diode characteristic.

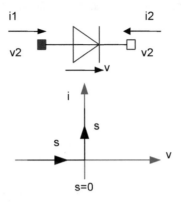

Fig. 3.24: Ideal diode characteristic.

Since the voltage level of the ideal diode would go to infinity in an ordinary voltage-current diagram, a parameterised description is more appropriate, where both the voltage v and the current i, same as i1, are functions of the parameter s. When the diode is off no current flows and the voltage is negative, whereas when it is on there is no voltage drop over the diode and the current flows.

```modelica
model Diode "Ideal diode"
  extends TwoPin;
  Real s;
  Boolean off;
equation
  off = s < 0;
  if off
    then v=s;
    else v=0;    // conditional equations
  end if;
  i = if off then 0 else s;
end Diode;
```

When-equations have been introduced in Modelica to express instantaneous equations, i.e., equations that are valid only at certain points in time that, for example, occur at discontinuities when specific conditions become true, so-called *events*. The syntax of when-equations for the case of a vector of conditions is shown below. The equations in the when-equation are activated when at least one of the conditions becomes true, and remain activated only for a time instant of zero duration. A single condition is also possible.

```modelica
when {condition1, condition2, ...} then
    <equations>
end when;
```

A bouncing ball is a good example of a hybrid system for which the when-equation is appropriate when modelled. The motion of the ball is characterised by the variable height above the ground and the vertical velocity. The ball moves continuously between bounces, whereas discrete changes occur at bounce times, as depicted in Figure 3.25. When the ball bounces against the ground its velocity is reversed. An ideal ball would have an elasticity coefficient of1 and would not lose any energy at a bounce. A more realistic ball, as the one modelled below, has an elasticity coefficient of 0.9, making it keep 90 percent of its speed after the bounce.

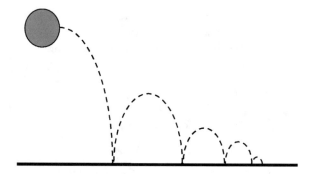

Fig. 3.25: Trajectory of bouncing ball.

The bouncing ball model contains the two basic equations of motion relating height and velocity as well as the acceleration caused by the gravitational force. At the bounce instant the velocity is suddenly reversed and slightly decreased, i.e., velocity (after bounce) = -c*velocity (before bounce), which is accomplished by the special reinit syntactic form of instantaneous equation for reinitialisation: reinit(velocity,-c*pre(velocity)), which in this case reinitialises the velocity variable.

```
model BouncingBall "Simple model of a bouncing ball"
  constant Real g = 9.81 "Gravity constant";
  parameter Real c = 0.9 "Coefficient of restitution";
  parameter Real radius=0.1 "Radius of the ball";
  Real height(start = 1)    "Height of the ball center";
  Real velocity(start = 0)  "Velocity of the ball";
equation
  der(height) = velocity;
  der(velocity) = -g;
  when height <= radius then
    reinit(velocity,-c*pre(velocity));
  end when;
end BouncingBall;
```

Note that the equations within a when-equation are active only during the instant in time when the condition(s) of the when-equation become true, whereas the conditional equations within an if-equation are active as long as the condition of the if-equation is true.

If we simulate this model long enough, the ball will fall through the ground. This strange behaviour of the simulation, called shattering, or the Zeno effect is due to the limited precision of floating point numbers together with the event detection mechanism of the simulator, and occurs for some (unphysical) models where events may occur infinitely close to each other. The real problem in this case is that the model of the impact is not realistic the law new_velocity = -c * velocity does not hold for very small velocities. A simple fix is to state a condition when the ball falls through the ground and then switch to an equation stating that the ball is lying on the ground. A better but more complicated solution is to switch to a more realistic material model.

Synchronous Clocks and State Machines

Starting from Modelica version 3.3 two main groups of discrete-time and hybrid modelling features are available in the language: built-in synchronous clocks with constructs for clock-based modelling and synchronous clock-based state-machines.

As depicted in Figure 3.22, clocks and clocked discrete-time variable values are only defined at clock tick events for an associated clock and are undefined between clock ticks.

Synchronous clock-based modelling gives additional advantages for real-time and embedded system modelling in terms of higher performance and avoidance and/or detection of certain kinds of errors.

To give a very simple example, in the model below the clock variable clk is defined with an interval of 0.2 seconds. This clock is used for sampling the built-in time variable, with result put in x and defined only at clock ticks. Finally, the variable y is defined also between clock ticks by the use of the hold operator.

```
model BasicClockedSynchronous
  Clock clk "A clock variable";
  discrete Real x "Clocked discrete-time variable";
  Real y "Unclocked discrete-time variable";
equation
  clk = Clock(0.2); // Clock with an interval of 0.2 seconds
  x = sample(time,clk); // Sample the continuous time at clock ticks
  y = hold(x)+1; // Hold value is defined also between clock ticks
end BasicClockedSynchronous;
```

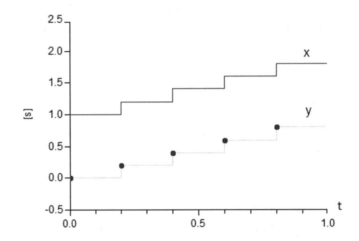

Fig. 3.26: Simulation of model BasicClockedSynchronous showing clocked discrete-time variable x and unclocked discrete-time variable y.

3.10 Modularity Facilites

3.10.1 Packages

Name conflicts are a major problem when developing reusable code, for example, libraries of reusable Modelica classes and functions for various application domains. No matter how carefully names are chosen for classes and variables it is likely that someone else is using some name for a different purpose. This problem gets worse if we are using short descriptive names since such names are easy to use and therefore quite popular, making them quite likely to be used in another person's code.

A common solution to avoid name collisions is to attach a short prefix to a set of related names, which are grouped into a package. For example, all names in the X-Windows toolkit have the prefix Xt, and most definitions in the Qt GUI library are prefixed with Q. This works reasonably well for a small number of packages, but the likelihood of name collisions increases as the number of packages grows.

Many programming languages, e.g., Java and Ada as well as Modelica provide a safer and more systematic way of avoiding name collisions through the concept of *package*. A package is simply a container or name space for names of classes, functions, constants, and other allowed definitions. The package name is prefixed to all definitions in the package using standard dot notation. Definitions can be *imported* into the name space of a package.

Modelica has defined the package concept as a restriction and enhancement of the class concept. Thus, inheritance could be used for importing definitions into the name space of another package. However, this gives conceptual modelling problems since inheritance for import is not really a package specialisation. Instead, an import language construct is provided for Modelica packages. The type name Voltage together with all other definitions in Modelica.SIunits is imported in the example below, which makes it possible to use it without prefix for declaration of the variable v. By contrast, the declaration of the variable i uses the fully qualified nameModelica.SIunits.Ampere of the type Ampere, even though the short version also would have been possible. The fully qualified long name for Ampere can be used since it is found using the standard nested lookup of the Modelica standard library placed in a conceptual top-level package.

```
package MyPack
  import MyTypes.*;
  import Modelica.SIunits.*;

  class Foo;
    Voltage v;
    Modelica.SIunits.Ampere i;
  end Foo;
end MyPack;
```

Importing definitions from one package into another class as in the above example has the drawback that the introduction of new definitions into a package may cause name clashes with definitions in packages using that package. For example, if a type definition named Voltage is introduced into the package MyTypes, a compilation error would arise in the package MyPack since it is also defined in the package Modelica.SIunits.

An alternative solution to the short-name problem that does not have the drawback of possible compilation errors when new definitions are added to libraries, is introducing short convenient name aliases for prefixes instead of long package prefixes. This is possible using the renaming form of import statement as in the package MyPack below, where the package name SI is introduced instead of the much longer Modelica.SIunits.

Another disadvantage with the above package is that the Ampere type is referred to using standard nested lookup and not via an explicit import statement. Thus, in the worst case we may have to do the following in order to find all such dependencies and the declarations they refer to:

1. Visually scan the whole source code of the current package, which might be large.
2. Search through all packages containing the current package, i.e., higher up in the package hierarchy, since standard nested lookup allows used types and other definitions to be declared anywhere above the current position in the hierarchy.

Instead, a *well-designed package* should state all its dependencies *explicitly* through import statements which are easy to find. We can create such a package, e.g., the package MyPack below, by adding the prefix encapsulated in front of the package keyword. This prevents nested lookup outside the package boundary (unfortunately with one exception, names starting with a dot ., ensuring that all dependencies on other packages outside the current package have to be explicitly stated as import statements. This kind of encapsulated package represents an independent unit of code and corresponds more closely to the package concept found in many other programming languages, e.g., Java or Ada.

```
encapsulated package MyPack
  import SI = Modelica.SIunits;
  import Modelica;

  class Foo;
   SI.Voltage v;
    Modelica.SIunits.Ampere i;
  end Foo;
  ...
end MyPack;
```

3.10.2 Annotations

A Modelica annotation is extra information associated with a Modelica model. This additional information is used by Modelica environments, e.g., for supporting documentation or graphical model editing. Most annotations do not influence the execution of a simulation, i.e., the same results should be obtained even if the annotations are removed–but there are exceptions to this rule. The syntax of an annotation is as follows:

```
annotation (annotation_elements)
```

where annotation_elements is a comma-separated list of annotation elements that can be any kind of expression compatible with the Modelica syntax. The following is a resistor class with its associated annotation for the icon representation of the resistor used in the graphical model editor:

```
model Resistor
  ...
  annotation(Icon(coordinateSystem(preserveAspectRatio = true,
  extent = {{-100,-100},{100,100}}, grid = {2,2}),
  graphics = {
    Rectangle(extent = {{-70,30},{70,-30}}, lineColor = {0,0,255},
    fillColor = {255,255,255},
      fillPattern = FillPattern.Solid),
    Line(points = {{-90,0},{-70,0}}, color = {0,0,255}),
    Line(points = {{70,0},{90,0}}, color ={0,0,255}),
    Text(extent = {{-144,-40},{142,-72}}, lineColor = {0,0,0},
    textString = "R=\%R"),
  ...
  ));
end Resistor;
```

Another example is the predefined annotation choices used to generate menus for the graphical user interface:

```
annotation(choices(choice=1 "P", choice=2 "PI", choice=3 "PID"));
```

The annotation Documentation is used for model documentation, as in this example for the Capacitor model:

```
  annotation(
    Documentation(info = "<HTML>
    <p>
      The linear capacitor connects the branch voltage <i>v</i> with
        the
      branch current <i>i</i> by <i>i = C * dv/dt</i>.
      The Capacitance <i>C</i> is allowed to be positive, zero, or
        negative.
    </p>
    </HTML>
  )
```

The external function annotation arrayLayout can be used to explicitly give the layout of arrays, e.g., if it deviates from the defaults rowMajor and columnMajor order for the external languages C and Fortran 77 respectively.

This is one of the rare cases of an annotation influencing the simulation results, since the wrong array layout annotation obviously will have consequences for matrix computations. An example:

```
annotation(arrayLayout = "columnMajor");
```

3.10.3 Naming Conventions

You may have noticed a certain style of naming classes and variables in the examples in this chapter. In fact, certain naming conventions, described below, are being adhered to. These naming conventions have been adopted in the Modelica standard library, making the code more readable and somewhat reducing the risk for name conflicts. The naming conventions are largely followed in the examples in this book and are recommended for Modelica code in general:

- Type and class names (but usually not functions) always start with an uppercase letter, e.g., Voltage.
- Variable names start with a lowercase letter, e.g., body, with the exception of some one-letter names such as T for temperature.
- Names consisting of several words have each word capitalised, with the initial word subject to the above rules, e.g., ElectricCurrent and bodyPart.
- The underscore character is only used at the end of a name, or at the end of a word within a name, often to characterise lower or upper indices, e.g., body_low_up.
- Preferred names for connector instances in (partial) models are p and n for positive and negative connectors in electrical components, and name variants containing a and b, e.g., flange_a and flange_b, for other kinds of otherwise-identical connectors often occurring in two-sided components.

For more details on naming conventions see the documentation inside the package.

3.11 Modelica Standard Library

Much of the power of modelling with Modelica comes from the ease of reusing model classes. Related classes in particular areas are grouped into packages to make them easier to find.

A special package, called Modelica, is a standardised predefined package that together with the Modelica Language is developed and maintained by the Modelica Association. This package is also known as the *Modelica Standard Library*, abbreviated MSL. It provides constants, types, connector classes, partial models, and model classes of components from various application areas, which are grouped into sub-packages of the Modelica package, known as the Modelica standard libraries.

The following is a subset of the growing set of Modelica standard libraries currently available:

- Modelica.Constants – Common constants from mathematics, physics, etc.
- Modelica.Icons – Graphical layout of icon definitions used in several packages.
- Modelica.Math – Definitions of common mathematical functions.
- Modelica.SIUnits – Type definitions with SI standard names and units.
- Modelica.Electrical – Common electrical component models.
- Modelica.Blocks – Input/output blocks for use in block diagrams.
- Modelica.Mechanics.Translational – 1D mechanical translational components.
- Modelica.Mechanics.Rotational – 1D mechanical rotational components.
- Modelica.Mechanics.MultiBody – MBS library–3D mechanical multibody models.
- Modelica.Thermal – Thermal phenomena, heat flow, etc. components.
- ...

Additional libraries are available in application areas such as thermodynamics, hydraulics, power systems, data communication, etc.

The Modelica Standard Library can be used freely under an open source license for both noncommercial and commercial purposes. The full documentation as well as the source code of these libraries appear at the Modelica web site.

So far the models presented have been constructed of components from single-application domains. However, one of the main advantages with Modelica is the ease of constructing multidomain models simply by connecting components from different application domain libraries. The DC (direct current) motor depicted in Figure 3.27 is one of the simplest examples illustrating this capability.

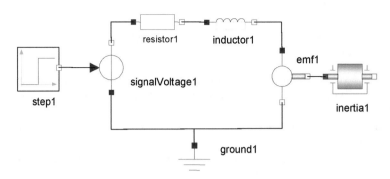

Fig. 3.27: A multidomain DCMotorCircuit model with mechanical, electrical, and signal block components.

This particular model contains components from the three domains, mechanical, electrical, and signal blocks, corresponding to the libraries Modelica.Mechanics, Modelica.Electrical, and Modelica.Blocks.

Model classes from libraries are particularly easy to use and combine when using a graphical model editor, as depicted in Figure 3.28, where the DC-motor model is being constructed. The left window shows theModelica.Mechanics.Rotational library, from which icons can be dragged and dropped into the central window when performing graphic design of the model.

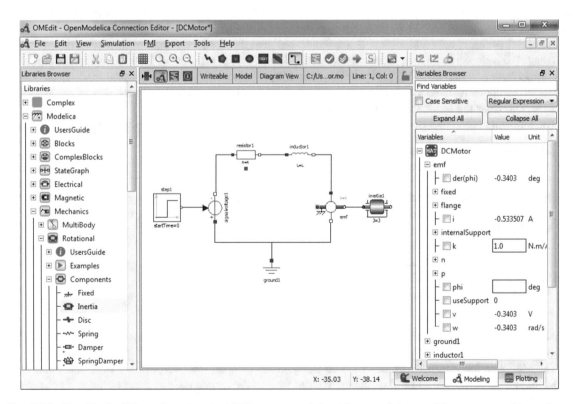

Fig. 3.28: Graphical editing of an electrical DC-motor model, with part of the model component hierarchy at the right and icons of the Modelica standard library in the left window.

3.12 Implementation and Execution of Modelica

In order to gain a better understanding of how Modelica works it is useful to take a look at the typical process of translation and execution of a Modelica model, which is sketched in Figure 3.29. Note that this is the typical process used by most Modelica tools. There are alternative ways of executing Modelica, some of which are currently at the research stage, that often keep more of the object-oriented structure at the later stages of translation and execution.

First the Modelica source code is parsed and converted into an internal representation, usually an abstract syntax tree. This representation is analysed, type checking is done, classes are inherited and expanded, modifications and instantiations are performed, connect-equations are converted to standard equations, etc. The result of this analysis and translation process is a flat set of equations, constants, variables, and function definitions. No trace of the object-oriented structure remains apart from the dot notation within names.

After flattening, all of the equations are topologically sorted according to the data-flow dependencies between the equations. In the case of general differential algebraic equations (DAEs), this is not just sorting, but also manipulation of the equations to convert the coefficient matrix into block lower triangular form, a so-called BLT transformation. Then an optimiser module containing algebraic simplification algorithms, symbolic index reduction methods, etc., eliminates most equations, keeping only a minimal set that eventually will be solved numerically, see also the chapter in this book about numerical approximation methods.

Then independent equations in explicit form are converted to assignment statements. This is possible since the equations have been sorted and an execution order has been established for evaluation of the equations in conjunction with the iteration steps of the numeric solver. If a strongly connected set of equations appears, this set is transformed by a symbolic solver, which performs a number of algebraic transformations to simplify the dependencies between the variables. It can sometimes solve a system of differential equations if it has a symbolic solution. Finally, C code is generated, and linked with a numeric equation solver that solves the remaining, drastically reduced, equation system.

The approximations to initial values are taken from the model definition or are interactively specified by the user. If necessary, the user also specifies the parameter values. A numeric solver for differential-algebraic equations (or in simple cases for ordinary differential equations) computes the values of the variables during the specified simulation interval $[t_0, t_1]$. The result of the dynamic system simulation is a set of functions of time, such as R2.v(t) in the simple circuit model. Those functions can be displayed as graphs and/or saved in a file.

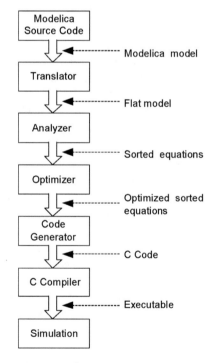

Fig. 3.29: The typical stages of translating and executing a Modelica model.

In most cases (but not always) the performance of generated simulation code (including the solver) is similar to handwritten C code. Often Modelica is more efficient than straightforwardly written C code, because additional opportunities for symbolic optimization are used by the system, compared to what a human programmer can manually handle.

3.12.1 Hand Translation of the Simple Circuit Model

Let us return once more to the simple circuit model, previously depicted in Figure 3.7, but for the reader's convenience also shown below in Figure 3.30. It is instructive to translate this model by hand, in order to understand the process.

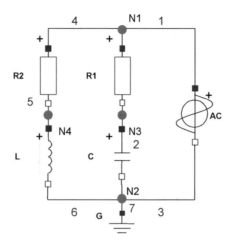

Fig. 3.30: The SimpleCircuit model once more, with explicitly labeled connection nodes N1, N2, N3, N4, and wires 1 to 7.

Classes, instances and equations are translated into a flat set of equations, constants, and variables (see the equations in Table 3.1), according to the following rules:

1. For each class instance, add one copy of all equations of this instance to the total differential algebraic equation (DAE) system or ordinary differential equation system (ODE)–both alternatives can be possible, since a DAE in a number of cases can be transformed into an ODE.
2. For each connection between instances within the model, add connection equations to the DAE system so that potential variables are set equal and flow variables are summed to zero.

The equation v=p.v-n.v is defined by the class TwoPin. The Resistor class inherits the TwoPin class, including this equation. The SimpleCircuit class contains a variable R1 of type Resistor. Therefore, we include this equation instantiated for R1 as R1.v=R1.p.v-R1.n.v into the system of equations.

The wire labeled 1 is represented in the model as connect(AC.p, R1.p). The variables AC.p and R1.p have type Pin. The variable v is a *potential* variable representing voltage potential. Therefore, the equality equation R1.p.v=AC.p.v is generated. Equality equations are always generated when potential variables are connected.

Notice that another wire (labeled 4) is attached to the same pin, R1.p. This is represented by an additional connect-equation: connect(R1.p.R2.p). The variable i is declared as a flow variable. Thus, the equation AC.p.i+R1.p.i+R2.p.i=0 is generated. Zero-sum equations are always generated when connecting flow variables, corresponding to Kirchhoff's second law.

The complete set of equations (see Table 3.1) generated from the SimpleCircuit class consists of 32 differential-algebraic equations. These include 32 variables, as well as time and several parameters and constants.

AC	0 = AC.p.i+AC.n.i AC.v = Ac.p.v-AC.n.v AC.i = AC.p.i AC.v = AC.VA∗ sin(2∗ AC.PI∗ AC.f∗ time);	L	0 = L.p.i+L.n.i L.v = L.p.v-L.n.v L.i = L.p.i L.v = L.L∗ der(L.i)
R1	0 = R1.p.i+R1.n.i R1.v = R1.p.v-R1.n.v R1.i = R1.p.i R1.v = R1.R∗ R1.i	G	G.p.v = 0

R2	0 = R2.p.i+R2.n.i
	R2.v = R2.p.v-R2.n.v
	R2.i = R2.p.i
	R2.v = R2.R∗ R2.i
		wires	R1.p.v = AC.p.v // wire 1
			C.p.v = R1.n.v // wire 2
			AC.n.v = C.n.v // wire 3
			R2.p.v = R1.p.v // wire 4
			L.p.v = R2.n.v // wire 5
			L.n.v = C.n.v // wire 6
			G.p.v = AC.n.v // wire 7

C	0 = C.p.i+C.n.i
	C.v = C.p.v-C.n.v
	C.i = C.p.i
	C.i = C.C∗ der(C.v)
		flow
		at
		node
			0 = AC.p.i+R1.p.i+R2.p.i // N1
			0 = C.n.i+G.p.i+AC.n.i+L.n.i // N2
			0 = R1.n.i + C.p.i // N3
			0 = R2.n.i + L.p.i // N4

Table 3.1: The equations extracted from the simple circuit model–an implicit DAE system.

Table 3.2 gives the 32 variables in the system of equations, of which 30 are algebraic variables since their derivatives do not appear. Two variables, C.v and L.i, are dynamic variables since their derivatives occur in the equations. In this simple example the dynamic variables are state variables, since the DAE reduces to an ODE.

R1.p.i	R1.n.i	R1.p.v	R1.n.v	R1.v
R1.i	R2.p.i	R2.n.i	R2.p.v	R2.n.v
R2.v	R2.i	C.p.i	C.n.i	C.p.v
C.n.v	C.v	C.i	L.p.i	L.n.i
L.p.v	L.n.v	L.v	L.i	AC.p.i
AC.n.i	AC.p.v	AC.n.v	AC.v	AC.i
G.p.i	G.p.v			

Table 3.2: The variables extracted from the simple circuit model.

3.12.2 Transformation to State Space Form

The implicit differential algebraic system of equations (DAE system) in Table 3.1 should be further transformed and simplified before applying a numerical solver. The next step is to identify the kind of variables in the DAE system. We have the following four groups:

1. All constant variables which are model parameters, thus easily modified between simulation runs and declared with the prefixed keyword parameter, are collected into a parameter vector p. All other constants can be replaced by their values, thus disappearing as named constants.
2. Variables declared with the input attribute, i.e., prefixed by the input keyword, that appears in instances at the highest hierarchical level, are collected into an input vector u.
3. Variables whose derivatives appear in the model (dynamic variables), i.e., the der( ) operator is applied to those variables, are collected into a state vector x.
4. All other variables are collected into a vector y of algebraic variables, i.e., their derivatives do not appear in the model.

For our simple circuit model these four groups of variables are the following:

p = { R1.R, R2.R, C.C, L.L, AC.VA, AC.f }

u = { AC.v }

$x = \{\ \text{C.v, L.i}\ \}$

$y = \{\ \text{R1.p.i, R1.n.i, R1.p.v, R1.n.v, R1.v, R1.i, R2.p.i, R2.n.i, R2.p.v, R2.n.v, R2.v, R2.i, C.p.i, C.n.i, C.p.v, C.n.v,}$
$\text{C.i, L.n.i, L.p.v, L.n.v, L.v, AC.p.i, AC.n.i, AC.p.v, AC.n.v, AC.i, AC.v, G.p.i, G.p.v}\ \}$

We would like to express the problem as the smallest possible ordinary differential equation (ODE) system (in the general case a DAE system) and compute the values of all other variables from the solution of this minimal problem. The system of equations should preferably be in an explicit state space form as below.

$$\dot{x} = f(x,t) \tag{3.12}$$

That is, the derivative $\dot{x}$ with respect to time of the state vector x is equal to a function of the state vector x and time. Using an iterative numerical solution method for this ordinary differential equation system, at each iteration step, the derivative of the state vector is computed from the state vector at the current point in time.

For the simple circuit model we have the following:

$$\dot{x} = \{C.v, L.i\}, u = \{AC.v\} \quad \text{(with constants: R1.R, R2.R, C.C, L.L, AC.VA, AC.f, AC.PI)} \tag{3.13}$$

$$\dot{x} = \{\mathbf{der}(C.v), \mathbf{der}(L.i)\} \tag{3.14}$$

3.12.3 Solution Method

We will use an iterative numerical solution method. First, assume that an estimated value of the state vector $x = \{C.v, L.i\}$ is available at $t = 0$ when the simulation starts. Use a numerical approximation for the derivative $\dot{x}$ (i.e. $\mathbf{der}(x)$) at time t, e.g.:

$$\mathbf{der}(x) = (x_{t+h} - x_t)/h \tag{3.15}$$

giving an approximation of x at time $t + h$:

$$x_{t+h} = x_t + \mathbf{der}(x) \times h \tag{3.16}$$

In this way, the value of the state vector x is computed one step ahead in time for each iteration, provided der(x) can be computed at the current point in simulated time. However, the derivative der(x) of the state vector can be computed from $\dot{x} = f(x,t)$, i.e., by selecting the equations involving der(x), and algebraically extracting the variables in the vector x in terms of other variables, as below:

$$\mathbf{der}(C.v) = C.i/C.C \tag{3.17}$$

$$\mathbf{der}(L.i) = L.v/L.L \tag{3.18}$$

Other equations in the DAE system are needed to calculate the unknowns C.i and L.v in the above equations. Starting with C.i, using a number of different equations together with simple substitutions and algebraic manipulations, we derive equations (3.19) through (3.21) below.

$C.i = R1.v/R1.R$
$$\text{using: } C.i = C.p.i = -R1.n.i = R1.p.i = R1.i = R1.v/R1.R \tag{3.19}$$

$R1.v = R1.p.v - R1.n.v = R1.p.v - C.v$
$$\text{using: } R1.n.v = C.p.v = C.v + C.n.v = C.v + AC.n.v = C.v + G.p.v = C.v + 0 = C.v \tag{3.20}$$

$R1.p.v = AC.p.v = AC.VA * sin(2 * AC.f * AC.PI * t)$

using: $AC.p.v = AC.v + AC.n.v = AC.v + G.p.v = AC.VA \times sin(2 \times AC.f \times AC.PI \times t) + 0$ (3.21)

In a similar fashion, we derive equations (3.22) and (3.23) below:

$L.v = L.p.v - L.n.v = R1.p.v - R2.v$

using: $L.p.v = R2.n.v = R1.p.v - R2.v$ and $L.n.v = C.n.v = AC.n.v = G.p.v = 0$ (3.22)

$R2.v = R2.R \times L.p.i$

using: $R2.v = R2.R \times R2.i = R2.R \times R2.p.i = R2.R \times (-R2.n.i) = R2.R \times L.p.i = R2.R \times L.i$ (3.23)

Collecting the five equations together:

$$C.i = R1.v/R1.R$$
$$R1.v = R1.p.v - C.v$$
$$R1.p.v = AC.VA \times sin(2 \times AC.f \times AC.PI \times t)$$
$$L.v = R1.p.v - R2.v$$
$$R2.v = R2.R \times L.i$$

(3.24)

By sorting the equations Eqs. 3.24 in data-dependency order, and converting the equations to assignment statements–this is possible since all variable values can now be computed in order–we arrive at the following set of assignment statements to be computed at each iteration, given values of C.v, L.i, and t at the same iteration:

```
R2.v   := R2.R * L.i;
R1.p.v := AC.VA * sin(2 * AC.f * AC.PI * time);
L.v    := R1.p.v - R2.v;
R1.v   := R1.p.v - C.v;
C.i    := R1.v/R1.R;

der(L.i) := L.v/L.L;
der(C.v) := C.i/C.C;
```

These assignment statements can be subsequently converted to code in some programming language, e.g., C, and executed together with an appropriate ODE solver, usually using better approximations to derivatives and more sophisticated forward-stepping schemes than the simple method described above, which, by the way, is called the *Euler integration* method. The algebraic transformations and sorting procedure that we somewhat painfully performed by hand on the simple circuit example can be performed completely automatically, and is known as *BLT partitioning*, i.e., conversion of the equation system coefficient matrix into block lower triangular form (Table 3.3).

	R2.v	R1.p.v	L.v	R1.v	C.i	L.i	C.v
R2.v = R2.R* L.i	1	0	0	0	0	0	0
R1.p.v = AC.VA* sin(2* AC.f* AC.PI* time)	0	1	0	0	0	0	0
L.v = R1.p.v - R2.v	1	1	1	0	0	0	0
R1.v = R1.p.v - C.v	0	1	0	1	0	0	0
C.i = R1.v/R1.R	0	0	0	0	1	0	0
der(L.i) = L.v/L.L	0	0	1	0	0	1	0
der(C.v) = C.i/C.C	0	0	0	0	1	0	1

Table 3.3: TBlock lower triangular form of the SimpleCircuit example.

The remaining 26 algebraic variables in the equation system of the simple circuit model that are not part of the minimal 7-variable kernel ODE system solved above can be computed at leisure for those iterations where their values are desired–this is not necessary for solving the kernel ODE system.

It should be emphasised that the simple circuit example is trivial. Realistic simulation models often contain tens of thousands of equations, nonlinear equations, hybrid models, etc. The symbolic transformations and reductions of equation systems performed by a real Modelica compiler are much more complicated than what has been shown in this example, e.g., including index reduction of equations and tearing of subsystems of equations.

simulate(SimpleCircuit,stopTime=5)]

plot(C.v, xrange={0,5})

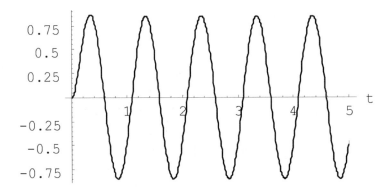

Fig. 3.31: Simulation of the SimpleCircuit model with a plot of the voltage C.v over the capacitor.

3.13 Tool Interoperability through Functional Mockup Interface

Even though Modelica is a universal modelling language that covers most application domains, there exist many established existing modelling and simulation tools in various application domains. There is a strong industrial need for interoperability and model interchange between tools. The Functional Mockup Interface standard (FMI, www.fmi-standard.org) has been developed to fulfil this need. It allows exporting models compiled into C-code combined with XML-based descriptors, which can be imported and simulated in other tools, both Modelica and non-Modelica tools. The standard also supports co-simulation between tools, and is already supported by a number of tool vendors.

3.14 Summary

This chapter has given a quick overview of the most important concepts and language constructs in Modelica. We have also defined important concepts such as object oriented mathematical modelling and acausal physical modelling, and briefly presented the concepts and Modelica language constructs for defining components, connections, and connectors. The chapter concludes with an in-depth example of the translation and execution of a simple model.

3.15 Literature and Further Reading

This chapter has given a short introduction to Modelica and equation-based modelling. A much more complete presentation of Modelica together with applications and the technology behind Modelica simulation tools can be found in [13].

The recent history of mathematical modelling languages is described in some detail in[14], whereas bits and pieces of the ancient human history of the invention and use of equations can be found in [122], and the picture on Newton's second law in latin in [144]. Early work on combined continuous/discrete simulation is described in [233] followed by [65]. A very comprehensive overview of numeric and symbolic methods for simulation is available in [68]. This author's first simulation work involving solution of the Schrödinger equation for a particular case is described in [113].

Current versions of several Modelica tools are described at www.modelica.org, including OpenModelica, Wolfram SystemModeler (before 2012 called MathModelica), and Dymola meaning the Dynamic modelling Laboratory. Several of the predecessors of the Modelica language are described in the following publications including Dymola meaning the Dynamic modelling Language [88, 89], Omola [203, 9], ObjectMath [119, 299, 120], NMF [248], Smile [93], etc.

Speed-Up, the earliest equation-based simulation tool, is presented in [251], whereas Simula-67–the first object-oriented programming language–is described in [35]. The early CSSL language specification is described in [258] whereas the ACSL system is described in [209]. The Hibliz system for an hierarchical graphical approach to modelling is presented in [89]. Software component systems are presented in [12, 260].

The Simulink system for block oriented modelling is described in [201], whereas the MATLAB language and tool are described in [202].

The DrModelica electronic notebook with the examples and exercises from this book has been inspired by DrScheme [102] and DrJava [8], as well as by Mathematica [306], a related electronic book for teaching mathematics [80], and the MathModelica (in 2012 renamed to Wolfram SystemModeler) environment [117, 229, 110]. The first version of DrModelica is described in [185, 186].

The OpenModelica version of notebook called OMNotebook is described in [103, 11]. Applications and extensions of this notebook concept are presented in [250, 210, 271].

General Modelica articles and books: [204, 117, 90], a series of 17 articles (in German) of which [225](Otter 1999) is the first, [270, 114, 91, 107, 111, 112].

The proceedings from the following conferences, as well as several conferences not listed here, which contain a number of Modelica related papers: the International Modelica Conference: [105, 226, 106, 255, 172, 18, 64, 75] and the Equation-Based Object-Oriented Languages and Tools (EOOLT) workshop series: [116, 115, 118, 66, 217]. the Scandinavian Simulation Conference, e.g.: [104] and some later conferences in that series, two special conferences with focus on biomedical [109] and safety issues [108], respectively.

3.16 Self-assessment

1. Create a class Add that calculates the sum of two parameters, which are Integer numbers with given values.
2. Create an instance of class Dog that overrides the default name with "Timmy"

```
class Dog
  constant Real legs = 4;
  parameter String name =  "Scott";
end Dog;
```

3. What do the terms **partial, class,** and **extends** stand for?
4. Consider the Bicycle class below.

```
record Bicycle
  Boolean hasWheels = true;
  Integer nrOfWheels = 2;
end
```

Define a record `ChildrensBike` that inherits from the class `Bicycle` and is meant for kids. Give the variables values.

5. Write a class, `Birthyear`, which calculates the year of birth from this year together with a person's age. Point out the declaration equations and the normal equations. Write an instance of the class `Birthyear` above. The class, let's call it `MartinsBirthyear`, shall calculate Martin's year of birth, call the variable `martinsBirthyear`, who is a 29-year-old. Point out the modification equation. Check your answer, e.g. by writing as below:[7]

```
val(martinsBirthday.birthYear,0)
```

6. Consider the class `Ptest` below:

```
class Ptest
   parameter Real x;
   parameter Real y;
   Real z;
   Real w;
equation
   x + y = z;
end Ptest;
```

What is wrong with this class? Is there something missing?

7. Create the class `Average` that calculates the average between two integers, using an `algorithm` section. Make an instance of the class and send in some values. Simulate and then test the result of the instance class by writing:

```
instanceVariable.classVariable.
```

8. Write a class `AverageExtended` that calculates the average of 4 variables (a, b, c, and d). Make an instance of the class and send in some values. Simulate and then test the result of the instance class as suggested in the question above.

9. Using an *if-equation*, write a class `Lights` that sets the variable switch (integer) to one if the lights are on and zero if the lights are off.

10. Using a *when-equation*, write a class `LightSwitch` that is initially switched off and switched on at time 5. *Tip:* `sample(start, interval)` returns `true` and triggers time events at time instants and `rem(x, y)` returns the integer remainder of x/y, such that `div(x,y) * y + rem(x, y) = x`.

Acknowledgements

This work has been supported by Vinnova in the ITEA3 OPENCPS project and in the RTISIM project. Support from the Swedish Government has been received from the ELLIIT project, as well as from the European Union in the H2020 INTO-CPS project. The OpenModelica development is supported by the Open Source Modelica Consortium.

---

[7] In OpenModelica, the expression val(martinsBirthday.birthYear,0) means the birthYear value at time=0, at the beginning of the simulation. Interpolated values at other simulated time points can also be obtained. Constant expressions such as 33+5, and function calls, can also be evaluated interactively.

Chapter 4
Causal-Block Diagrams: A Family of Languages for Causal Modelling of Cyber-Physical Systems

Cláudio Gomes, Joachim Denil, and Hans Vangheluwe

Abstract The description of a complex system in terms of constituent components and their interaction is one of the most natural and intuitive ways of decomposition. Causal Block Diagram (CBD) models combine subsystem blocks in a network of relationships between input signals and output signals. Popular modelling and simulation tools such as Matlab/Simulink® implement different variants from the family of Causal Block Diagram formalisms. This chapter gives an overview of modelling and simulation of systems with software and physical components using Causal Block Diagrams. It describes the syntax and - both declarative and operational - semantics of CBDs incrementally. Starting from simple algebraic models (no notion of time), we introduce, first a discrete notion of time (leading to discrete-time CBDs) and subsequently, a continuous notion of time (leading to continuous-time CBDs). Each new variant builds on the previous ones. Because of the heavy dependency of CBDs on numerical techniques, we give an intuitive introduction to this important field, pointing out main solutions as well as pitfalls.

Learning Objectives

After reading this chapter, we expect you to be able to:

- Judge when to employ each of the variants of the CBD formalism (algebraic, discrete time, or continuous time);
- Understand the corresponding syntax and semantics of each of the formalisms
- Identify the main issues such as numerical accuracy encountered when simulating physical systems and mitigate them

4.1 Introduction

The design process of complex systems, aided by the technology advances in the last century, is rapidly shifting from small scale development of isolated systems, to large scale development of integrated systems [15].

The graphical representation of such systems using blocks and arrows is one of the first methods used to represent systems. One of the benefits of this notation is that complex systems can be hierarchically decomposed

Cláudio Gomes
University of Antwerp, Belgium
e-mail: claudio.gomes@uantwerp.be

Joachim Denil
Flanders Make, Belgium
e-mail: joachim.denil@uantwerp.be

Hans Vangheluwe
McGill University, Canada
e-mail: hans.vangheluwe@uantwerp.be

© The Author(s) 2020
P. Carreira et al. (eds.), *Foundations of Multi-Paradigm Modelling for Cyber-Physical Systems*,
https://doi.org/10.1007/978-3-030-43946-0_4

97

into sub-systems, thus providing a way to deal with complexity. Causal Block Diagrams (CBD) is a formalisation of this intuitive graphical notation.

Originally, CBDs were used to represent and simulate analog circuits [15, 280, 28, 128], with the most commonly used blocks illustrated in Table 4.1. Nowadays, this formalism is widely used in the modelling and simulation of systems that comprise a physical and a controller component, as depicted in Figure 4.1.

In this kind of system, the controller, often implemented in software, monitors the activity of physical processes by means of sensors, takes appropriate decisions, and influences the physical processes through actuators. This architecture can be generalised to networked software and many other processes, not necessarily physical. Examples include cruise controllers, thermostats, robotic arms, and insulin pumps.

For the purposes of introducing the CBD formalism, we hold on to the traditional view of a physical process being controlled by a software component, also known as a feedback control system [16].

Table 4.1: Block representation for analog circuits. Reproduced from [45].

ELEMENT TYPE	LANGUAGE SYMBOL	DIAGRAMMATIC SYMBOL	DESCRIPTION
BANG-BANG	B	e_i — B n — e_o	e_o, +1, e_i, -1
DEAD SPACE	D	e_i — D n — e_o	e_o, e_i, P_2 P_1
FUNCTION GENERATOR	F	e_i — F n — e_o	e_o, P_2 e_i P_1
GAIN	G	e_i — (n) P_1 — e_o	$e_o = P_1 e_i$
HALF POWER	H	e_i — H n — e_o	$e_o = \sqrt{e_i}$ SQUARE ROOT
INTEGRATOR	I	e_1 P_1 e_2 P_2 I — e_o e_3 P_3	$e_o = P_1 + \int (e_1 + e_2 P_2 + e_3 P_3) \, dt$
JITTER	J	J n — e_o	RANDOM NUMBER GENERATOR BETWEEN ±1
CONSTANT	K	(n) P_1 — e_o	$e_o = P_1$
LIMITER	L	e_i — L n — e_o	e_o P_2, e_i, P_1

n REPRESENTS THE BLOCK NUMBER

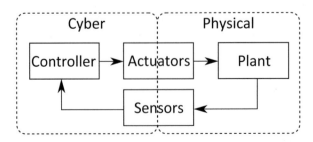

Fig. 4.1: Generic embedded system structure.

In the next section, a simple running example will be introduced, as well as some necessary background concepts. In the remaining sections, CBDs are introduced gradually in three different flavors: algebraic, discrete time, and continuous time CBDs. These are distinguished by the class of blocks at the disposal of the modeler. The gradual presentation allows for a deeper understanding of all the concepts related to modelling and simulation of CBDs. The last few sections of the chapter deal with advanced concepts, related to the simulation of CBDs.

4.2 Background

A dynamical system is characterised by a state and a notion of evolution rules. The state is a set of point values in a state space. The evolution rules describe how the state evolves over an independent variable, usually time.

The domain of the time variable dictates the kind of dynamical system: if time ranges over a continuous set, usually $\mathbb{R}$, then the dynamical system is continuous time; if time ranges over a countable set, usually $\mathbb{N}$, then the dynamical system is discrete time.

Cruise Control System

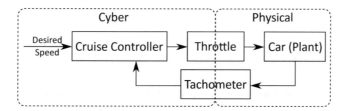

Fig. 4.2: Cruise control system.

Consider the cruise control system depicted in Figure 4.2. The car is a dynamical system whose state (e.g., the velocity) evolves continuously in time. The state of software controller, on the other hand, evolves discretely through time, as any software variable does, by consequence of assignment instructions.

The system in Figure 4.2 is an example of a feedback control system: the physical system – the car – is actuated by control inputs generated by control software – the cruise controller. The sensors are the tachometers that translate wheel revolutions per minute into instantaneous velocity. The actuators are the motor throttle and brakes. The cruise controller software decides, based on current speed of the car, which amount of thrust (throttle or brakes) should be applied to restore the car to a desired speed despite drag, weight, and other factors.

Obviously, it is not desirable to wait for a car prototype to be built, in order to test the control software. This motivates the use of two dynamical systems: *(a)* A continuous dynamical system which acts as a mock-up of a real car, and *(b)* A discrete time dynamical system which acts as a mock-up of the software unit.

With these models, early evaluation of many different control strategies can be performed, at a much lower cost. Furthermore, the chosen controller model can then be used to automatically generate the software code.

4.2.1 Models of Physical Systems

Physical systems are inherently continuous: their state evolves continuously through time. Ordinary Differential Equations (ODE) describe how physical quantities change continuously in time, and are thus ideal models to describe the behaviour of physical systems. Initial value problems are ODEs that have a constraint on the initial state. Together with the ODEs, extra equations can be added to represent the outputs of the system. Formally, we consider models of the form:

$$x'(t) = F(x, u, t),$$
$$y(t) = G(x, u, t), \tag{4.1}$$
$$x(0) = x_0,$$

where $x(t), x'(t), y(t), u(t)$ are vectors, $x(t) = [x_1(t), \dots, x_n(t)]^T$ represents the state vector, $u(t)$ represents the input, and x_0 is the initial state. The state here denotes the minimal information required to determine, together with the input $u(t)$ and function $F(x, u, t)$, the complete future states of the system. Function G is the output function.

Dynamic behavior model

A possible model of the dynamic behavior of the car, to be used in the design of the cruise controller in the Cruise Controller example is:

$$v' = \frac{1}{m}(T - kv)$$
$$y = v \tag{4.2}$$

where v is the velocity, v', T is the thrust force and k depends on the air density and car shape. This model assumes that the car moves in a straight line and neglects any effects that gravity might induce. Reasonable assumptions for early experimentation.

4.2.2 Discrete Time Models

While the solution of ODEs is continuous, the state of the software unit, in the cruise control system presented in Example 4.2, can only evolve discretely, by the nature of the digital computer on which it runs. For mathematical treatment (e.g., proving that the cruise controller is correct), differential equations may be used for an early specification of the control software. However, when it comes to simulating those models in a digital computer, the only available option is to use discrete time models.

First order difference equations, and a constraint on their initial state, allow the specification of such models. They take the form

$$x^{[s+1]} = F(x^{[s]}, u^{[s+1]}),$$
$$y^{[s]} = G(x^{[s]}, u^{[s]}), \tag{4.3}$$
$$x^{[0]} = x_0,$$

where s denotes the step, $x^{[s]}$ is the state vector at step s, $u^{[s+1]}$ is the input vector, $y^{[s]}$ the output vector, and x_0 the initial value of the state vector. The new vector $x^{[s+1]}$ is computed from the old one $x^{[s]}$ and input $u^{[s+1]}$, according to the specification function F. The output function G allows values to be read from the dynamical system. The repeated application of functions F and G yields the discrete evolution of the state and output vectors. Difference equations are sometimes written as

$$x^{[s]} = F(x^{[s-1]}, u^{[s]})$$
$$y^{[s]} = G(x^{[s]}, u^{[s]})$$
$$x^{(0)} = x_0.$$

4.2.2.1 Discrete Control Systems

Software Controller

The cruise control software can be described by the following difference equation:

$$e^{[s+1]} = e^{[s]} + h \left(v_d^{[s+1]} - v^{[s+1]} \right)$$
$$T^{[s]} = K_p \left(v_d^{[s]} - v^{[s]} \right) + K_i e^{[s]}$$

(4.4)

where v_d is the velocity that the car should be kept at (input); v is the actual velocity (input) of the car; $\left(v_d^{[s+1]} - v^{[s+1]} \right)$ is the instantaneous error; $e^{[s]}$ denotes the accumulated error (state); $T^{[s]}$ is the thrust force to be transmitted to the car (output). Finally, K_p and K_i are constant parameters of the controller.

The controller in Example 4.2.2.1 gets the car velocity input $v^{[s]}$ from the readings of the tachometer (recall Figure 4.2). If the differential equation Eq. 4.2 is used to model the car, then $v(t)$ is a continuous quantity and so we can relate it to the input of the controller by $v^{[s]} = v(s \times \Delta t)$, where Δt denotes the constant interval of time between two successive tachometer readings. From now on, assume that Eq. 4.2 is coupled to Eq. 4.4 in this manner.

Intuitively, the thrust force $T^{[s]}$ is proportional to the instantaneous error $v_d^{[s]} - v^{[s]}$ and to the accumulation of errors $e^{[s]}$ from previous steps.

The following two paragraphs give an intuitive rationale for each component of Eq. 4.4.

$K_p \left(v_d^{[s]} - v^{[s]} \right)$ component.

When the instantaneous error is large, the current velocity is far away from the desired one, so the thrust force should be large in order to ensure that the car quickly accelerates/brakes toward the desired speed. When the instantaneous error is small, the car is almost at the desired speed, so the thrust force should be smaller to avoid causing discomfort to the driver. For now, neglect the $K_i e^{[s]}$ component in the thrust force calculation. After a while, if the thrust force given by $T^{[s]} = K_p \left(v_d^{[s]} - v^{[s]} \right)$ becomes symmetric (same magnitude, opposite direction) with the drag force in Eq. 4.2, the car acceleration will be null and its speed will be kept constant. However, that speed will not be exactly equal to the desired speed, because $-kv \neq 0 \implies \left(v_d^{[s]} - v^{[s]} \right) \neq 0$. This is where the $K_i e^{[s]}$ component, neglected until now, has its use.

$K_i e^{[s]}$ component.

This component accumulates the instantaneous error over time, and contributes to the thrust force accordingly. Suppose that the thrust force is counteracted by the drag force and the car is kept at a constant speed, below the desired velocity, just like in the previous paragraph. Then, the accumulated error will keep growing, ensuring that the second component continues to increase the thrust force until it overcomes the drag force. Notice that this might cause the car to overshoot the desired speed, which can be dangerous. The accumulated error will start decreasing once that happens, decreasing the thrust force. The choice of parameters K_p and K_i is an important part of tuning the controller.

4.2.2.2 Discretisation of Differential Equations

The example introduced in this section represents a typical feedback control system, the majority of which are developed using CBDs. In the following sections it will become clear how this is achieved.

4.3 Algebraic Causal Block Diagrams

Algebraic CBDs are CBDs where the only atomic blocks permitted are algebraic ones: summation, negation, inversion, product, raise to power and roots. These can be used to represent systems where there is no notion of time and no evolving state. In other words, the time is a constant: *now*.

While it may seem restricted, these kind of systems arise in the study of the steady state behaviour of dynamic systems, and are used to represent algebraic functions.

Steady State

Consider the car dynamics in Eq. 4.2. The steady state behaviour of the car happens when it is not accelerating. That is, for known constants T, k:

$$0 = \frac{1}{m}(T - kv)$$

The equation gives insight about the torque required to keep the car at the same speed: $T = kv$. The larger the drag force, the larger the torque, and the more energy is required.

4.3.1 Syntax

The main constituents of a CBD are blocks and connections between blocks. Blocks can be atomic or composite. Composite blocks stand for an external CBD, specified elsewhere. These blocks will be drawn with a dashed contour. In algebraic CBDs, atomic blocks can be summation, negation, inversion, product, raise to power, and roots. These will be denoted with the appropriate mathematical symbol.

Since a block can have more than one input and more than one output, the notion of ports is essential to distinguish between inputs and between outputs. Atomic blocks have up to two input ports - depending on the operation - and one output port. Composite blocks can have any number of input and output ports.

Figure 4.3a shows an example of an Algebraic CBD, that calculates the drag force d affecting a car, as it moves with a velocity v, given as input. The composite block c refers to a CBD that calculates the drag coefficient, detailed in Figure 4.3b.

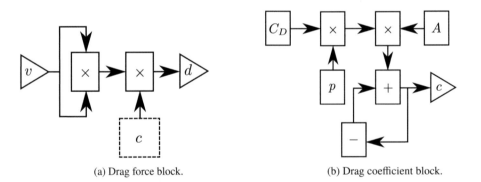

(a) Drag force block.   (b) Drag coefficient block.

Fig. 4.3: Example Algebraic CBDs.

The ports associated with a block will not be drawn explicitly but they are part of the CBD and have identifiers (ids). The directed connections will make clear which input ports and output ports are associated with a block. When there is a need, the input port id is shown at the border of the associated block. In the specification of a composite block, the input and output ports are represented as triangles. Whether the port is input or output is clear from the context.

Blocks are also referred to by ids. The id of a port is comprised of the name of the port, and the id of the parent block. The id of a block is formed by its name, along with the id of its parent CBD. Note the recursive definition. While the id is almost never visible in the graphic representation, it is always defined.

Some of the uses of ids are: the unambiguous description of connections between ports and the unambiguous identification of individual blocks after the flattening process (Section 4.3.2).

More often, the names of the blocks will be depicted in the graphical representations, to enhance their readability. Names are not ids, they are a part of the id. For instance, the product blocks in Figure 4.3a have two ports with distinct ids, even though these are not depicted in the graphical representation. In the same picture, the name v of the input and the name d of the output port are shown. Similarly, the name c of the composite block is shown. Notice that in Figure 4.3b the same name c also denotes the output port. There is no ambiguity as the composite block and the output port have distinct ids, even though they have the same name.

Whenever a composite block is used, all its internal blocks adopt different ids, based on the id of the CBD where the composite block is used. For instance, the fact that the composite block c is used in the CBD of Figure 4.3a means that, when processing that CBD, the ids of the inner blocks/ports of c (detailed in Figure 4.3b) include the id of c. This has two important consequences:

1. the id of any element depends ultimately on where it is being used, or where any of its parents are being used;
2. if the composite blocks are replaced by their specification in a *flattening* process, there will be no two ids alike, thus ensuring the well formedness of the CBD.

4.3.2 Semantics

The meaning of an algebraic CBD is an association of a value to each of the ports in the CBD. It can be conveyed in two general ways: by writing the mathematical equations that correspond to the CBD (translational semantics), or by giving an algorithm which computes the value associated with each input/output port (operational semantics).

To simplify both these approaches, it is assumed that all composite blocks are replaced by their specification in a *flattening* process. This process is done recursively until all composite blocks have been replaced by their specification [232]. The following aspects are important to ensure the well-formedness of the flattened CBD:

1. After replacing a composite block, their input/output ports (e.g., the triangular ones in Figure 4.3b) will be connected from both sides. These are redundant ports and hence substituted by a single connection.
2. The id of the replaced composite block is still part of the ids of its inner blocks/ports. This ensures uniqueness among ids after the flattening process is complete.

Figure 4.4 shows the result of replacing the composite block c with its specification (in Figure 4.3b). The ids, shown explicitly in the picture, contain the ids of the composite block replaced. During the process, the port c of the composite block became redundant and thus was removed.

The meaning of a flattened CBD is the same as the original CBD. Every block/port has a unique id and every connection is between ports. The semantics of CBDs can be thus explained assuming they have been flattened.

4.3.2.1 Translational Semantics

Given a flattened algebraic CBD, the equations that it represents can be written down using the rules shown in Table 4.2.

The system of equations that results from applying the following rules to each port, block and connection, of an algebraic CBD, can then be solved for the unknowns to get the values associated with each port in the CBD. The flattened algebraic CBD depicted in Figure 4.4 is translated into the following set of algebraic equations:

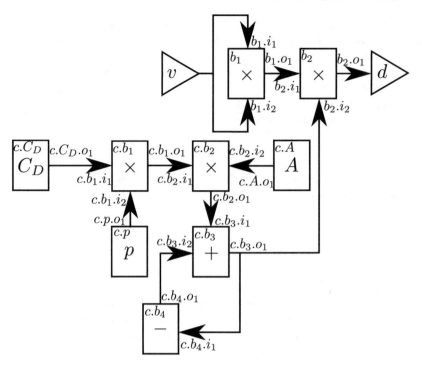

Fig. 4.4: Flattened version of algebraic CBD depicted in Figure 4.3a.

Table 4.2: Translational semantics of a flattened algebraic CBD.

1. Assign a unique mathematical variable to the id of each port in the CBD.
2. Let (p,q) denote a connection from port id p to port id q, and let $\mathrm{var}(p)$ and $\mathrm{var}(q)$ denote the mathematical variables corresponding to p and q, according to the assignment made in Rule 1. Then, the equation associated with the connection (p,q) is $\mathrm{var}(p) = \mathrm{var}(q)$.
3. For each atomic block, let the sequence $p_1, p_2, \ldots$ denote the list of ids of its inputs ports, and let q denote id of the output port:

   a. If the block is a constant with value c, then it has no input ports and the resulting equation is $\mathrm{var}(q) = c$;
   b. If it is a summation, then the resulting equation is $\mathrm{var}(q) = \mathrm{var}(p_1) + \mathrm{var}(p_2)$;
   c. If it is a product, then the resulting equation is $\mathrm{var}(q) = \mathrm{var}(p_1) \times \mathrm{var}(p_2)$;
   d. If it is a negation, then the resulting equation is $\mathrm{var}(q) = -\mathrm{var}(p_1)$;
   e. If it is an inversion, the resulting equation is $\mathrm{var}(q) = \frac{1}{\mathrm{var}(p_1)}$;
   f. If it is a raise-to-power, the resulting equation is $\mathrm{var}(q) = \mathrm{var}(p_1)^{\mathrm{var}(p_2)}$;
   g. If it is a root, the resulting equation is $\mathrm{var}(q) = \mathrm{var}(p_1)^{\frac{1}{\mathrm{var}(p_2)}}$;

$$\begin{aligned}
\mathrm{var}(v) &= \mathrm{var}(b_1.i_1) \\
\mathrm{var}(v) &= \mathrm{var}(b_1.i_2) \\
\mathrm{var}(b_1.o_1) &= \mathrm{var}(b_2.i_1) \\
\mathrm{var}(b_2.o_1) &= \mathrm{var}(d) \\
\mathrm{var}(c.b_3.o_1) &= \mathrm{var}(b_2.i_2) \\
\mathrm{var}(c.C_d.o_1) &= \mathrm{var}(c.b_1.i_1) \\
\mathrm{var}(c.b_1.o_1) &= \mathrm{var}(c.b_2.i_1) \\
\mathrm{var}(c.p.o_1) &= \mathrm{var}(c.b_1.i_2) \\
\mathrm{var}(c.A.o_1) &= \mathrm{var}(c.b_2.i_2) \\
\mathrm{var}(c.b_2.o_1) &= \mathrm{var}(c.b_3.i_1) \\
\mathrm{var}(c.b_3.o_1) &= \mathrm{var}(c.b_4.i_1) \\
\mathrm{var}(c.b_4.o_1) &= \mathrm{var}(c.b_3.i_2) \\
\mathrm{var}(c.C_d.o_1) &= C_d
\end{aligned}$$

$$\begin{aligned}
\mathrm{var}(b_1.o_1) &= \mathrm{var}(b_1.i_1) \times \mathrm{var}(b_1.i_2) \\
\mathrm{var}(b_2.o_1) &= \mathrm{var}(b_2.i_1) \times \mathrm{var}(b_2.i_2) \\
\mathrm{var}(c.p.o_1) &= p \\
\mathrm{var}(c.b_1.o_1) &= \mathrm{var}(c.b_1.i_1) \times \mathrm{var}(c.b_1.i_2) \\
\mathrm{var}(c.A.o_1) &= A \\
\mathrm{var}(c.b_2.o_1) &= \mathrm{var}(c.b_2.i_1) \times \mathrm{var}(c.b_2.i_2) \\
\mathrm{var}(c.b_3.o_1) &= \mathrm{var}(c.b_3.i_1) + \mathrm{var}(c.b_3.i_2) \\
\mathrm{var}(c.b_4.o_1) &= -\mathrm{var}(c.b_4.i_1)
\end{aligned}$$

(4.5)

The system in Eq. 4.5 can be simplified to a quadratic drag force $\text{var}(b_2.o_1) = \text{var}(v)^2 \times \frac{1}{2} \times C_d \times p \times A$, where p is the air density, C_D the drag coefficient, and A the cross sectional area of the car. Obviously, the value of the input port v has to be known in order to solve for the value of the output port $b_2.o_1$.

Any system of algebraic equations that uses operations supported by the atomic blocks of algebraic CBDs can be represented as an algebraic CBD. The CBD in Figure 4.3b can be drawn directly from the equation

$$c = C_D \times p \times A - c \tag{4.6}$$

where c is the output, and C_D, A, p constants.

4.3.2.2 Operational Semantics

Instead of deferring the responsibility of computing the values associated with each port, to an equation solver, it is possible to do so directly. Two such algorithms are presented.

Algorithm 1 presents the dataflow version of the operational semantics. A list of atomic blocks to be computed is revisited iteratively until no blocks remain. A block can be computed only after all the blocks it depends on have been computed. The algorithm terminates after $O((\#\text{atomic blocks in } D)^2)$ iterations.

The inefficiency of this algorithm lies in not taking advantage of the dependencies between blocks to come up with an optimal execution order for blocks. An improved algorithm will be presented later, after formalizing the dependency information between blocks.

Algorithm 1 Data-flow algorithm to evaluate an Algebraic CBD D.

---

function EVALALGEBRAICCBD(D, $v_1, \ldots, v_n$)
    Let val(p) be the computed value associated with port identified by p.
    Let $i_1, \ldots, i_n$ be the ids of the input ports associated with the CBD D.
    Let $o_1, \ldots, o_m$ be the ids of the output ports associated with the CBD D.
    Then, val(i_1) := $v_1, \ldots,$ val(i_n) := v_n are the values associated with each input ports of D.
    Let B denote the set of atomic blocks of D not yet computed.
    Initially, $B :=$ all atomic blocks in D.
    while $B \neq \{\}$ **do**
        for $b_i \in$ B **do**
            Let p denote the single output port of b_i.
            Let $P = \{p_1, p_2, \ldots\}$ denote the inputs ports of b_i.
            Let $Q = \{q_1, q_2, \ldots\}$ denote the output ports connected to each input port $p_j \in P$, respectively.
            Let $\mathcal{B} = \text{block}(q_1) \cup \text{block}(q_2) \cup \ldots$ be the set of blocks that b_i depends on, where block(q_j) is the block associated
with port q_j or the empty set, if no such block exists.
            if $\mathcal{B} \cap B = \{\}$ **then**
                Remark: val(q_1), val(q_2), $\ldots$ have been computed before.
                val(p) :=COMPUTEBLOCK(b_i, val(q_1), val(q_2), $\ldots$)
                Let $\mathcal{P} = \{\rho_1, \rho_2, \ldots\}$ be the set of ports to which port p connects to.
                val(ρ_j) := val(p), for $\rho_1 \in \mathcal{P}$
                $B := B \setminus \{b_i\}$
            end if
        end for
    end while
    $P = \{\}$
    return val(o_1), $\ldots,$ val(o_m)
end function
function COMPUTEBLOCK(b, val(q_1), val(q_2), $\ldots$)
    if b is a summation block **then**
        return val(q_1) + val(q_2)
    end if
    if b is a Constant Block with value v **then**
        return v
    end if
    $\ldots$
end function

The advantage of Algorithm 1 is its simplicity. It represents the execution model of the dataflow paradigm and, provided that no algebraic loops exist, it always finds the values associated with every port of a flattened CBD.

An algebraic loop arises when a block depends indirectly on itself. It is thus natural to think of the CBD in terms of a dependency graph and identify the cycles thereof. In the CBD of Figure 4.4, blocks $c.b_4$ and $c.b_3$ are part of one algebraic loop. Algebraic loops also happen in algebraic systems of equations. For example, in Eq. 4.6, the c variable depends on itself.

Both these loops where introduced artificially for the purposes of illustration. They can easily be removed by reformulating the mathematical expression that the CBD represents. However, in general, not all algebraic loops can be removed by this method and a way to detect them is required.

Given a flattened CBD, its corresponding dependency graph can be created applying the rules in Table 4.3. For example, Figure 4.5 shows the dependency graph of the flattened CBD shown in Figure 4.4.

Table 4.3: Rules for constructing the dependency graph.

1. For each block identified by b, create a unique node v. Let node(b) denote the corresponding node.
2. For each connection (p,q) from port id p to port id q, let b_p and b_q denote the block ids associated with ports p and q, respectively. If p or q have no associated blocks, then ignore this connection and proceed to the next one. Create a directed edge (node(b_q), node(b_p)) in the dependency graph, to mark that fact that b_q depends on b_p.

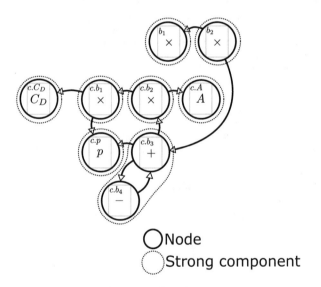

Fig. 4.5: Dependency graph and strong components.

The dependency graph makes the detection of algebraic loops a simple matter of detecting the strong components in the graph. Formally, a strong component $S = \{n_1, n_2, \ldots\}$ of a graph G is a set of nodes where, between every $n_i, n_j \in S$, there are two different paths: $p_1 : n_i \overset{*}{\Rightarrow} n_j$ and $p_2 : n_j \overset{*}{\Rightarrow} n_i$. This implies that every node in a strong component is either the only node in that strong component, or depends on itself, through some other node, also in the same strong component. Figure 4.5 illustrates the strong components of the dependency graph. As expected, the blocks $c.b_4$ and $c.b_3$ are part of the same strong component.

Tarjan's algorithm [262] accepts a graph and outputs a sorted list of strong components. The sort order of the strong components is a topological order according to the dependencies between strong components. For the example in Figure 4.5, one possible topological order is:

$$\{c.C_D\}, \{c.p\}, \{c.A\}, \{c.b1\}, \{c.b2\}, \{c.b_3, c.b_4\}, \{b_1\}, \{b_2\}$$

If no algebraic loops exist in the flattened graph, then the sorted list of strong components returned by the algorithm is just the topological sort of the nodes in dependency graph. In this list, a singleton strong component always appears after the nodes it depends on.

In the case where algebraic loops exist, all nodes belonging to the same algebraic loop will be in the same strong component. Regarding the sort order, non-singleton strong components appear after all components on which it depends on. A strong component depends on other if any one of its comprising nodes depends on at least one of the other's nodes.

These two facts about the sorted strong component list, given by Tarjan's algorithm [262], provide a basis for an improved algebraic CBD operational semantics, that not only can compute the values associated with all output ports much faster than Algorithm 1, but also detects algebraic loops. Algorithm 2 summarizes the steps for computing the values of all ports of a flattened CBD, under the improved algorithm.

Algorithm 2 Evaluation an Algebraic CBD D with support for algebraic loops.

---

function EVALALGEBRAICCBD($D, v_1, \ldots, v_n$)
    Let val(p) be the computed value associated with port identified by p.
    Let $i_1, \ldots, i_n$ be the ids of the input ports associated with the CBD D.
    Let $o_1, \ldots, o_m$ be the ids of the output ports associated with the CBD D.
    Then, val(i_1) := $v_1, \ldots,$ val(i_n) := v_n are the values associated with each input ports of D.
    Let G denote the dependency graph induced by D.
    Let $SC = (S_1, S_2, \ldots)$ denote the sorted list of strong components obtained with Tarjan's algorithm.
    for $S_i \in SC$ **do**
        if $S_i = \{n\}$ **then**
            Let b denote the id of the unique block such that node(b) = n.
            Let p denote the id of the output port associated with b.
            Let $\{q_1, q_2, \ldots\}$ denote the ids of the input ports of b.
            Remark: val(q_1), val(q_2), $\ldots$ have been computed.
            val(p) :=COMPUTEBLOCK(b, val(q_1), val(q_2), $\ldots$)
            Let $\mathcal{P} = \{\rho_1, \rho_2, \ldots\}$ be the ports that port p connects to.
            val(ρ_j) := val(p), for $p_j \in \mathcal{P}$
        else if $S_i = \{n_1, n_2, \ldots\}$ **then**
            Let $b_1, b_2, \ldots$ be the unique blocks such that node(b_1) = n_1, node(b_2) = $n_2, \ldots$
            Let $p_1, p_2, \ldots$ denote the ids of the outputs ports of $b_1, b_2, \ldots$ respectively.
            Let $Q_1, Q_2, \ldots$ denote the sets of ids of the inputs ports of $b_1, b_2, \ldots$ respectively, where $Q_i = \left\{q_1^{(i)}, q_2^{(i)}, \ldots\right\}$.
            For each Q_i there might be input ports whose value is unknown, because these are connected to unknown output
ports. Let $\bar{Q}_i = \left\{\bar{q}_1^{(i)}, \bar{q}_2^{(i)}, \ldots\right\} \subseteq Q_i$ denote the set of input ports whose value is known.
            (val(p_1), val(p_2), $\ldots$) :=SOLVELOOP(b, val($\bar{q}_1^{(1)}$), val($\bar{q}_2^{(1)}$), $\ldots,$ val($\bar{q}_1^{(2)}$), $\ldots$)
            for $p_i \in p_1, p_2, \ldots$ **do**
                Let $\mathcal{P}_i = \left\{\rho_1^{(i)}, \rho_2^{(i)}, \ldots\right\}$ be the ports that port p_i connects to.
                val($\rho_j^{(i)}$) := val(p_i), for $\rho_j^{(i)} \in \mathcal{P}_i$
            end for
        end if
    end for
    return val(o_1), $\ldots,$ val(o_m)
end function

The SOLVELOOP function computes the values of all unknown ports (whose value is unknown) associated with the blocks in the loop. A input port is unknown when an unknown output port is connected to it. An output port is unknown when the block it is associated with belongs to the strong component. Equivalently, an input port is known when a known output port is connected to it. A known output port is associated with a block that does not belong to the strong component.

Essentially, solving an algebraic loop amounts to computing the solution of a matrix equation of the form $X = F(X, U)$:

$$
\underbrace{\begin{bmatrix} \mathrm{val}(p_1) \\ \mathrm{val}(p_2) \\ \cdots \end{bmatrix}}_{X} = \underbrace{\begin{bmatrix} F_1(\mathrm{val}(p_1), \mathrm{val}(p_2), \ldots, \mathrm{val}(\bar{q}_1^{(1)}), \mathrm{val}(\bar{q}_2^{(1)}), \ldots, \mathrm{val}(\bar{q}_1^{(2)}), \ldots) \\ F_2(\mathrm{val}(p_1), \mathrm{val}(p_2), \ldots, \mathrm{val}(\bar{q}_1^{(1)}), \mathrm{val}(\bar{q}_2^{(1)}), \ldots, \mathrm{val}(\bar{q}_1^{(2)}), \ldots) \\ \cdots \end{bmatrix}}_{F(X,U)}
\tag{4.7}
$$

Where $X = [\mathrm{val}(p_1), \mathrm{val}(p_2), \ldots]^T$ denotes the unknown values of the output ports of the strong component, and $U = \left[\mathrm{val}(\bar{q}_1^{(1)}), \mathrm{val}(\bar{q}_2^{(1)}), \ldots, \mathrm{val}(\bar{q}_1^{(2)}), \ldots\right]^T$ denote the known values of the input ports.

In Eq. 4.7, the unknown input ports are not considered because these depend directly, by algebraic equality, on the output ports connected to them. So finding the values of the unknown output ports is enough to be able to find the values of all unknown ports of the strong component.

The definition of F depends on the atomic blocks that belong to the strong component. If F is linear, then the above equation can be written in the form $AX = BU$ and solved with any technique suitable to solve linear systems of equations (see [60, Chapter 6&7]). Matrices A and B depend on the blocks in the strong component, and the product BU is known.

If F is non-linear, successive substitution techniques ((see [60, Chapter 10])) can be used in an attempt to find X.

Caution has to be taken when non-linear loops are solved, as they might not have a solution, or a unique solution. The iterative methods require initial guess values to be provided for X, and depending on those initial guesses, different solutions might be attained. Both the initial guesses, and the solutions attained have to be physically meaningful, as the equations often represent the characteristics of physical systems (e.g., drag forces, concentrations, etc. . .).

For the algebraic loop containing bocks $c.b_4$ and $c.b_3$ in Figure 4.5, the resulting linear system of equations and its analytical solution is:

$$
\begin{cases}
\mathrm{val}(c.b_3.o_1) = \mathrm{val}(c.b_3.i_1) + \mathrm{val}(c.b_3.i_2) \\
\mathrm{val}(c.b_3.i_2) = \mathrm{val}(c.b_4.o_1) \\
\mathrm{val}(c.b_4.o_1) = -\mathrm{val}(c.b_4.i_1) \\
\mathrm{val}(c.b_4.i_1) = \mathrm{val}(c.b_3.o_1)
\end{cases} \leftrightarrow
$$

$$
\begin{cases}
\mathrm{val}(c.b_3.o_1) - \mathrm{val}(c.b_4.o_1) = \mathrm{val}(c.b_3.i_1) \\
\mathrm{val}(c.b_3.o_1) + \mathrm{val}(c.b_4.o_1) = 0
\end{cases} \leftrightarrow
$$

$$
\underbrace{\begin{bmatrix} 1 & -1 \\ 1 & 1 \end{bmatrix}}_{A} \underbrace{\begin{bmatrix} \mathrm{val}(c.b_3.o_1) \\ \mathrm{val}(c.b_4.o_1) \end{bmatrix}}_{X} = \underbrace{\begin{bmatrix} 1 \\ 0 \end{bmatrix}}_{B} \underbrace{\begin{bmatrix} \mathrm{val}(c.b_3.i_1) \end{bmatrix}}_{U} \leftrightarrow
$$

$$
\begin{bmatrix} \mathrm{val}(c.b_3.o_1) \\ \mathrm{val}(c.b_4.o_1) \end{bmatrix} = \begin{bmatrix} \frac{1}{2}\mathrm{val}(c.b_3.i_1) \\ -\frac{1}{2}\mathrm{val}(c.b_3.i_1) \end{bmatrix}
$$

In this section, the syntax and semantics of Algebraic CBDs were described. These are used in systems where there is no notion of evolving state and time. Any algebraic system of equations can be written as an algebraic CBD.

We described two equivalent approaches to obtain the meaning of an algebraic CBD, summarised in Figure 4.6. The solutions are only approximately equal because, in the presence of algebraic loops, these may have to be solved iteratively to get an approximate solution.

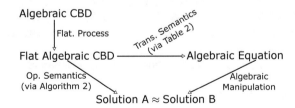

Fig. 4.6: Algebraic CBDs semantic equivalence (approximate).

In the next section, we expand the available atomic blocks to introduce the notion of evolving state via discrete jumps in time.

4.4 Discrete-time CBDs

In this section, the Discrete time (DT) CBDs are presented. Syntactically, the only difference to the Algebraic CBDs, is that the DT CBDs allow the modeler to use not only algebraic blocks, but also a step delay block. Because the Delay block has a state, which gets updated whenever the block is computed, the other blocks in a DT CBD no longer have static outputs (as in the algebraic CBDs case), but instead change whenever they are computed. DT CBDs share many similarities with discrete time dynamical systems, presented in Section 4.2.2.

4.4.1 Syntax

The step delay block has two inputs i_1, i_c and one output o_1. It is represented with a $\mathcal{D}$ symbol, as highlighted in the DT CBD of Figure 4.7. The input port i_c is called the initial condition and is distinguished by its subscript.

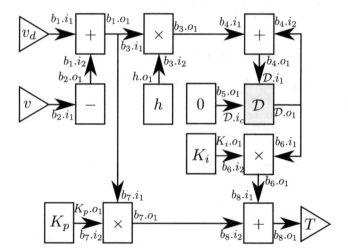

Fig. 4.7: Discrete-time CBD of the cruise controller with an highlighted Delay block $\mathcal{D}$. Equivalent to Eq. 4.4.

4.4.2 Semantics

The fact that the output of the delay block changes whenever it is computed means that the output of any other block that depends on the delay will also be dynamic. To formalize the multiple different values that any single port can assume, the notion of step is necessary. A step is a natural index that allows the distinction between the different outputs of each block. It is no different than the index used in difference equations, presented in Section 4.2.2.

4.4.2.1 Translational

The output of the delay block is defined in terms of the input provided at the previous step. This is the essence of difference equations, where the current values are calculated from the previous ones. It is then natural that the meaning of a DT CBD is a set of difference equations.

Similarly to the algebraic CBD case, the flattening process ensures that only atomic blocks remain in the DT CBD. Given a flattened DT CBD, the difference equations that it represents can be written following the rules specified in Table 4.4.

Table 4.4: Translational semantics of a flattened DT CBD.

1. Assign a unique mathematical variable to the identifier of each port in the CBD.
2. Let (p, q) denote a connection from port identified by p to port identified by q, and let var(p) and var(q) denote the mathematical variables corresponding to p and q, following the assignment made in Rule 1. Then, the equation associated with the connection (p, q) is
   $$\text{var}(p)^{[s+1]} = \text{var}(q)^{[s+1]}.$$
3. Let $p_1, p_2, \ldots$ denote the list of ids of inputs ports of an atomic block, and let q denote id of its single output port:

   a. If the block is a delay block, then the resulting equations are
      $$\text{var}(q)^{[s+1]} = \text{var}(p_1)^{[s]} \text{ and}$$
      $$\text{var}(q)^{[0]} = \text{var}(p_c)^{[0]};$$
   b. If the block is a constant block with value c, then the resulting equation is
      $$\text{var}(q)^{[s+1]} = c;$$
   c. If the block is a summation block, then the resulting equation is
      $$\text{var}(q)^{[s+1]} = \text{var}(p_1)^{[s+1]} + \text{var}(p_2)^{[s+1]};$$
   d. If the block is a product block, then the resulting equation is
      $$\text{var}(q)^{[s+1]} = \text{var}(p_1)^{[s+1]} \times \text{var}(p_2)^{[s+1]};$$
   e. If the block is a negation block, then the resulting equation is
      $$\text{var}(q)^{[s+1]} = -\text{var}(p_1)^{[s+1]};$$
   f. If the block is an inversion block, then the resulting equation is
      $$\text{var}(q)^{[s+1]} = \frac{1}{\text{var}(p_1)^{[s+1]}};$$
   g. If the block is a raise-to-power block, then the resulting equation is
      $$\text{var}(q)^{[s+1]} = \left(\text{var}(p_1)^{[s+1]}\right)^{\text{var}(p_2)^{[s+1]}};$$
   h. If the block is a root block, then the resulting equation is
      $$\text{var}(q)^{[s+1]} = \left(\text{var}(p_1)^{[s+1]}\right)^{\frac{1}{\text{var}(p_2)^{[s+1]}}};$$

The result is a set of difference equations, along with initial conditions (see Rule 3(a) of Table 4.4), that can be solved, either to obtain a closed-form solution, or simulated, by an independent solver. As an example, the DT CBD represented in Figure 4.7 corresponds to, after simplification and renaming of variables, the software controller of Eq. 4.4.

Conversely, any difference equation written in the form of Eq. 4.3 can be represented as a DT CBD. This is illustrated in Figure 4.8.

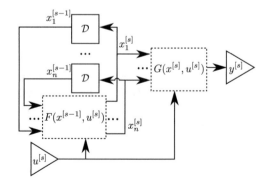

Fig. 4.8: Difference equation (written in the form of Eq. 4.3) can be represented in a DT CBD. The i-th component of the vector x is represented as x_i.

4.4.2.2 Operational

When compared to the algebraic CBDs operational semantics, in Algorithm 2, any algorithm that simulates DT CBDs has to compute not single values for variables, but discrete signals. A discrete signal is an ordered list of values, indexed by the step.

The operational meaning of the DT CBDs is thus the computation of the discrete time signal associated with each port. That can be done by fixing the step at 0, then computing all values in the CBD, as if it were an algebraic CBD. Then, step is incremented to 1, and the evaluation of all values is repeated, and so on. The fact that, within the same step, the DT CBD is evaluated as if it were an algebraic CBD, allows us to reuse the EVALALGEBRAICCBD function, defined in Algorithm 2, with some minor changes:

1. A parameter s is added to the COMPUTEBLOCK function, denoting the current step.
2. All values are indexed by the current step. For example, the instruction $\mathrm{val}(p) := \text{COMPUTEBLOCK}(b, \mathrm{val}(q_1), \mathrm{val}(q_2), \ldots)$
   becomes
   $\mathrm{val}(p)^{(s)} := \text{COMPUTEBLOCK}(b, \mathrm{val}(q_1)^{(s)}, \mathrm{val}(q_2)^{(s)}, \ldots, s);$

Algorithm 3 summarizes the operational semantics. The definition of the COMPUTEBLOCK function is included, to specify the computation of the delay block. The computations of the remaining atomic blocks are trivial.

Algorithm 3 Operational Semantics of an DT CBD D.

---

function EVALDISCRETETIMECBD($D, v_1, \ldots, v_n, N$)
    Let $\mathrm{val}(p)$ be the computed value associated with port identified by p.
    Let $i_1, \ldots, i_n$ be the ids of the input ports associated with the CBD D.
    Let $o_1, \ldots, o_m$ be the ids of the output ports associated with the CBD D.
    Then, $\mathrm{val}(i_1) := v_1, \ldots, \mathrm{val}(i_n) = v_n$ are the values associated with each input ports of D.
    $s := 0$
    while $n \leq N$ **do**
        EVALALGEBRAICCBD($D, v_1^{(s)}, \ldots, v_n^{(s)}, s$)
        $s := s + 1$
    end while
    return $\mathrm{val}(o_1), \ldots, \mathrm{val}(o_m)$
end function
function COMPUTEBLOCK($b, \mathrm{val}(q_1), \ldots, s$)
    if b is a delay block **then**
        if $s = 0$ **then**
            return $\mathrm{val}(q_c)^{(0)}$, where q_c is the id of the initial condition port.
        else
            return $\mathrm{val}(q_c)^{(s-1)}$.
        end if
    end if
    if b is a summation block **then**
        return $\mathrm{val}(q_1)^{(s)} + \mathrm{val}(q_2)^{(s)}$
    end if
    $\ldots$
end function

If there are algebraic loops in the DT CBD, they are handled in the same as way in the algebraic CBDs. A note has to be made, however, about the dependencies of the delay block. At the first step ($s = 0$), the output of the delay block is equal to the input associated with its initial condition port ($\mathrm{val}(q_c)^{(0)}$ in Algorithm 3). At any other step $s > 0$, the output is computed from the *previous* step $s - 1$. This means that, except for $s = 0$, the delay block has no algebraic dependencies. And at $s = 0$ it depends on whatever block is connected to its initial condition port. As a result, Rule 2 of Table 4.3 has to be adapted specifically for the Delay block and the current step being computed.

Following the same structure as Section 4.3, this section presented the syntax and semantics, both translational and operational, of DT CBDs. As with algebraic CBDs, the two semantics commute.

4.5 Continuous-time CBDs

The meaning of algebraic CBDs is a set of algebraic equations, and the meaning of Discrete time CBDs is a set of difference equations.

As shown in Section 4.2.1, differential equations are ideal to represent physical systems, whose state evolves continuously in time. By the end of this section, it will be clear that Continuous Time (CT) CBDs too, are suited to model these systems.

4.5.1 Syntax

Syntactically, CT CBDs include the standard algebraic blocks, a derivative, and an integral block. The Delay block is not included.

The derivative and integral blocks have two inputs i_1, i_c, and one output o_1. The input subscripted by c denotes the initial condition. Both blocks will be denoted by the appropriate mathematical symbol: $\frac{d}{dt}$ and $\int$.

Continuous-time CBD

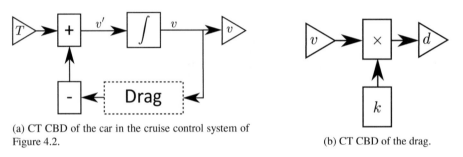

(a) CT CBD of the car in the cruise control system of
Figure 4.2.

(b) CT CBD of the drag.

Fig. 4.9: Example CT CBDs.

Figure 4.9a shows a CT CBD example, with the Drag block specified in Figure 4.9b.

4.5.2 Semantics

4.5.2.1 Translational Semantics to Differential Equations

The meaning of a flattened CT CBD is a system of Ordinary Differential Equations (ODEs). Table 4.5 shows the rules that build such system. The meaning of Figure 4.9a is represented, after simplification and renaming the variables, in Eq. 4.2.

Furthermore, any ODE written in the form of Eq. 4.1 can be translated to a CT CBDs as illustrated in Figure 4.10.

4.5.2.2 Basics of ODE Discretization

In many ODEs arising in science and engineering, and this includes, by the translational semantics, CT CBDs, a closed-form solution cannot be found. One of the possible ways to get insight into the solution is via simulation. Since most simulations are performed in a digital computer, solutions to ODE's obtained via simulation can only be approximate.

Table 4.5: Translational semantics of a flattened CT CBD.

---

1. Assign a unique mathematical variable to the identifier of each port in the CBD.
2. Let (p, q) denote a connection from port identified by p to port identified by q, and let $\text{var}(p)$ and $\text{var}(q)$ denote the mathematical variables corresponding to p and q, following the assignment made in Rule 1. Then, the equation associated with the connection (p, q) is $\text{var}(p)(t) = \text{var}(q)(t)$.
3. Let $p_1, p_2, \ldots$ denote the list of ids of inputs ports of an atomic block, and let q denote id of its single output port:

   a. If the block is an Integral block, then the resulting equation is $\text{var}(q)(t) = \int_0^t \text{var}(p_1)(\tau)d\tau + \text{var}(p_c)(0)$;
   b. If the block is a Derivative block, then the resulting equations are $\text{var}(q)(t) = \text{var}(p_1)'(t)$ and $\text{var}(q)(0) = \text{var}(p_c)(0)$;
   c. If the block is a summation block, then the resulting equation is $\text{var}(q)(t) = \text{var}(p_1)(t) + \text{var}(p_2)(t)$;
   d. ...

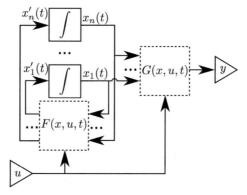

Fig. 4.10: First order ODE, written in the form of Eq. 4.1, can represented as a CT CBD. The i-th component of the vector x is represented as x_i.

In the following paragraphs, we show how to translate ODEs into difference equations, whose solution, obtained via simulation, approximate the solution of the ODEs. See [?, 60] for a more detailed exposition on numerical approximation methods. The method explained is then used in Section 4.5.2.3 as the basis to describe how a CT CBD is translated into discrete time CBDs, which approximate the solution to the original.

Consider a first order ODE without input:

$$x'(t) = F(x, t)$$
$$x(0) = x_0 \tag{4.8}$$

where
- $x(t) = [x_1(t), \ldots, x_n(t)]^T$ is the state vector, and
- $F(x, t) = [F_1(x, t), \ldots, F_n(x, t)]^T$ is the state derivative function, and x_0 the initial value of x.

Let $x_i(t)$ denote the i-th state trajectory and $x_i'(t) = F_i(x, t)$ the i-th state derivative. Assuming that $x_i(t)$ and any of its derivatives are smooth, it can be approximated around any point t^* by the Taylor series [302]:

$$x_i\left(t^* + \Delta t\right) = x_i\left(t^*\right) + x_i'\left(x\left(t^*\right), t^*\right)\Delta t + x_i^{(2)}\left(t^*\right)\frac{\Delta t^2}{2!} + \ldots \tag{4.9}$$

Using Taylor's theorem, it is possible to write the Taylor series expansion in the finite form of a polynomial and a residual in Lagrange form [60]:

$$x_i(t^* + \Delta t) = x_i(t^*) + x_i'(x(t^*), t^*)\Delta t + \ldots + x_i^{(n)}(t^*)\frac{\Delta t^n}{n!} + x_i^{(n+1)}(\xi(t^*))\frac{\Delta t^{n+1}}{(n + 1)!}$$

where $\xi(t^*)$ is an unknown number between t^* and $t^* + \Delta t$. The residual term $x_i^{(n+1)}(\xi(t^*))\frac{\Delta t^{n+1}}{(n+1)!}$ denotes the truncation error. It cannot be computed directly but, since any of the derivatives of x_i are smooth in all points between t^* and $t^* + \Delta t$, there exists a maximum constant $K < \infty$, such that, for any t^* and any n,

$$\frac{x_i^{(n+1)}(\xi(t^*))}{(n+1)!} \leq K$$

An upper bound can then be written for the remainder term in the Big O notation:

$$\lim_{\Delta t \to 0} \frac{x_i^{(n+1)}(\xi(t^*))}{(n+1)!}\Delta t^{n+1} \leq K\Delta t^{n+1} = O(\Delta t^{n+1})$$

Notice that the Big O notation $O(\Delta t^{n+1})$ highlights the dominant term as $\Delta t \to 0$.

Taylor theorem allows us to write the Taylor series taking into account the ODE of Eq. 4.8 and replacing the residual term by its order:

$$x_i\,(t^* + \Delta t) = x_i\,(t^*) + F_i\,(x\,(t^*)\,,t^*)\,\Delta t + O(\Delta t^2) \tag{4.10}$$

For small $\Delta t < 1$ we can neglect the $O\left(\Delta t^2\right)$ term and approximate $x_i\,(t^* + \Delta t)$ by:

$$x_i\,(t^* + \Delta t) \approx x_i\,(t^*) + F_i\,(x\,(t^*)\,,t^*)\,\Delta t \tag{4.11}$$

Going back to the vector case, this suggests that we can approximate the solution vector $x\,(t)$ by the following algorithm:

$$\begin{aligned}
x(\Delta t) &:\approx x(0) + F(x(0),0)\Delta t \\
x(\Delta t + \Delta t) &:\approx x(\Delta t) + F(x(\Delta t), \Delta t)\Delta t \\
x(2\Delta t + \Delta t) &:\approx x(2\Delta t) + F(x(2\Delta t), 2\Delta t)\Delta t
\end{aligned} \tag{4.12}$$

$$\cdots$$

Let $x^{[s]} = x(s\Delta t)$, we get the Forward Euler method to numerically approximate Eq. 4.8:

$$x^{[s+1]} = x^{[s]} + F(x^{[s]}, s\Delta t)\Delta t \tag{4.13}$$

The Taylor series, introduced in Eq. 4.9 also works backward from any point, including the point $x_i\,(t^* + \Delta t)$:

$$x((t^* + \Delta t) - \Delta t) = x(t^* + \Delta t) - x'(x(t^* + \Delta t), t^*)\Delta t + O(\Delta t^2) \tag{4.14}$$

Replacing the derivative by F, from Eq. 4.8, neglecting the residual term, and simplifying gives the Newton's Difference Quotient:

$$\frac{x(t^* + \Delta t) - x(t^*)}{\Delta t} \approx F(x(t^* + \Delta t)) \tag{4.15}$$

Rewritten as a difference equation gives:

$$x^{[s+1]} = x^{[s]} + F(x^{[s+1]})\Delta t \tag{4.16}$$

Contrary to the Forward Euler, it is not possible to get an iterative algorithm immediately out of this method: the vector term $x^{[s+1]}$ depends on itself. This is an algebraic loop (recall Section 4.3.2.2). It requires that the matrix equation be solved for $x^{[s+1]}$. The presence of these loops distinguishes implicit (with loops) from explicit (without loops) methods. The important point is that, as shown in Section 4.3.2.2, these loops can be solved at each simulation step.

4.5.2.3 Translational Semantincs to Discrete-time CBDs

As explained in the previous section, differential equations can be discretised to difference equations by means of numerical approximation techniques, which can then be easily simulated.

Since any CT CBD can be translated into an ODE (by Table 4.5), and since the meaning of a discrete time CBD is a system of difference equations, it is natural to wonder whether a discrete time CBD can be transformed directly into a discrete time CBD, which realizes the approximation. The only blocks that need to be approximated are the derivative and the integral. All the other blocks are algebraic. The integral block is left as an exercise.

The derivative block outputs the derivative of its input u, at time t:

$$y(t) = u'(t)$$

except at time $t = 0$, where the output $y(0)$ is given by the input initial condition u_c, i.e., $y(0) = u_c(0)$.

Applying the Newton's Difference Quotient, from Eq. 4.15, yields:

$$\frac{u(t + \Delta t) - u(t)}{\Delta t} \approx y(t + \Delta t)$$

Solving for the output $y(t + \Delta t)$ and writing as a difference equation gives:

$$y^{[s]} :\approx \frac{u^{[s]} - u^{[s-1]}}{\Delta t} \tag{4.17}$$

Since the input is not differentiable at time $t = 0$, the initial condition of the derivative block is provided with an initial value $y(0) = u_c(0)$.

It is easy to build a discrete time CBD equivalent to Eq. 4.17 using a delay and algebraic blocks. The delay block will ensure the delayed signal of the input ($u^{[s-1]}$) can be obtained. However, at the initial step, $s = 0$, the delay block has to have an initial condition defined because the value $u^{[-1]}$, in Eq. 4.17, is unknown. Let u_{-1} denote this unknown value. u_{-1} cannot be equal to $y^{[0]}$. That does not satisfy the initial condition of the derivative, expressed as:

$$y^{[0]} \approx \frac{u^{[0]} - u_{-1}}{\Delta t} = u_c^{[0]}$$

To find out the initial condition of the delay, one can rearrange the above equation to get $u_{-1} = u^{[0]} - \Delta t \cdot u_c^{[0]}$, which defines the initial condition of the delay. Figure 4.11 shows the transformation rule.

In this section, CT CBDs where introduced, and its meaning given as a translation to discrete time CBDs.

Until now, we have introduced the minimal set of concepts that allow one to use and understand the semantics of CBDs. We skipped over a few details which, to become a proficient user of CT CBDs, have to be covered in the next section. For example, we took for granted that the solution computed by the Forward Euler method is accurate. Furthermore, we have not introduced the operational semantics of continuous time CBDs. Such algorithm can be easily devised for the approximations of the integral and derivative blocks already given – Forward Euler and Newton's Difference Quotient – with special attention to the fact that the approximation of the derivative may introduce algebraic loops in the CBD. However, if those approximation methods are used, the algorithm will be hardly useful: the translation to discrete time CBDs is equivalent and already deals with the algebraic loops problem for free.

4.6 Advanced Concepts and Extensions

This section allows you to devise smarter approximation methods, that minimising the error in the approximation. We focus on numerical integration methods, that is, approximations of the integrator block, because these are the most commonly used when modelling physical systems (see Figure 4.10). Finally, we introduce an extension that is widely used in CBDs: logic blocks. These allow higher level reasoning to be used in CBDs, conveying more expressive power to the modeller, but introducing other interesting challenges when it comes to simulation.

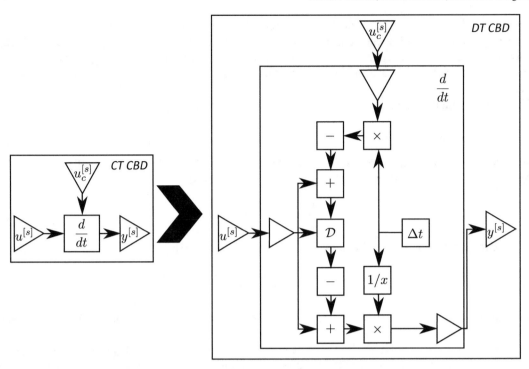

Fig. 4.11: Sample CT CBD with a Derivative block (on the left) and the corresponding discrete time CBD (on the right).

4.6.1 Approximation Error

Consider the Forward Euler method in Eq. 4.13. To derive it, the term $O(\Delta t^2)$ of the Taylor series was neglected (recall Eqs. 4.9, 4.10, and 4.11).

Let $x(t)$ denote the solution to Eq. 4.8 approximated with Forward Euler, and let $\hat{x}$ denote the real solution. The first term $(x(0) = \hat{x}(0))$ is known from the initial condition of Eq. 4.8. The second term $(x(\Delta t) = x(0) + F(x(0))\Delta t)$ deviates from the true solution $\hat{x}(\Delta t)$ by an order $O(\Delta t^2)$, which is the residual term ignored in the Taylor series (recall Eq. 4.10 and Eq. 4.11). Formally, that is,

$$\|\hat{x}(\Delta t) - x(\Delta t)\| = \left\| \hat{x}(0) + F(\hat{x}(0))\Delta t + O(\Delta t^2) - x(0) - F(x(0)) \right\| = O(\Delta t^2)$$

The third term $x(2\Delta t)$ deviates further from the true solution not only because of the residual – of order $O(\Delta t^2)$– but also because $F(x(\Delta t), \Delta t)$ is evaluated with the approximated term $x(\Delta t)$ (most likely $F(x(\Delta t)) \neq F(\hat{x}(\Delta t))$). The iteration continues and it is easy to see that the error accumulates over the iterations.

In order to analyse the accumulation of errors, it is best to distinguish two kinds of errors: the local truncation error, due to the ignored residual term, and the derivative error, due to evaluating the derivative F at approximated points x. Both these errors contribute to the accumulation of error over time, that is, the global error.

The local truncation error denotes the deviation made by a single step of the numerical method, starting from accurate information, i.e., with no previously accumulated error.

Let

$$\hat{x}((s + 1) \cdot \Delta t) = \hat{x}(s \cdot \Delta t) + F(\hat{x}(s \cdot \Delta t)) \Delta t + O\left(\Delta t^2\right) \tag{4.18}$$

denote the real solution expanded with the infinite Taylor series, and let

$$x((s + 1) \cdot \Delta t) \approx \hat{x}(s \cdot \Delta t) + F(\hat{x}(s \cdot \Delta t)) \Delta t$$

denote the solution computed across one step of the Forward Euler method, starting from accurate information. The local truncation error is thus given as

$$\|\hat{x}((s+1)\cdot \Delta t) - x((s+1)\cdot \Delta t)\| = O(\Delta t^2) \tag{4.19}$$

Studying the global error is more difficult as it depends on the derivative error, which, for a generic analysis, can be any function F. If any error in the parameter of F gets amplified, then the global error will grow faster. If it gets contracted, then the global error will grow in a slower fashion. To formalize, suppose that we know that F satisfies:

$$\|F(\hat{x}(t)) - F(x(t))\| \le K_f \|\hat{x}(t) - x(t)\|, \text{ for all } t \in \mathbb{R}, \tag{4.20}$$

where $0 \le K_f < \infty$ is a constant. Then, the error at the second step of the Forward Euler can be derived as follows:

$$
\begin{aligned}
&\|\hat{x}(2\Delta t) - x(2\Delta t)\| \\
&= \left\| \hat{x}(\Delta t) + F(\hat{x}(\Delta t))\Delta t + O(\Delta t^2) - (x(\Delta t) + F(x(\Delta t))\Delta t) \right\| \\
&= \left\| \hat{x}(\Delta t) - x(\Delta t) + (F(\hat{x}(\Delta t)) - F(x(\Delta t)))\Delta t + O(\Delta t^2) \right\| \\
&\le \|\hat{x}(\Delta t) - x(\Delta t)\| + \|F(\hat{x}(\Delta t)) - F(x(\Delta t))\| \Delta t + O(\Delta t^2) \\
&\le \|\hat{x}(\Delta t) - x(\Delta t)\| + K_f \|\hat{x}(\Delta t) - x(\Delta t)\| \Delta t + O(\Delta t^2) \\
&= \left(2 + K_f \Delta t\right) O(\Delta t^2) = O(2\Delta t^2)
\end{aligned}
\tag{4.21}
$$

Notice that, as $\Delta t \to 0$, the big O definition implies that

$$\left(2 + K_f \Delta t\right) O(\Delta t^2) = O(2\Delta t^2 + K_f \Delta t^3) = O(2\Delta t^2).$$

Similarly, for the third step: $\|\hat{x}(3\Delta t) - x(3\Delta t)\| = O(3\Delta t^2)$. And after s steps, we have $\|\hat{x}(s\Delta t) - x(s\Delta t)\| = O(s\Delta t^2)$.

To run the simulation up to time t_f the Forward Euler method performs $t_f/\Delta t$ steps, which gives

$$\left\| \hat{x}(t_f) - x(t_f) \right\| = O(\frac{t_f}{\Delta t}\Delta t^2) = O(\Delta t)$$

Which says that the global error will be approximately linear in the size of Δt, as $\Delta t \to 0$. For a more accurate expression of the global error of the Forward Euler method, see [60, 61].

An important conclusion is that, provided that function F obeys the condition in Eq. 4.20, the global error can be minimised by simply taking smaller Δt at each step of the simulation using the Forward Euler method. This is called convergence, a property that any useful numerical method should satisfy.

Since there is a limit to how small Δt can be made in a digital computer, a numerical method which has an higher order of accuracy than $O(\Delta t)$, for example, $O(\Delta t^2)$, will allow for larger steps to be taken. In the next section, we introduce other numerical methods.

4.6.2 Other Numerical Methods

4.6.2.1 Backward Euler Method

The Taylor series (Eq. 4.10) also works backward from any point, including the point $(t^* + \Delta t)$, as was done in Eq. 4.14. Neglecting the residual term, we get:

$$x_i((t^* + \Delta t) - \Delta t) \approx x_i(t^* + \Delta t) - x_i'(x(t^* + \Delta t))\Delta t$$

After some simplifications we get a method that resembles the Forward Euler method:

$$x(t^* + \Delta t) \approx x_i(t^*) + F(x(t^* + \Delta t))\Delta t \tag{4.22}$$

This leads to the Backward Euler method:

$$x^{[s+1]} = x^{[s]} + F(x^{[s+1]})\Delta t \tag{4.23}$$

When compared to explicit methods, the backward Euler requires the solution to an algebraic loop, so it will incur some extra computation at each simulation step. Furthermore, the global and local errors of the backward Euler are of the same order as the Forward Euler method. Their difference lies in the fact that the derivative used to make the estimation of $x^{[s+1]}$ is the closest to it. In the Forward Euler, the derivative is an *out-dated* one. This has benefits when dealing with stiff systems. See [61, 60, 69] for more details.

4.6.2.2 Second Order Taylor Method

Until now we have always neglected the term $O(\Delta t^2)$ of the Taylor series. Let us see what happens when we neglect higher order terms. For example, the Taylor series, after neglecting the term $O(\Delta t^3)$, becomes:

$$x(t^* + \Delta t) \approx x(t^*) + F(x(t^*),t^*)\Delta t + \frac{dF(x(t^*),t^*)}{dt}\frac{\Delta t^2}{2!}$$

The derivative $\frac{dF(x(t^*),t^*)}{dt}$ can be expanded with the chain rule [1]:

$$\frac{dF(x(t^*),t^*)}{dt} = \frac{\partial F(x(t^*),t^*)}{\partial x}F(x(t^*),t^*) + \frac{\partial F(x(t^*),t^*)}{\partial t}$$

The second order Taylor series method then becomes:

$$x^{[s+1]} = x^{[s]} + F(x^{[s]}, s \cdot \Delta t)\Delta t$$
$$+ \left(\frac{\partial F(x^{[s]}, s \cdot \Delta t)}{\partial x}F(x^{[s]}, s \cdot \Delta t) + \frac{\partial F(x^{[s]}, s \cdot \Delta t)}{\partial t}\right)\frac{\Delta t^2}{2!} \qquad (4.24)$$

The local truncation error of this method is the neglected term $O(\Delta t^3)$, better than the Euler methods. The disadvantage of this method is that it requires the calculation (symbolically or numerically) of the partial derivatives of F – a costly operation. The global error is in the order of $O(\Delta t^2)$.

Higher order Taylor methods require even more derivative calculations, making them impractical. There are methods that offer that same global error order with far less computation at each step. We show one next.

4.6.2.3 Midpoint Method

The backward Euler method makes use of the most up-to-date derivative to estimate the solution at $t^* + \Delta t$ with the disadvantage that it requires more computation to solve the implicit equation. To avoid this, but still trying to be better than Forward Euler, we can try to estimate the derivative at halfway between t^* and $t^* + \Delta t$ and use that derivative to compute $x(t^* + \Delta t)$:

$$x(t^* + \Delta t) = x(t^*) + F(x(t^* + \frac{\Delta t}{2}))\Delta t.$$

However, we do not know the value of $x(t^* + \frac{\Delta t}{2})$. We can use Taylor series again to get

$$x(t^* + \frac{\Delta t}{2}) = x(t^*) + F(x(t^*))\frac{\Delta t}{2}$$

Thus we arrive at the midpoint method:

$$x^{[s+1]} = x^{[s]} + F\left[x^{[s]} + F\left(x^{[s]}, s \cdot \Delta t\right)\frac{\Delta t}{2}, \left(s + \frac{1}{2}\right) \cdot \Delta t\right]\Delta t \qquad (4.25)$$

The midpoint method, Eq. 4.25, can be generalised to

---

[1] Notice that, to be general, we represent the derivative $F(x(t^*),t^*)$ as a function that depends directly on the time. If this is not the case, then $\frac{\partial F(x(t^*),t^*)}{\partial t} = 0$.

$$x_C^{[s+1]} = x^{[s]} + \beta_{C1} \cdot \Delta t \cdot F^{[s]} + \beta_{C2} \cdot \Delta t \cdot F\left(x^{[s]} + \beta_p \cdot F^{[s]} \cdot \Delta t, \left(s + \alpha_p\right) \cdot \Delta t\right)$$

where $F^{[s]} = F\left(x^{[s]}, s \cdot \Delta t\right)$, $\beta_p = \alpha_p = \frac{1}{2}$, $\beta_{C1} = 0$, and $\beta_{C2} = 1$.

Expanding $F\left(x^{[s]} + \beta_p \cdot F^{[s]} \cdot \Delta t, \left(s + \alpha_p\right) \cdot \Delta t\right)$ with the multi-variate version of the Taylor series, we get:

$$F\left(x^{[s]} + \beta_p \cdot F^{[s]} \cdot \Delta t, \left(s + \alpha_p\right) \cdot \Delta t\right)$$
$$\approx F^{[s]} + \beta_p \cdot \frac{\partial F^{[s]}}{\partial x} \cdot F^{[s]} \cdot \Delta t + \alpha_p \cdot \frac{\partial F^{[s]}}{\partial t} \cdot \Delta t$$

Where the quadratic term was neglected. Plugging it into the previous equation gives:

$$x_C^{[s+1]} = x^{[s]} + \beta_{C1} \cdot F^{[s]} \cdot \Delta t +$$
$$\beta_{C2} \cdot \left[F^{[s]} + \beta_p \cdot \frac{\partial F^{[s]}}{\partial x} \cdot F^{[s]} \cdot \Delta t + \alpha_p \cdot \frac{\partial F^{[s]}}{\partial t} \cdot \Delta t\right] \cdot \Delta t$$
$$= x^{[s]} + (\beta_{C1} + \beta_{C2}) F^{[s]} \cdot \Delta t +$$
$$\beta_{C2}\left[\beta_p \cdot \frac{\partial F^{[s]}}{\partial x} \cdot F^{[s]} + \alpha_p \cdot \frac{\partial F^{[s]}}{\partial t}\right] \cdot \Delta t^2$$

To find the local truncation error, let us find the Taylor series expansion of the true solution and then compare it to the previous equation. The true solution can be expanded as:

$$\hat{x}^{[s+1]} = \hat{x}^{[s]} + F^{[s]} \cdot \Delta t + \frac{1}{2}\frac{\partial F^{[s]}}{\partial x}\Delta t^2 + O(\Delta t^3)$$

Applying the chain rule to the derivative yields:

$$\hat{x}^{[s+1]} = \hat{x}^{[s]} + F^{[s]} \cdot \Delta t + \frac{1}{2} \cdot \left[\frac{\partial F^{[s]}}{\partial x}F^{[s]} + \frac{\partial F^{[s]}}{\partial t}\right] \cdot \Delta t^2 + O(\Delta t^3)$$

Comparing $\hat{x}^{[s+1]}$ with $x_C^{[s+1]}$, and assuming that these start from a true solution $\hat{x}^{[s]}$ gives:

$$x_C^{[s+1]} = \hat{x}^{[s+1]} \leftrightarrow$$
$$\hat{x}^{[s]} + (\beta_{C1} + \beta_{C2}) F^{[s]} \cdot \Delta t + \beta_{C2}\left[\beta_p \cdot \frac{\partial F^{[s]}}{\partial x} \cdot F^{[s]} + \alpha_p \cdot \frac{\partial F^{[s]}}{\partial t}\right] \cdot \Delta t^2$$
$$= \hat{x}^{[s]} + F^{[s]} \cdot \Delta t + \frac{1}{2} \cdot \left[\frac{\partial F^{[s]}}{\partial x}F^{[s]} + \frac{\partial F^{[s]}}{\partial t}\right] \cdot \Delta t^2 + O(\Delta t^3)$$

When solved for the parameters, the above equation gives:

$$\begin{cases} \beta_{C1} + \beta_{C2} = 1 \\ 2\beta_{C2}\beta_p = 1 \\ 2\beta_{C2}\alpha_p = 1 \end{cases}$$

As long as the parameters $\beta_p, \alpha_p, \beta_{C1}, \beta_{C2}$ obey the above system of equations, the generic method will have a local truncation error of order $O(\Delta t^3))$, without having to compute any derivative of F. This also shows that the midpoint method is but an element of a family of methods, all with different sets of parameters, called the two stage Runge Kutta methods [69].

By the same argument as the Forward Euler (in Section 4.6.1), we conclude that the global error of the two stage Runge Kutta method is of order $O(\Delta t^2)$.

4.6.3 Adaptive-Step Size

The numerical integration schemes introduced until now use a step size Δt assumed to be constant throughout the simulation process. These numerical algorithms are computationally expensive in systems where the dynamic behaviour changes slowly except in some limited intervals of time.

Recall that the order of growth of the global error ultimately depends on the Lipschitz constant K_f, in Eq. 4.20. This constant represents the worst case deviation of function F as a response to deviations in its parameters[2], for all possible values of $\hat{x}(t)$.

A larger K_f indicates that the global error *may* grow faster, which means that the step size Δt should be smaller. To clarify: if a system has a large K_f, it means that there is at least one pair of values $x(t)$ and $\hat{x}(t)$ for which $\|F(\hat{x}(t)) - F(x(t))\|$ is large. This does not imply the deviations of F are large for all possible pairs of values $x(t)$ and $\hat{x}(t)$. Furthermore, it does not imply that, if the system were to be simulated in a bounded region (e.g, for $0 < t < t_f$), the Lipschitz constant in that region would be smaller. A smaller Lipschitz constant means that the Δt can be larger.

For a given derivative F, it is hard to find the proper K_f in order to pick the right Δt.

An algorithm that can change the Δt throughout the simulation, not only leverages the features of each region in the state space to improve the run-time performance of the simulation but also frees the user from the burden of picking an appropriate Δt. All of this without sacrificing accuracy.

The change of the Δt has to be triggered by some estimate of the error being committed at each simulation step. Assuming the estimate is available, Δt is increased if the error becomes too small and decreased if the error is too large.

The challenge is to come up with a good estimate of the error being committed. Suppose we are given two methods, with local truncation errors $O(c\Delta t^v)$ and $O(c'\Delta t^{v'})$, respectively, with c, c', v, v' positive constants. Formally, let $x(t)$ be the solution computed by the first method, $\tilde{x}(t)$ by the second, and $\hat{x}(t)$ be the real solution. Then, after one inaccurate step, solutions $x(t + \Delta t)$ and $\tilde{x}(t + \Delta t)$ can be written as:

$$x(t + \Delta t) = \hat{x}(t + \Delta t) + O(c\Delta t^v)$$
$$\tilde{x}(t + \Delta t) = \hat{x}(t + \Delta t) + O(c'\Delta t^{v'}) \tag{4.26}$$

Comparing $x(t + \Delta t)$ with $\tilde{x}(t + \Delta t)$ yields

$$\|x(t + \Delta t) - \tilde{x}(t + \Delta t)\| = \left\|O(c\Delta t^v) - O(c'\Delta t^{v'})\right\|$$

The big O notation tells that there exist constants K_1 and K_2 such that, in the limit $\Delta t \to 0$,

$$\left\|O(c\Delta t^v) - O(c'\Delta t^{v'})\right\| = \left\|K_1 c\Delta t^v - K_2 c'\Delta t^{v'}\right\|$$

Assuming that $c' > c$ and that $v' < v$ (the other cases are similar) we have, as $\Delta t \to 0$,

$$\|x(t + \Delta t) - \tilde{x}(t + \Delta t)\| = O(c'\Delta t^{v'}) = \|\tilde{x}(t + \Delta t) - \hat{x}(t + \Delta t)\|,$$

thus proving that comparing the solutions of the two methods gives an estimate of the error in the same order as the local truncation error of the least accurate method.

From the previous sections, there are two approaches to affect the accuracy of a method: (a) use a smaller step-size and (b) use an higher order approximation method (e.g., the midpoint).

The approach (a) is straightforward: simply take any existing numerical method, compute the solution twice (once with two half steps, and once with a single step), and compare the two estimates.

For an example of approach (b), use the midpoint method to compute the solution $x(t)$, and, at each step, compare it with the result $\tilde{x}(t)$ of the Forward Euler method. It is easy to see that there is some redundant computation in this approach. Fortunately, higher order Runge-Kutta methods can be combined, reusing most of the redundant computation. These are called the Runge-Kutta Fehlberg methods.

---

[2] $F(x(t)) = F(\hat{x}(t) + e(t))$, with e being the approximation error.

4.6.4 Logic Blocks

Decision blocks are widely used in CBDs to increase the expressiveness of the language. The most common decision block is the switch block. The switch block, shown in Figure 4.12, outputs the value $u(t)$ or $v(t)$ depending on the value of $c(t)$. If $c(t) \geq 0$, $u(t)$ is the output, otherwise $v(t)$. The translational semantics are:

$$y(t) = \begin{cases} u(t), & \text{if } c(t) \geq 0 \\ v(t), & \text{otherwise} \end{cases}$$

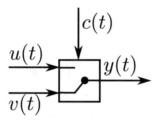

Fig. 4.12: Switch block

As will be presented shortly, the operational semantics of this block introduce interesting challenges.

4.6.4.1 Discontinuity Handling – Zero-Crossing Location

Recall that the simulation of continuous time CBDs can only be performed approximately in a digital computer. See Section 4.5.2. This means that the simulation of a continuous CBDs is actually a discrete set of points

$$x(0), x(1\Delta t_1), x(2\Delta t_2), \ldots$$

computed with an adaptive step size method (see Section 4.6.3).

Suppose the time is t and the simulator is going to compute the solution to the output of the switch block $y(t + \Delta t)$. Furthermore, assume that $y(t) = u(t)$, that is, $c(t) \geq 0$. If $c(t + \Delta t) \leq 0$, then two issues can be identified:

1. $y(t + \Delta t) = v(t + \Delta t)$ may be very different than $y(t)$, because $v(t + \Delta t) \neq u(t + \Delta t)$.
2. $t + \Delta t$ may not represent the exact time at which the signal $c(t)$ crossed the zero. That is $c(t + \Delta t) = 0 - \delta$, for some $\epsilon > 0$.

The second issue implies that, by the intermediate value theorem, there exists at least one point $t^* \in [t, t + \Delta t]$, at which $c(t + \Delta t) = 0$. Ideally, Δt should be picked in a way such that $t + \Delta t \approx t^*$, thus minimizing δ, for two reasons:

1. Accuracy is improved since all the blocks that depend on the solution $x(t + \Delta t)$ will produce outputs that are close to the switching point t^*;
2. Integrator blocks, which apply the numerical methods presented in Section 4.6.2 may need to be aware of the discontinuity in their inputs, caused by the discontinuity of y around the point t^* (issue 1 above).

To see why this can be a problem, consider the abstract CBD shown in Figure 4.13. It can be written as a differential equation $x'(t) = F(x(t))$ (recall Figure 4.10). At the time of the discontinuity t^*, in the limit $\epsilon \to 0$, $x(t^* - \epsilon) = x(t^* + \epsilon)$, but $F(x(t^* - \epsilon)) \neq F(x(t^* + \epsilon))$ because of the switch block. This causes a fundamental assumption about the behavior of F—the condition in Eq. 4.20—to be violated. Formally,

$$\|F(x(t^* + \epsilon)) - F(x(t^* - \epsilon))\| \leq K_f \|x(t^* + \epsilon) - x(t^* - \epsilon)\| \leq 0$$

is a contradiction.

Without the Lipschitz condition assumption, it is hard to guarantee an order for the growth of the global error. There are multiple ways to address this problem, once the exact time of the discontinuity is located (see [211, 310]). We focus here in the location of the time of the discontinuity (also called root-finding, or zero crossing location in the literature).

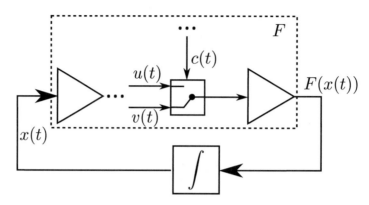

Fig. 4.13: Abstract CBD which may violate the condition in Eq. 4.20.

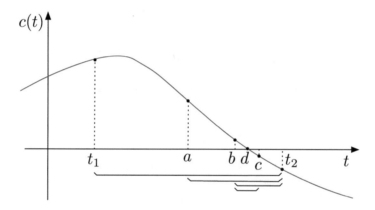

Fig. 4.14: The bisection method

Different algorithms have been proposed over the years (see [60, 69]). The essence is always the same: locate t^* in an interval $[t, t + \Delta t]$ such that the condition of the switch satisfies $c(t + \Delta t) = 0$.

A robust yet simple algorithm is the bisection method. As the name hints, the method works by iteratively bisecting the interval. At each iteration it selects the subinterval where the zero-crossing is present to search for the zero location. The algorithm is illustrated in Figure 4.14. The initial steps detects a zero-crossing in the interval between t_1 and t_2. The iterative procedure evaluates first point a, then point b, then point c and finally point d that is within the tolerance bounds.

Other algorithms are described in the literature [60].

4.7 Global Error Euler Method

The global error measures the accumulated deviation of the numerical method from the true solution, across any number of steps. We need to assume that $F(x, t)$ is Lipschitz continuous.

At the initial step, the global error is the same as the local truncation error: $e^{(1)} = x^{(1)} - \tilde{x}^{(1)} = x^{(0)} + F(x^{(0)}, k\Delta t)\Delta t + O\left(\Delta t^2\right) - \left(x^{(k)} + F(x^{(k)}, k\Delta t)\Delta t\right) = O\left(\Delta t^2\right)$.

At the second step:

$$e^{(2)} = x^{(2)} - \tilde{x}^{(2)} =$$

$$x^{(1)} + F(x^{(1)}, \Delta t)\Delta t + O\left(\Delta t^2\right) - \left(\tilde{x}^{(1)} + F(\tilde{x}^{(1)}, \Delta t)\Delta t\right) =$$

$$e^{(1)} + F(x^{(1)}, \Delta t)\Delta t - F(\tilde{x}^{(1)}, \Delta t)\Delta t + O\left(\Delta t^2\right) =$$

$$\left[F(x^{(1)}, \Delta t) - F(\tilde{x}^{(1)}, \Delta t)\right]\Delta t + e^{(1)} + O\left(\Delta t^2\right)$$

We know that, for some constant K, $\tilde{x}^{(1)} = x^{(1)} - e^{(1)}$, we can write:

$$\left[F(x^{(1)}, \Delta t) - F(x^{(1)} - e^{(1)}, \Delta t)\right]\Delta t + e^{(1)} + O\left(\Delta t^2\right)$$

The Lipschitz continuity condition tells us that:

$$\left|F(x^{(1)}, \Delta t) - F(x^{(1)} - e^{(1)}, \Delta t)\right| \leq Ke^{(1)},$$

Hence, we can give an upper bound on the above error:

$$e^{(2)} = \left[F(x^{(1)}, k\Delta t) - F(\tilde{x}^{(1)}, k\Delta t)\right]\Delta t + e^{(1)} + O\left(\Delta t^2\right) \leq$$

$$Ke^{(1)}\Delta t + e^{(1)} + O\left(\Delta t^2\right) \leq Ke^{(1)}\Delta t + e^{(1)} + O\left(\Delta t^2\right) =$$

$$(K\Delta t + 1)\,e^{(1)} + O\left(\Delta t^2\right).$$

In general,

$$e^{(k+1)} =$$

$$x^{(k+1)} - \tilde{x}^{(k+1)} =$$

$$x^{(k)} + F(x^{(k)}, k\Delta t)\Delta t + O\left(\Delta t^2\right) - \left(\tilde{x}^{(k)} + F(\tilde{x}^{(k)}, k\Delta t)\Delta t\right) =$$

$$e^{(k)} + \left[F(x^{(k)}, k\Delta t) - F(\tilde{x}^{(k)}, k\Delta t)\right]\Delta t + O\left(\Delta t^2\right) \leq$$

$$K\Delta t e^{(k)} + e^{(k)} + O\left(\Delta t^2\right) \leq O\left(\Delta t^2\right) + (K+1)\,e^{(k)}.$$

Expanding recursively, and assuming that $\Delta t < 1$ we get:

$$e^{(1)} = O\left(\Delta t^2\right)$$

$$e^{(2)} \leq O\left(\Delta t^2\right) + (K+1)\,O\left(\Delta t^2\right) = (K+2)\,O\left(\Delta t^2\right)$$

$$e^{(3)} \leq O\left(\Delta t^2\right) + (K+2)\,O\left(\Delta t^2\right) = (K+3)\,O\left(\Delta t^2\right)$$

$$\cdots$$

$$e^{(k)} \leq O\left(\Delta t^2\right) + (K+k-1)\,O\left(\Delta t^2\right) = (K+k)\,O\left(\Delta t^2\right).$$

To simulate the system from 0 to t_f, we require $\frac{t_f}{\Delta t}$ steps. Thus, the error will be:

$$e^{(\frac{t_f}{\Delta t})} \leq \left(K + \frac{t_f}{\Delta t}\right)O\left(\Delta t^2\right) =$$

$$KO\left(\Delta t^2\right) + \frac{t_f}{\Delta t}O\left(\Delta t^2\right) \leq \left(K + t_f\right)O\left(\Delta t\right) = O\left(\Delta t\right)$$

This leads us to conclude that the order of global error of the forward Euler is $O\left(\Delta t\right)$.

4.8 Summary

Causal Block Diagrams represent a formalisation of the intuitive graphical notation of blocks and arrows. This chapter introduced the different variants of this formalism, in a gradual manner.

The most typical uses for these formalisms are: *(i)* Algebraic CBDs to study the steady state behaviour of systems; *(ii)* Discrete time CBDs to represent computation and software components; and *(iii)* Continuous time CBDs to model physical systems. To connect these three variants, a running example of a cruise control system was used.

Algebraic CBDs represent algebraic systems where there is no notion of passing time. Discrete time CBDs mixes in the passage of time, although at discrete points. These are analogous to difference equations. Finally, continuous time CBDs, were time is a continuum, correspond to differential equations. x

The advantage of CBDs over plain difference/differential equations is the natural support for hierarchical descriptions of complex systems, providing a way to manage complexity.

The disadvantage is in the ability to reuse models of physical components, represented as CBDs. Physical objects do not have a notion of inputs and outputs. They are best modeled with equations where any variable can be an input/output, depending on whether it is known (see Acausal modelling chapter). This way, the same component can be reused in many different settings, with its input/outputs defined upon instantiation. In CBDs, the modeler is forced to think of the possible instantiations of the model, and define the inputs/outputs accordingly.

CBDs are widely used in the development of embedded systems. Understanding their semantics and the numerical techniques employed are a stepping stone into understanding other modelling languages.

4.9 Literature and Further Reading

Among the references already cited, we highlight: [69] provides an extensive overview of the simulation of continuous systems. [279] gives a good introduction to continuous system modelling and simulation, for someone with a background in Computer Science. [60] provides a mathematically oriented description of multiple numerical techniques. Last but not least, [211] and [310] provide an overview of the the challenges involved in hybrid system simulation, of which CBDs with logic blocks are part of.

4.10 Self Assessment

1. What does it mean for the control software to be correct?
2. What is the role of the component $K_p \left(v_d^{[s]} - v^{[s]} \right)$ in Eq. 4.4? And what is the role of $K_i e^{[s]}$ in the same equation? (*hint: write Eq. 4.4 without the $K_i e^{[s]}$ component, and solve it together with Eq. 4.2 for a constant velocity of the car*).
3. What the meaning of an algebraic equation, like the one shown in Example 4.3.
4. What is the general procedure to draw a CBD from a given algebraic equation?
5. What is the worst case CBD for Algorithm 1.
6. How can the integral block be approximated with the Forward Euler method, in Eq. 4.13?
7. Construct a CT CBD for the initial value problem $x''(t) = -x; x(0) = 1; x'(0) = 1$ using only integrator blocks. Is it possible to use only derivative blocks?
8. Transform the CT CBDs to DT CBDs using the approximation method taught here, and check whether there are algebraic loops.
9. Apply the backward Euler method to approximate the integral block, as in Question 6 above.
10. In case of approach (b), which of the two approximated solutions should be displayed to the user?

Acknowledgements

This work was partially funded with PhD fellowship from the Agency for Innovation by Science and Technology in Flanders (IWT), and the COST Action MPM4CPS. Partial support by the Flanders Make strategic research center for the manufacturing industry is also gratefully acknowledged.

Chapter 5
DEVS: Discrete-Event Modelling and Simulation for Performance Analysis of Resource-Constrained Systems

Yentl Van Tendeloo and Hans Vangheluwe

Abstract DEVS is a popular formalism for modelling complex dynamic systems using a discrete-event abstraction. At this abstraction level, a timed sequence of pertinent "events" input to a system (or internal, in the case of timeouts) cause instantaneous changes to the state of the system. Between events, the state does not change, resulting in a piecewise constant state trajectory. Main advantages of DEVS are its rigorous formal definition, and its support for modular composition. This chapter introduces the Classic DEVS formalism in a bottom-up fashion, using a simple traffic light example. The syntax and operational semantics of Atomic (i.e., non-hierarchical) models are introduced first. The semantics of Coupled (hierarchical) models is then given by translation into Atomic DEVS models. As this formal "flattening" is not efficient, a modular abstract simulator which operates directly on the coupled model is also presented. This is the common basis for subsequent efficient implementations. We continue to actual applications of DEVS modelling and simulation, as seen in performance analysis for queueing systems. Finally, we present some of the shortcomings in the Classic DEVS formalism, and show solutions to them in the form of variants of the original formalism.

Learning Objectives

After reading this chapter, we expect you to be able to:

- Understand the difference between DEVS and other (similar) formalisms
- xplain the semantics of a given DEVS model
- Understand the relation and difference between a DEVS model and its simulator
- Apply DEVS to simple queueing problems
- Understand the major limitations of DEVS and existing extensions

5.1 Introduction

DEVS [309] is a popular formalism for modelling complex dynamic systems using a discrete-event abstraction. At this abstraction level, a timed sequence of pertinent "events" input to a system cause instantaneous changes to the state of the system. These events can be generated externally (*i.e.*, by another model) or internally (*i.e.*, by the model itself due to timeouts). The next state of the system is defined based on the previous state of the system and the event. Between events, the state does not change, resulting in a piecewise constant state

Yentl Van Tendeloo
University of Antwerp, Belgium
e-mail: Yentl.VanTendeloo@uantwerpen.be

Hans Vangheluwe
University of Antwerp - Flanders Make, Belgium; McGill University, Canada
e-mail: hans.vangheluwe@uantwerpen.be

© The Author(s) 2020
P. Carreira et al. (eds.), *Foundations of Multi-Paradigm Modelling for Cyber-Physical Systems*,
https://doi.org/10.1007/978-3-030-43946-0_5

trajectory. Simulation kernels must only consider states at which events occur, skipping over all intermediate points in time. This is in contrast with discrete time models, where time is incremented with a fixed increment, and only at these times is the state updated. Discrete event models have the advantage that their time granularity can become (theoretically) unbounded, whereas time granularity is fixed in discrete time models. Nonetheless, the added complexity makes it unsuited for systems that naturally have a fixed time step.

Main advantages of DEVS compared to other discrete event formalisms are its rigorous formal definition, and its support for modular composition. Comparable discrete event formalisms are DES and Statecharts, though significant differences exist.

Compared to DES, DEVS offers modularity which makes it possible to nest models inside of components, thus generating a hierarchy of models. This hierarchy necessitates couplings and (optionally) ports, which are used for all communication between two components. In contrast, DES models can directly access other models and send them events in the form of method invocation. Additionally, DEVS offers a cleaner split between the simulation model and the simulation kernel.

Compared to Statecharts, DEVS offers a more rigorous formal definition and a different kind of modularity. Statecharts leaves a lot of room for interpretation, resulting in a wide variety of interpretations [140, 138]. In contrast, DEVS is completely formally defined, and there is a reference algorithm (*i.e.*, an abstract simulator). While both DEVS and Statecharts are modular formalisms, Statecharts creates hierarchies through composite states, whereas DEVS uses composite models for this purpose. Both have their distinct advantages, making both variants useful in practice.

This chapter provides an introductory text to DEVS (often referred to as Classic DEVS nowadays) through the use of a simple example model in the domain of traffic lights. We start from a simple autonomous traffic light, which is incrementally extended up to a traffic light with policeman interaction. Each increment serves to introduce a new component of the DEVS formalism and the corresponding (informal) semantics. We start with atomic (*i.e.*, non-hierarchical) models in Section 5.2, and introduce coupled (*i.e.*, hierarchical) models in Section 5.3. An abstract simulator, defining the reference algorithm, is covered in Section 5.4. Section 5.5 moves away from the traffic light example and presents a model of a simple queueing system. Even though DEVS certainly has its applications, several variants have spawned to tackle some of its shortcomings. These variants, together with a rationale and the differences, are discussed in Section 5.6. Finally, Section 5.7 summarises the chapter.

5.2 Atomic DEVS models

We commence our explanation of DEVS with the atomic models. As their name suggests, atomic models are the indivisible building blocks of a model.

Throughout this section, we build up the complete formal specification of an atomic model, introducing new concepts as they become required. In each intermediate step, we show and explain the concepts we introduce, how they are present in the running example, and how this influences the semantics. The domain we will use as a running example throughout this chapter is a simple traffic light.

5.2.1 Autonomous Model

The simplest form of a traffic light is an autonomous traffic light. Looking at it from the outside, we expect to see a trace similar to that of Figure 5.1. Visually, Figure 5.2 presents an intuitive representation of a model that could generate this kind of trace.

Trying to formally describe Figure 5.2, we distinguish these elements:

1. **State Set (*S*)** The most obvious aspect of the traffic light is the state it is in, which is indicated by the three different colours it can have. These states are *sequential*: the traffic light can only be in one of these states at the same time[1]. The set of states is not limited to enumeration style as presented here, but can contain an arbitrary number of attributes.

---

[1] In contrast to, say, Statecharts.

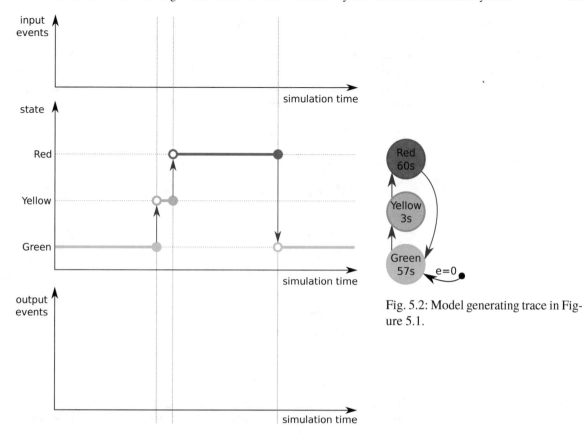

Fig. 5.2: Model generating trace in Figure 5.1.

Fig. 5.1: Trace of the autonomous traffic light.

$$S : \times_{i=1}^{n} S_i$$

2. **Time Advance** (ta) For each of the states just defined, we notice the timeout in them. Clearly, some states take longer to process than others. For example, whereas we will stay in green and red a relatively long time, the time in the yellow state is only brief. This function needs to be defined for each and every element of the state set, and needs to deterministically return a duration. The duration can be any positive real number, including zero and infinity. A negative time is disallowed, as this would require simulation to go back in time. DEVS allows a time advance of exactly zero, even though this is impossible in real life. Two use cases for this exist: the delay might be very small and irrelevant to the problem we are modelling, or the state is an artificial state, without any real-world equivalent (*e.g.*, as part of a design pattern). Note that DEVS does not consider time bases, despite the use of seconds in our visualisation. Simulation time is just a real number, and the interpretation given to it is up to the user. Whether these units indicate seconds, years, or even π seconds, is completely up to the users, as long as it is fixed throughout the simulation.

$$ta : S \rightarrow \mathbb{R}^+_{0,+\infty}$$

3. **Internal Transition** (δ_{int}) With the states and timeouts defined, the final part is the definition of which is the next state from a given state. This is the job of the internal transition function, which gives the next state for each and every state. As it is a function, every state has at most one next state, preventing any possible ambiguity. Note that the function does not necessarily have to be total, nor injective: some states might not have a next state (*i.e.*, if the time advance was specified as $+\infty$), and some states have the same state as next state. Up to now, only the internal transition function is described as changing the state. Therefore, it is not allowed for other functions (*e.g.*, time advance) to modify the state: their state access is read-only.

$$\delta_{int} : S \rightarrow S$$

4. **Initial Total State** (q_{init}) We also need to define the initial state of the system. While this is not present in the original specification of the DEVS formalism, we include it here as it is a vital element of the model [285]. But note that, instead of being an "initial state (s_{init})", it is a total state. This means that we not only select the initial state of the system, but also define how long we are already in this state. Elapsed time is therefore added to the definition of the initial total state, to allow more flexibility when modelling a system. To the simulator, it will seem as if the model has already been in the initial state for some time.

$$q_{init} : (s, e)|s \in S, 0 \le e \le ta(s)$$

We describe the model in Figure 5.2 as a 4-tuple of these three elements.

$$\langle S, q_{init}, \delta_{int}, ta \rangle$$

$$S = \{\text{GREEN, YELLOW, RED}\}$$
$$q_{init} = (\text{GREEN}, 0.0)$$
$$\delta_{int} = \{\text{GREEN} \rightarrow \text{YELLOW},$$
$$\text{YELLOW} \rightarrow \text{RED},$$
$$\text{RED} \rightarrow \text{GREEN}\}$$
$$ta = \{\text{GREEN} \rightarrow delay_{green},$$
$$\text{YELLOW} \rightarrow delay_{yellow},$$
$$\text{RED} \rightarrow delay_{red}\}$$

For this simple formalism, we define the semantics as in Algorithm 4. The model is initialized with simulation time set to 0, and the state set to the initial state (*e.g.*, GREEN). Simulation updates the time with the return value of the time advance function, and executes the internal transition function on the current state to get the new state.

Algorithm 4 DEVS simulation pseudo-code for autonomous models.

$time \leftarrow 0$
$current\_state \leftarrow initial\_state$
$last\_time \leftarrow -initial\_elapsed$
while not termination_condition() **do**
    $time \leftarrow last\_time + ta(current\_state)$
    $current\_state \leftarrow \delta_{int}(current\_state)$
    $last\_time \leftarrow time$
end while

5.2.2 Autonomous Model with Output

Recall that DEVS is a modular formalism, with only the atomic model having access to its internal state. This naturally raises a problem for our traffic light: others have no way of knowing its current state (*i.e.*, its colour).

We therefore want the traffic light to output its colour, in this case in the form of a string (and not as the element of an enumeration). For now, the output is tightly linked to the set of state, but this does not need to be the case: the possible values to output can be completely distinct from the set of states. Our desired trace is shown in Figure 5.3. We see that we now output events indicating the start of the specified period. Recall, also, that DEVS is a discrete event formalism: the output is only a single event indicating the time and is not a continuous signal. The receiver of the event thus would have to store the event to know the current state of the traffic light at any given point in time. Visually, the model is updated to Figure 5.4, using the exclamation mark on a transition to indicate output generation.

Analysing the updated model, we see that two more concepts are required to allow for output.

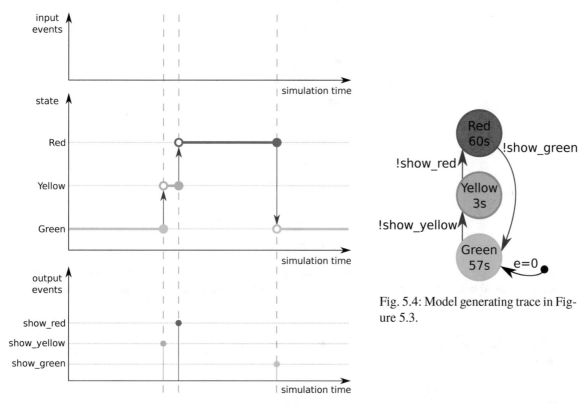

Fig. 5.4: Model generating trace in Figure 5.3.

Fig. 5.3: Trace of the autonomous traffic light with output.

1. **Output Set (Y)**

   Similarly to defining the set of allowable states, we should also define the set of allowable outputs. This set serves as an interface to other components, defining the events they expect to receive. Events can have complex attributes as well, though we again limit ourself to simple events for now. If ports are used, each port has its own output set.

   $$Y : \times_{i=1}^{l} Y_i$$

2. **Output Function (λ)**

   With the set of allowable events defined, we still need a function to actually generate the events. Similar to the other functions, the output function is defined on the state, and deterministically returns an event (or no event). As seen in the Figure of the model, the event is generated *before* the new state is reached. This means that instead of the new state, the output function still uses the old state (*i.e.*, the one that is being left). For this reason, the output function needs to be invoked right before the internal transition function. In the case of our traffic light, the output function needs to return the name of the *next* state, instead of the current state. For example, if the output function receives the GREEN state as input, it needs to generate a *show_yellow* event.

   Similar to the time advance function, this function does not output a new state, and therefore state access is read-only. This might require some workarounds: outputting an event often has some repercussions on the model state, such as removing the event from a queue or increasing a counter. Since the state cannot be written to, these changes need to be remembered and executed as soon as the internal transition is executed. Note that it is possible for the output function not to return any output, in which case it returns ϕ.

   $$\lambda : S \rightarrow Y \cup \{\phi\}$$

   The model can be described as a 6-tuple.

   $$\langle Y, S, q_{init}, \delta_{int}, \lambda, ta \rangle$$

$$Y = \{show\_green, show\_yellow, show\_red\}$$
$$S = \{\text{GREEN}, \text{YELLOW}, \text{RED}\}$$
$$q_{init} = (\text{GREEN}, 0.0)$$
$$\delta_{int} = \{\text{GREEN} \to \text{YELLOW},$$
$$\text{YELLOW} \to \text{RED},$$
$$\text{RED} \to \text{GREEN}\}$$
$$\lambda = \{\text{GREEN} \to show\_yellow,$$
$$\text{YELLOW} \to show\_red,$$
$$\text{RED} \to show\_green\}$$
$$ta = \{\text{GREEN} \to delay_{green},$$
$$\text{YELLOW} \to delay_{yellow},$$
$$\text{RED} \to delay_{red}\}$$

The pseudo-code is slightly altered to include output generation, as shown in Algorithm 5. Recall that output is generated before the internal transition is executed, so the method invocation happens right before the transition.

Algorithm 5 DEVS simulation pseudo-code for autonomous models with output.

```
time ← 0
current_state ← initial_state
last_time ← −initial_elapsed
while not termination_condition() do
    time ← last_time + ta(current_state)
    output(λ(current_state))
    current_state ← δ_int(current_state)
    last_time ← time
end while
```

5.2.3 Interruptable Model

Our current traffic light specification is still completely autonomous. While this is fine in most circumstances, police might want to temporarily shut down the traffic lights, when they are managing traffic manually. To allow for this, our traffic light must process externally generated incoming events; such as events from a policeman to shutdown or startup again. Figure 5.5 shows the trace we wish to obtain. A model generating this trace is shown in Figure 5.6, using a question mark to indicate event reception.

We once more require two additional elements in the DEVS specification.

1. **Input Set** (X)

   Similar to the output set, we need to define the events we expect to receive. This is again a definition of the interface, such that others know which events are understood by this model.

$$X = \times_{i=1}^{m} X_i$$

2. **External Transition** (δ_{ext})

   Similar to the internal transition function, the external transition function is allowed to define the new state as well. First and foremost, the external transition function is still dependent on the current state, just like the internal transition function. The external transition function has access to two more values; elapsed time and the input event. The *elapsed time* indicates how long it has been for this atomic model since the last transition (either internal or external). Whereas this number was implicitly known in the internal transition function (*i.e.*, the value of the time advance function), here it needs to be passed explicitly. Elapsed time

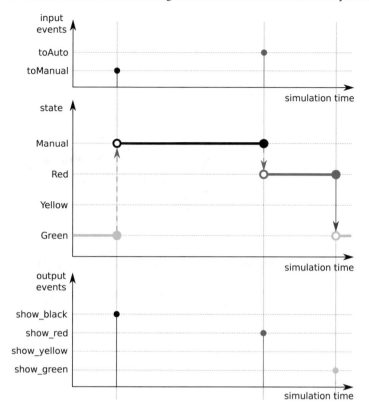

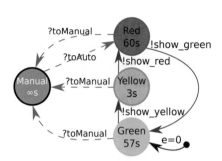

Fig. 5.6: Naive model that should generate the trace in Figure 5.5 (but does not).

Fig. 5.5: Trace of the autonomous traffic light.

is a number in the range $[0, ta(s)]$, with s being the current state of the model. Note that it is inclusive of both 0 and $ta(s)$: it is possible to receive an event exactly after a transition happened, or exactly before an internal transition happens. The combination of the current state and the elapsed time is often called the *total state* (Q) of the model. We have previously seen the total state, in the context of the initial total state. The received event is the final parameter to this function. A new state is deterministically defined through the combination of these three parameters. Since the external transition function takes multiple parameters, multiple external transitions might be defined for a single state.

$$\delta_{ext} : Q \times X \to S$$
$$Q = \{(s, e) | s \in S, 0 \le e \le ta(s)\}$$

While we now have all elements of the DEVS specification for atomic models, we are not done yet. When we include the additional state MANUAL, we also need to send out an output message indicating that the traffic light is off. But recall that an output function was only invoked before an internal transition function, so not before an external transition function. To have an output nonetheless, we need to make sure that an internal transition happens before we actually reach the MANUAL state. This can be done through the introduction of an artificial intermediate state, which times out immediately, and sends out the *turn_off* event. Instead of going to MANUAL upon reception of the *toManual* event, we go to the artificial state GOING_MANUAL. The time advance of this state is set to 0, since it is only an artificial state without any meaning in the domain under study. Its output function will be triggered immediately due to the time advance of zero, and the *turn_off* output is generated while transferring to MANUAL. Similarly, when we receive the *toAuto* event, we need to go to an artificial GOING_AUTO state to generate the *show_red* event. A visualization of the corrected trace and corresponding model is shown in Figure 5.7 and Figure 5.8 respectively.

Finally, we give the full specification of the traffic light as an atomic DEVS model, defined by a 8-tuple.

$$\langle X, Y, S, q_{init}, \delta_{int}, \delta_{ext}, \lambda, ta \rangle$$

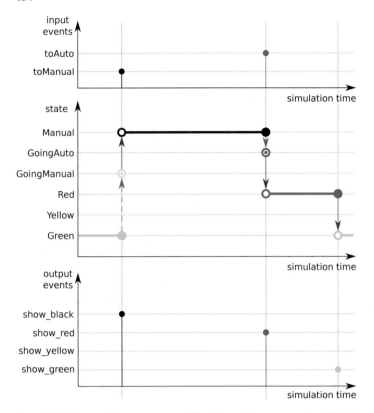

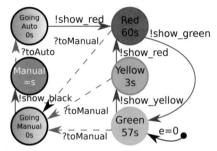

Fig. 5.8: Model generating trace in Figure 5.7.

Fig. 5.7: Trace of the interrupt traffic light with corrected artificial states.

$$X = \{toAuto, toManual\}$$

$$Y = \{show\_green, show\_yellow, show\_red, turn\_off\}$$

$$S = \{\text{GREEN, YELLOW, RED, GOING\_MANUAL, GOING\_AUTO, MANUAL}\}$$

$$q_{init} = (\text{GREEN}, 0.0)$$

$$\delta_{int} = \{\text{GREEN} \rightarrow \text{YELLOW},$$
$$\text{YELLOW} \rightarrow \text{RED},$$
$$\text{RED} \rightarrow \text{GREEN},$$
$$\text{GOING\_MANUAL} \rightarrow \text{MANUAL},$$
$$\text{GOING\_AUTO} \rightarrow \text{RED}\}$$

$$\delta_{ext} = \{(\text{GREEN}, \_, toManual) \rightarrow \text{GOING\_MANUAL}$$
$$(\text{YELLOW}, \_, toManual) \rightarrow \text{GOING\_MANUAL}$$
$$(\text{RED}, \_, toManual) \rightarrow \text{GOING\_MANUAL}$$
$$(\text{MANUAL}, \_, toAuto) \rightarrow \text{GOING\_AUTO}\}$$

$$\lambda = \{\text{GREEN} \rightarrow show\_yellow,$$
$$\text{YELLOW} \rightarrow show\_red,$$
$$\text{RED} \rightarrow show\_green,$$
$$\text{GOING\_MANUAL} \rightarrow turn\_off,$$
$$\text{GOING\_AUTO} \rightarrow show\_red\}$$

$$ta = \{\text{GREEN} \rightarrow delay_{green},$$
$$\text{YELLOW} \rightarrow delay_{yellow},$$
$$\text{RED} \rightarrow delay_{red},$$
$$\text{MANUAL} \rightarrow +\infty,$$
$$\text{GOING\_MANUAL} \rightarrow 0,$$
$$\text{GOING\_AUTO} \rightarrow 0\}$$

Algorithm 6 presents the complete semantics of an atomic model in pseudo-code. Similar to before, we still have the same simulation loop, but now we can be interrupted externally. At each time step, we need to determine whether an external interrupt is scheduled before the internal interrupt. If that is not the case, we simply continue like before, by executing the internal transition. If there is an external event that must go first, we execute the external transition.

Algorithm 6 DEVS simulation pseudo-code for interruptable models.

$time \leftarrow 0$
$current\_state \leftarrow initial\_state$
$last\_time \leftarrow -initial\_elapsed$
while not termination_condition() **do**
   $next\_time \leftarrow last\_time + ta(current\_state)$
   if $time\_next\_event \leq next\_time$ **then**
      $elapsed \leftarrow time\_next\_event - last\_time$
      $current\_state \leftarrow \delta_{ext}((current\_state, elapsed), next\_event)$
      $time \leftarrow time\_next\_event$
   else
      $time \leftarrow next\_time$
      $output(\lambda(current\_state))$
      $current\_state \leftarrow \delta_{int}(current\_state)$
   end if
   $last\_time \leftarrow time$
end while

Reflective Question Why is the output function not called as well before the external transition is invoked?

5.3 Coupled DEVS Models

While our traffic light example is able to receive and output events, there are no other atomic models to communicate with. To combine different atomic models together and have them communicate, we now introduce coupled models. This will be done in the context of our previous traffic light, which is connected to a policeman. The details of the traffic light are exactly like before; the details of the policeman are irrelevant here, as long as it outputs *toAuto* and *toManual* events.

5.3.1 Basic Coupling

The first problem we encounter with coupling the traffic light and policeman together is the structure: how do we define a set of models and their interrelations? This is the core definition of a coupled model: it is merely a structural model that couples models together. Contrary to the atomic models, there is *no behaviour whatsoever* associated to a coupled model. Behaviour is the responsibility of atomic models, and structure that of coupled models.

To define the basic structure, we need three elements.

1. **Model instances** (D)
   The set of model instances defines which models are included within this coupled model.
2. **Model specifications** ($MS = \{M_i | i \in D\}$)
   Apart from defining the different instances of submodels, we must include the atomic model specification of these models. For each element defined in D, we include the 8-tuple specifying the atomic model. By definition, a submodel of the coupled DEVS model always needs to be an atomic model. Later on, we will see how this can be extended to support arbitrarily hierarchies.

$$MS = \{M_i | i \in D\} = \{\langle X_i, Y_i, S_i, q_{init,i}, \delta_{int,i}, \delta_{ext,i}, \lambda_i, ta_i\rangle | i \in D\}$$

3. **Model influencees** ($IS = \{I_i | i \in D \cup \{self\}\}$)
   Apart from defining the model instances and their specifications, we need to define the connections between them. Connections are defined through the use of influencee sets: for each atomic model instance, we define

the set of models influenced by that model. There are some limitations on couplings, to make sure that inconsistent models cannot be created. The following two constraints are imposed:

- *A model should not influence itself.* This constraint makes sense, as otherwise it would be possible for a model to influence itself directly. While there is no significant problem with this in itself, it would cause the model to trigger both its internal and external transition simultaneously. As it is undefined which one should go first, this situation is not allowed. In other words, a model should not be an element in its own set of influencees.

$$\forall i \in D : i \notin I_i$$

- *Only links within the coupled model are allowed.* This is another way of saying that connections should respect modularity. Models should not directly influence models outside of the current coupled model, nor models deeper inside of other submodels at this level. In other words, the influenced model should be a subset of the set of models in this coupled model.

$$\forall i \in D : I_i \subseteq D$$

Note that there is no explicit constraint on algebraic loops (*i.e.*, a loop of models that have a time advance equal to zero, preventing the progression of simulated time). If this situation is not resolved, it is possible for simulation to get stuck at that specific point in time. The situation is only problematic if the circular dependency never gets resolved, causing a livelock of the simulation.

A coupled model can thus be defined as a 3-tuple.

$$\langle D, MS, IS \rangle$$

5.3.2 Input and Output

Our coupled model now couples two atomic models together. And while it is now possible for the policeman to pass the event to the traffic light, we again lost the ability to send out the state of the traffic light. The events cannot reach outside of the current coupled model. Therefore, we need to augment the coupled model with input and output events, which serve as the interface to the coupled model. This adds the components X_{self} and Y_{self} to the tuple, respectively the set of input and output events, resulting in a 5-tuple.

$$\langle X_{self}, Y_{self}, D, MS, IS \rangle$$

The constraints on the couplings need to be relaxed to accommodate for the new capabilities of the coupled model: a model can be influenced by the input events of the coupled model, and likewise the models can also influence the output events of the coupled model. The previously defined constraints are relaxed to allow for *self*, the coupled model itself.

$$\forall i \in D \cup \{self\} : i \notin I_i$$
$$\forall i \in D \cup \{self\} : I_i \subseteq D \cup \{self\}$$

5.3.3 Tie-breaking

Recall that DEVS is considered a formal and precise formalism. But while all components are precisely defined, their interplay is not completely defined yet: what happens when the traffic light changes its state at exactly the same time as the policeman performs its transition? Would the traffic light switch on to the next state first and then process the policeman's interrupt, or would it directly respond to the interrupt, ignoring the internal event? While it is a minimal difference in this case, the state reached after the timeout might respond significantly different to the incoming event.

DEVS solves this problem by defining a **tie-breaking function** (*select*). This function takes all conflicting models and returns the one that gets priority over the others. After the execution of that internal transition, and possibly the external transitions that it caused elsewhere, it might be that the set of imminent models has changed. If multiple models are still imminent, we repeat the above procedure (potentially invoking the *select* function again with the new set of imminent models).

$$select : 2^D \rightarrow D$$

This new addition changes the coupled model to a 6-tuple.

$$\langle X_{self}, Y_{self}, D, MS, IS, select \rangle$$

5.3.4 Translation Functions

Finally, in this case we had full control over both atomic models that are combined. We might not always be that lucky, as it is possible to reuse atomic models defined elsewhere. Depending on the application domain of the reused models, they might work with different events. For example, if our policeman and traffic light were both predefined, with the policeman using *go_to_work* and *take_break* and the traffic light listening to *toAuto* and *toManual*, it would be impossible to directly couple them together. While it is possible to define wrapper blocks (*i.e.*, artificial atomic models that take an event as input and, with time advance equal to zero, output the translated version), DEVS provides a more elegant solution to this problem.

Connections are augmented with a **translation function** ($Z_{i,j}$), specifying how the event that enters the connection is translated before it is handed over to the endpoint of the connection. The function thus maps output events to input events, potentially modifying their content.

$$
\begin{aligned}
Z_{self,j} &: X_{self} \rightarrow X_j \;\; \forall j \in D \\
Z_{i,self} &: Y_i \rightarrow Y_{self} \;\; \forall i \in D \\
Z_{i,j} &: Z_i \rightarrow X_j \;\;\;\; \forall i, j \in D
\end{aligned}
$$

These translation functions are defined for each connection, including those between the coupled model's input and output events.

$$ZS = \{Z_{i,j} | i \in D \cup \{self\}, j \in I_i\}$$

The translation function is implicitly assumed to be the identity function if it is not defined. In case an event needs to traverse multiple connections, all translation functions are chained in order of traversal.

With the addition of this final element, we define a coupled model as a 7-tuple.

$$\langle X_{self}, Y_{self}, D, MS, IS, ZS, select \rangle$$

5.3.5 Closure Under Coupling

Similar to atomic models, we need to formally define the semantics of coupled models. But instead of explaining the semantics from scratch, by defining some pseudo-code, we map coupled models to equivalent atomic models. Semantics of a coupled model is thus defined in terms of an atomic model. In addition, this flattening removes the constraint of coupled models that their submodels should be atomic models: if a coupled model is a submodel, it can be flattened to an atomic model.

In essence, for any coupled model specified as

$$< X_{self}, Y_{self}, D, MS, IS, ZS, select >$$

we define an equivalent atomic model specified as

$$< X, Y, S, q_{init}, \delta_{int}, \delta_{ext}, \lambda, ta >$$

Therefore, we have to define all the elements of the atomic model in terms of elements of the coupled model. The input and output variables X and Y are easy, since they stay the same.

$$X = X_{self}$$
$$Y = Y_{self}$$

From an external point of view, the interface of the atomic and coupled model is identical: it has the same input and output events and expects the same kind of data on all of them.

The state S encompasses the parallel composition of the states of all the submodels, including their elapsed times (*i.e.*, the total state Q, as defined previously):

$$S = \times_{i \in D} Q_i$$

with the total states Q_i previously defined as:

$$Q_i = \{(s_i, e_i)|s_i \in S_i, 0 \le e_i \le ta_i(s_i)\}, \forall i \in D$$

The elapsed time is stored for each model separately, since the elapsed time of the new atomic model updates more frequently than each submodel's elapsed time.

The initial total state is again composed of two components: the initial state (s_{init}) and the initial elapsed time (e_{init}). First, we consider the elapsed time (e_{init}), which is intuitively equal to the time since the last transition of the flattened model, meaning that it is the minimum of all initial elapsed times, found in all atomic submodels. Then, the initial state (s_{init}) is to be specified, which is the composition of all initial total states of all atomic submodels. Note, however, that for each initial total state of the submodels, the minimum elapsed time (e_{init}) is to be decremented. Indeed, q_{init} specifies that we entered state s_{init} in the past, more specifically e_{init} time units ago.

$$q_{init} = (s_{init}, e_{init})$$
$$s_{init} = (..., (s_{init,i}, e_{init,i} - e_{init}), ...)$$
$$e_{init} = min_{i \in D}\{e_{init,i}\}$$

The time advance function ta then returns the minimum of all remaining times.

$$ta(s) = min\{\sigma_i = ta_i(s_i) - e_i | i \in D\}$$

The imminent component is chosen from the set of all models with the specified minimum remaining time (IMM). This set contains all models whose remaining time (σ_i) is identical to the time advance of the flattened model (ta). The *select* function is then used to reduce this set to a single element i^*.

$$IMM(s) = \{i \in D | \sigma_i = ta(s)\}$$
$$i^* = select(IMM(s))$$

The output function λ executes the output function of i^* and applies the translation function, but only if the model influences the flattened model directly (*i.e.*, if the output of i^* is routed to the coupled model's output). If there is no connection to the coupled model's output (*i.e.*, i^* is only coupled to other atomic models), no output function is invoked here. We will see later on that these events are still generated, but they are consumed internally elsewhere.

$$\lambda(s) = \begin{cases} Z_{i^*,self}(\lambda_{i^*}(s_{i^*})) & \text{if } self \in I_{i^*} \\ \phi & \text{otherwise} \end{cases}$$

The *internal transition function* is defined for each part of the state separately:

$$\delta_{int}(s) = (\ldots, (s_j', e_j'), \ldots)$$

With three kinds of models: (1) the model i^* itself, which just performs its internal transition function; (2) the models influenced by i^*, which perform their external transition based on the output generated by i^*; (3) models unrelated to i^*. In all cases, the elapsed time is updated.

$$(s'_j, e'_j) = \begin{cases} (\delta_{int,j}(s_j), 0) & \text{for } j = i^*, \\ (\delta_{ext,j}((s_j, e_j + ta(s)), Z_{i^*,j}(\lambda_{i^*}(s_{i^*}))), 0) & \text{for } j \in I_{i^*}, \\ (s_j, e_j + ta(s)) & \text{otherwise} \end{cases}$$

Note that the internal transition function includes external transition functions of submodels for those models influenced by i^*. As i^* outputs events that are consumed internally, this all happens internally.

The *external transition* function is similar to the internal transition function. Now two types are distinguished: (1) models directly connected to the input of the model, which perform their external transition; (2) models not directly connected to the input of the model, which only update their elapsed time.

$$\delta_{ext}((s, e), x) = (\dots, (s'_i, e'_i), \dots)$$

$$(s'_i, e'_i) = \begin{cases} (\delta_{ext,i}((s_i, e_i + e), Z_{self,i}(x)), 0) & \text{for } i \in I_{self} \\ (s_i, e_i + e) & \text{otherwise} \end{cases}$$

Reflective Question Is it possible to replace the translation function Z by an atomic DEVS model that takes input and puts the translated value on its output, while preserving semantics?

5.4 The DEVS Abstract Simulator

Up to now, the semantics of atomic models was defined through natural language and high-level pseudo-code. Coupled models were given semantics through a mapping to these atomic models. Both of these have their own problems. For atomic models, the pseudo-code is not sufficiently specific to create a compliant DEVS simulator: a lot of details of the algorithm are left unspecified (*e.g.*, where does the external event come from). For coupled models, the flattening procedure is elegant and formal, though it is highly inefficient to perform this flattening at run-time.

To counter these problems, we will define a more elaborate, and formal, simulation algorithm for both atomic and coupled models. Atomic models get a more specific definition with a clear interface, and coupled models get their own simulation algorithm without flattening. Coupled models are thus given "operational semantics" instead of "translational semantics".

This simulation algorithm, an *abstract simulator* forms the basis for more efficient simulation algorithms, and serves as a reference algorithm. Its goal is to formally define the semantics of both models in a concise way, without caring about performance or implementation issues. Adaptations are allowed, but the final result should be identical: simulation results are to be completely independent from the implementation. A direct implementation of the abstract simulator is inefficient, and actual implementations therefore vary significantly.

We now elaborate on the abstract simulator algorithm. For each atomic and coupled model, an instance is created of the respective algorithm.

Table 5.1 shows the different variables used, their type, and a brief explanation.

Table 5.1: Variables used in the abstract simulator.

name	type	explanation
t_l	time	simulation time of last transition
t_n	time	simulation time of next transition
t	time	current simulation time
e	time	elapsed time since last transition
s	state	current state of the atomic model
x	event	incoming event
y	event	outgoing event
from	model	source of the incoming message
parent	model	coupled model containing this model
self	model	current model

We furthermore distinguish five types of synchronization messages, as exchanged between the different abstract simulators. An overview of messages is shown in Table 5.2.

Table 5.2: Types of synchronization messages.

type	explanation
i	initialization of the simulation
$*$	transition in the model
x	input event for the model
y	output event from the model
$done$	computation finished for a model

First is the abstract simulation algorithm for atomic models, presented in Algorithm 7. This algorithm consists of a big conditional, depending on the message that is received. Atomic models only perform an operation upon reception of a message: there is no autonomous behaviour. This algorithm is invoked every time a synchronisation message is received. Messages consist of three components: the type of the message, the source of the message, and the simulation time. The conditional consists of three options: On the reception of an i message, we perform *initialization* of the simulation time.

Another option is the reception of a $*$ message, triggering a *transition*. The message consists of both a sender and the time at which the transition should happen. By definition, a transition can only happen at time t_n, so we assert this. After this check, we have to perform the following steps: (1) generate the output, (2) send it out to the sender of the $*$ message (our parent), (3) perform the internal transition, (4) update our time with the time advance, and (5) indicate to our parent that we finished processing the message, also passing along our time of next transition.

Finally, it is possible to receive an x message, indicating *external input*. This can happen anytime between our last transition (t_l), and our scheduled transition (t_n), so we again assert the simulation time. Note that these times are inclusive: due to the *select* function it is possible that another model comes right after or before our own scheduled transition. We perform the following steps: (1) compute the elapsed time (e) based on the provided simulation time (t), (2) perform the external transition, (3) update the simulation time of the next transition, and (4) indicate to our parent that we finished processing the message, also passing along our time of next transition.

Algorithm 7 DEVS atomic model abstract simulator.

---

if receive $(i, from, t)$ message **then**
    $t_l \leftarrow t - e$
    $t_n \leftarrow t_l + ta(s)$
    send $(done, self, t_n)$ to $parent$
else if receive $(*, from, t)$ message **then**
    if $t = t_n$ **then**
        $y \leftarrow \lambda(s)$
        if $y \neq \phi$ **then**
            send $(y, self, t)$ to $parent$
        end if
        $s \leftarrow \delta_{int}(s)$
        $t_l \leftarrow t$
        $t_n \leftarrow t_l + ta(S)$
        send $(done, self, t_n)$ to parent
    end if
else if receive $(x, from, t)$ message **then**
    if $t_l \leq t \leq t_n$ **then**
        $e \leftarrow t - t_l$
        $s \leftarrow \delta_{ext}((s, e), x)$
        $t_l \leftarrow t$
        $t_n \leftarrow t_l + ta(s)$
        send $(done, self, t_n)$ to $parent$
    else
        error: bad synchronization
    end if
end if

Recall that the abstract simulation algorithm did not have any autonomous behaviour. This indicates that there is another entity governing the progression of the simulation This simulation entity is the root coordinator, and it encodes the main simulation loop. Its algorithm is shown in Algorithm 8. As long as simulation needs to continue, it sends out a message to the topmost model in the hierarchy to perform transitions. When a reply is received, simulation time is progressed to the time indicated by the topmost model.

Algorithm 8 DEVS root coordinator.

---

send $(i, main, 0.0)$ to topmost coupled model top
wait for $(done, top, t_N)$
$t \leftarrow t_N$
while not $terminationCondition()$ **do**
    send $(*, main, t)$ to topmost coupled model top
    wait for $(done, top, t_N)$
    $t \leftarrow t_N$
end while

Finally, while not completely necessary due to the existence of the flattening algorithm, we also define a shortcut for the simulation of coupled models. The abstract simulation algorithm for coupled models is shown in Algorithm 9. Coupled models can receive all five different types of synchronisation messages.

First, the i message again indicates *initialization*. It merely forwards the message to all of its children and marks each child as active. Every coupled model has a t_l and t_n variable as well, which is defined as the maximum, respectively minimum, of its children. This is logical, as any transition of its children will also require an operation on the coupled model containing it. When a message is sent to a submodel, the submodel is marked as active. The use for this is shown in the processing of the *done* message.

Second, the $*$ message again indicates a *transition*. Contrary to the atomic models, a coupled model is unable to perform a transition itself. Instead, it forwards the message to the imminent submodel, found by executing the *select* function for all models that have that exact same t_n. Only a single model will be selected, and a $*$ message is sent to that model. Just like before, the model is marked as active to make sure that we wait for its computation to finish.

Third, a y message indicates an *output* message. The output message is output by the output function of a subcomponent, and needs to be routed through the coupled model. This part of the function is responsible for routing the message to the influencees of the model that sent out the message. Note that it is also possible that one of the influencees is $self$, indicating that the message needs to be routed externally (*i.e.*, to the output of the coupled model). In any case, the message needs to be translated using the translation function. The actual translation function that is invoked depends on the source and destination of the message.

Fourth, a x message can be received, indicating *input*. This is mostly identical to the output messages, only now can we also handle messages that were received from our own parent.

Finally, a *done* message can be received, indicating that a submodel has *finished* its computation. The submodel, which was marked as an active child, will now be unmarked. When *done* messages are received from all submodels (*i.e.*, all children are inactive), we determine our own t_l and t_n variables and send out the minimal t_n of all submodels. This time is then sent to the parent.

The abstract simulator for coupled models can work with any kind of submodel, not necessarily atomic models. In deep hierarchies, the *done* message always propagates the minimal t_n upwards in the hierarchy. In the end, the root coordinator will always receive the minimal t_n, which is the time of the earliest next internal transition.

Reflective Question Is it possible for an atomic DEVS model to do an internal and external transition at the same point in simulated time? Explain your answer.

5.5 Application to Queueing Systems

The usefulness of DEVS of course goes further than traffic lights. To present a more realistic model and highlight the potential for performance analysis, we present a simple queueing system next. While a lot has been done in queueing theory, we present simulation as an alternative to the mathematical solutions. Even though the

Algorithm 9 DEVS coupled model abstract simulator.

if receive $(i, from, t)$ message **then**
    for all d in D **do**
        send $(i, self, t)$ to d
        $active\_children \leftarrow active\_children \cup \{d\}$
    end for
else if receive $(*, from, t)$ message **then**
    if $t = t_n$ **then**
        $i* = select(\{M_i.t_n = t | i \in D\})$
        send $(*, self, t)$ to $i*$
        $active\_children \leftarrow active\_children \cup \{i^*\}$
    end if
else if receive $(y, from, t)$ message **then**
    for all $i \in I_{from} \setminus \{self\}$ **do**
        send $(Z_{from,i}(y), from, to)$ to i
        $active\_children \leftarrow active\_children \cup \{i\}$
    end for
    if $self \in I_{from}$ **then**
        send $(Z_{from,self}(y), self, t)$ to $parent$
    end if
else if receive $(x, from, t)$ message **then**
    if $t_l \leq t \leq t_n$ **then**
        for all $i \in I_{from}$ **do**
            send $(Z_{self,i}(x), self, t)$ to i
            $active\_children \leftarrow active\_children \cup \{i\}$
        end for
    end if
else if receive $(done, from, t)$ message **then**
    $active\_children \leftarrow active\_children \setminus \{from\}$
    if $active\_children = \phi$ **then**
        $t_l \leftarrow max\{t_{l,d} | d \in D\}$
        $t_n \leftarrow min\{t_{n,d} | d \in D\}$
        send $(done, self, t_n)$ to $parent$
    end if
end if

mathematical solutions have their advantages, simulation offers more flexibility and does not get that complex. It is, however, necessarily limited to "sampling": simulations will only take samples and will therefore generally not find rare and exceptional cases. Not taking them into account is fine in many situations, as it is now in our example model.

In this section, we present a simple queueing problem. Variations on this model — in either its behaviour, structure, or parameters — are easy to do.

5.5.1 Problem Description

In this example, we model the behaviour of a simple queue that gets served by multiple processors. Implementations of this queueing systems are widespread, such as for example at airport security. Our model is parametrisable in several ways: we can define the random distribution used for event generation times and event size, the number of processors, performance of each individual processor, and the scheduling policy of the queue when selecting a processor. Clearly, it is easier to implement this, and all its variants, in DEVS than it is to model it mathematically. For our performance analysis, we show the influence of the number of processors (*e.g.*, metal detectors) on the average and maximal queueing time of jobs (*e.g.*, travellers).

A model of this system can be shown in Figure 5.9. Events (people) are generated by a generator using some distribution function. They enter the queue, which decides the processor that they will be sent to. If multiple processors are available, it picks the processor that has been idle for the longest; if no processors are available, the event is queued until a processor becomes available. The queue works First-In-First-Out (FIFO) in case multiple events are queueing. For a processor to signal that it is available, it needs to signal the queue. The queue keeps track of available processors. When an event arrives at a processor, it is processed for some time, depending on the size of the event and the performance characteristics of the processor. After processing, the processor signals the queue and sends out the event that was being processed.

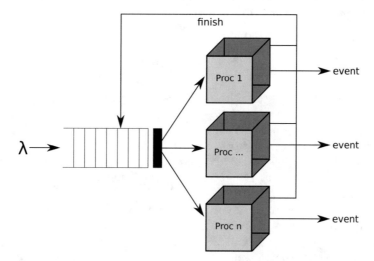

Fig. 5.9: Queue system with a single generator, single queue, and *n* processors.

5.5.2 Description in DEVS

While examples could be given purely in their formal description, they would not be executable and would introduce a significant amount of accidental complexity. We use the tool PythonPDEVS[2] [282, 283] to implement the DEVS model and perform simulations. In PythonPDEVS, DEVS models are implemented by defining methods that implement the different aspects of the tuple. All code within these methods is just normal Python code, though a minimal number of API calls is required in the case of a coupled DEVS model. Since most DEVS tools work similarly, these examples could easily be transposed to other DEVS simulation tools. An overview of popular DEVS simulation tools is shown in [284].

To specify this model, we first define the event exchanged between different models: the *Job*. A job is coded as a class Job. It has the attributes *size* (*i.e.*, indicative of processing time) and *creation time* (*i.e.*, for statistic gathering). The Job class definition is shown in Listing 1.

```
class Job:
    def __init__(self, size, creation_time):
        # Jobs have a size and creation_time parameter
        self.size = size
        self.creation_time = creation_time
```
Listing 1: PythonPDEVS code for the Job event.

We now focus on each atomic model separately, starting at the event generator.

The *generator* is defined as an atomic model using the class Generator, shown in Listing 2. Classes that represent an atomic model inherit from the AtomicDEVS class. They should implement methods that implement each of the DEVS components. Default implementations are provided for a passivated model, such that unused functions do not need to be defined. In the constructor, input and output ports are defined, as well as model parameters and the initial state. We see that the definition of the generator is very simple: we compute the time remaining until the next event (remaining), and decrement the number of events to send. The generator also keeps track of the current simulation time, in order to set the creation time of events. The time advance function returns the time remaining until the next internal transition. Finally, the output function returns a new customer

---

[2] Download: http://msdl.cs.mcgill.ca/projects/DEVS/PythonPDEVS

event with a randomly defined size. The job has an attribute containing the time at which it was generated. Recall, however, that the output function was invoked before the internal transition, so the current time has not yet been updated by the internal transition. Therefore, the output function also has to do this addition, without storing the result in the state (as it cannot write to the state).

```python
from pypdevs.DEVS import AtomicDEVS
from job import Job
import random

# Define the state of the generator as a structured object
class GeneratorState:
    def __init__(self, gen_num):
        # Current simulation time (statistics)
        self.current_time = 0.0
        # Remaining time until generation of new event
        self.remaining = 0.0
        # Counter on how many events to generate still
        self.to_generate = gen_num

class Generator(AtomicDEVS):
    def __init__(self, gen_param, size_param, gen_num):
        AtomicDEVS.__init__(self, "Generator")
        # Output port for the event
        self.out_event = self.addOutPort("out_event")
        # Define the state
        self.state = GeneratorState(gen_num)

        # Parameters defining the generator's behaviour
        self.gen_param = gen_param
        self.size_param = size_param

    def intTransition(self):
        # Update simulation time
        self.state.current_time += self.timeAdvance()
        # Update number of generated events
        self.state.to_generate -= 1
        if self.state.to_generate == 0:
            # Already generated enough events, so stop
            self.state.remaining = float('inf')
        else:
            # Still have to generate events, so sample for new duration
            self.state.remaining = random.expovariate(self.gen_param)
        return self.state

    def timeAdvance(self):
        # Return remaining time; infinity when generated enough
        return self.state.remaining

    def outputFnc(self):
        # Determine size of the event to generate
        size = max(1, int(random.gauss(self.size_param, 5)))
        # Calculate current time (note the addition!)
        creation = self.state.current_time + self.state.remaining
        # Output the new event on the output port
        return {self.out_event: Job(size, creation)}
```

Listing 2: PythonPDEVS code for the Generator atomic model.

Next up is the queue, which is the most interesting component of the simulation, as it is the part we wish to analyze. The Queue implementation is similar in structure to the Generator. Of course, the DEVS parts get a different specification, as shown in Listing 3. The queue takes a structural parameter, specifying the number of processors. This is needed since the queue has an output port for each processor. When an internal transition happens, the queue knows that it has just output an event to the first idle processor. It thus marks the first idle processor as busy, and removes the event it was currently processing. If there are events remaining in the queue, and a processor is available to process it, we process the first element from the queue and set the remaining_time counter. In the external transition, we check the port we received the event on. Either it is a signal of the processor to indicate that it has finished, or else it is a new event to queue. In the former case, we mark the processor that sent the event as idle, and potentially process a queued message. For this to work, the processor should include its ID in the event, as otherwise the queue has no idea who sent this message. In the latter case, we either process the event immediately if there are idle processors, or we store it in the queue. The time advance merely has to return the remaining_time counter that is managed in both transition functions. Finally in the output function, the model outputs the first queued event to the first available processor. Note that we can only read the events and processors, and cannot modify these lists: state modification is reserved for the transition functions. An important consideration in this model is the remaining_time counter, which indicates how much time remains before the event is processed. We cannot simply put the processing time of events in the time advance, as interrupts could happen during this time. When an interrupt happens (*e.g.*, another event arrives), the time advance is invoked again, and would return the total processing time, instead of the remaining time to process the event. To solve this problem, we maintain a counter that explicitly gets decremented when an external interrupt happens.

```python
from pypdevs.DEVS import AtomicDEVS

# Define the state of the queue as a structured object
class QueueState:
    def __init__(self, outputs):
        # Keep a list of all idle processors
        self.idle_procs = range(outputs)
        # Keep a list that is the actual queue data structure
        self.queue = []
        # Keep the process that is currently being processed
        self.processing = None
        # Time remaining for this event
        self.remaining_time = float("inf")


class Queue(AtomicDEVS):
    def __init__(self, outputs):
        AtomicDEVS.__init__(self, "Queue")
        # Fix the time needed to process a single event
        self.processing_time = 1.0
        self.state = QueueState(outputs)

        # Create 'outputs' output ports
        # 'outputs' is a structural parameter!
        self.out_proc = []
        for i in range(outputs):
            self.out_proc.append(self.addOutPort("proc_%i" % i))

        # Add the other ports: incoming events and finished event
        self.in_event = self.addInPort("in_event")
        self.in_finish = self.addInPort("in_finish")

    def intTransition(self):
        # Is only called when we are outputting an event
        # Pop the first idle processor and clear processing event
```

```
        self.state.idle_procs.pop(0)
        if self.state.queue and self.state.idle_procs:
            # There are still queued elements, so continue
            self.state.processing = self.state.queue.pop(0)
            self.state.remaining_time = self.processing_time
        else:
            # No events left to process, so become idle
            self.state.processing = None
            self.state.remaining_time = float("inf")
        return self.state

    def extTransition(self, inputs):
        # Update the remaining time of this job
        self.state.remaining_time -= self.elapsed
        # Several possibilities
        if self.in_finish in inputs:
            # Processing a "finished" event, so mark proc as idle
            self.state.idle_procs.append(inputs[self.in_finish])
            if not self.state.processing and self.state.queue:
                # Process first task in queue
                self.state.processing = self.state.queue.pop(0)
                self.state.remaining_time = self.processing_time
        elif self.in_event in inputs:
            # Processing an incoming event
            if self.state.idle_procs and not self.state.processing:
                # Process when idle processors
                self.state.processing = inputs[self.in_event]
                self.state.remaining_time = self.processing_time
            else:
                # No idle processors, so queue it
                self.state.queue.append(inputs[self.in_event])
        return self.state

    def timeAdvance(self):
        # Just return the remaining time for this event (or infinity else
            )
        return self.state.remaining_time

    def outputFnc(self):
        # Output the event to the processor
        port = self.out_proc[self.state.idle_procs[0]]
        return {port: self.state.processing}
```

Listing 3: PythonPDEVS code for the Queue atomic model.

The next atomic model is the Processor class, shown in Listing 4. It merely receives an incoming event and starts processing it. Processing time, computed upon receiving an event in the external transition, is dependent on the size of the task, but takes into account the processing speed and a minimum amount of processing that needs to be done. After the task is processed, we trigger our output function and internal transition function. We need to send out two events: one containing the job that was processed, and one to signal the queue that we have become available. For this, two different ports are used. Note that the definition of the processor would not be this simple in case there was no queue before it. We can now make the assumption that when we get an event, we are already idle and therefore don't need to queue new incoming events first.

```
from pypdevs.DEVS import AtomicDEVS

# Define the state of the processor as a structured object
```

```python
class ProcessorState(object):
    def __init__(self):
        # State only contains the current event
        self.evt = None

class Processor(AtomicDEVS):
    def __init__(self, nr, proc_param):
        AtomicDEVS.__init__(self, "Processor_%i" % nr)

        self.state = ProcessorState()
        self.in_proc = self.addInPort("in_proc")
        self.out_proc = self.addOutPort("out_proc")
        self.out_finished = self.addOutPort("out_finished")

        # Define the parameters of the model
        self.speed = proc_param
        self.nr = nr

    def intTransition(self):
        # Just clear processing event
        self.state.evt = None
        return self.state

    def extTransition(self, inputs):
        # Received a new event, so start processing it
        self.state.evt = inputs[self.in_proc]
        # Calculate how long it will be processed
        time = 20.0 + max(1.0, self.state.evt.size / self.speed)
        self.state.evt.processing_time = time
        return self.state

    def timeAdvance(self):
        if self.state.evt:
            # Currently processing, so wait for that
            return self.state.evt.processing_time
        else:
            # Idle, so don't do anything
            return float('inf')

    def outputFnc(self):
        # Output the processed event and signal as finished
        return {self.out_proc: self.state.evt,
                self.out_finished: self.nr}
```

Listing 4: PythonPDEVS code for the Processor atomic model.

The processor finally sends the task to the Collector class, shown in Listing 5. The collector is an artificial component that is not present in the system being modelled; it is only used for statistics gathering. For each job, it stores the time in the queue.

```python
from pypdevs.DEVS import AtomicDEVS

# Define the state of the collector as a structured object
class CollectorState(object):
    def __init__(self):
        # Contains received events and simulation time
        self.events = []
```

```
        self.current_time = 0.0
class Collector(AtomicDEVS):
    def __init__(self):
        AtomicDEVS.__init__(self, "Collector")
        self.state = CollectorState()
        # Has only one input port
        self.in_event = self.addInPort("in_event")

    def extTransition(self, inputs):
        # Update simulation time
        self.state.current_time += self.elapsed
        # Calculate time in queue
        evt = inputs[self.in_event]
        time = self.state.current_time - evt.creation_time - evt.
            processing_time
        inputs[self.in_event].queueing_time = max(0.0, time)
        # Add incoming event to received events
        self.state.events.append(inputs[self.in_event])
        return self.state

    # Don't define anything else, as we only store events.
    # Collector has no behaviour of its own.
```

Listing 5: PythonPDEVS code for the Collector atomic model.

With all atomic models defined, we only have to couple them together in a coupled model, as shown in Listing 6. In this system, we instantiate a generator, queue, and collector, as well as a variable number of processors. The number of processors is variable, but is still static during simulation. The couplings also depend on the number of processors, as each processor is connected to the queue and the collector.

```
from pypdevs.DEVS import CoupledDEVS

# Import all models to couple
from generator import Generator
from queue import Queue
from processor import Processor
from collector import Collector

class QueueSystem(CoupledDEVS):
    def __init__(self, mu, size, num, procs):
        CoupledDEVS.__init__(self, "QueueSystem")

        # Define all atomic submodels of which there are only one
        generator = self.addSubModel(Generator(mu, size, num))
        queue = self.addSubModel(Queue(len(procs)))
        collector = self.addSubModel(Collector())

        self.connectPorts(generator.out_event, queue.in_event)

        # Instantiate desired number of processors and connect
        processors = []
        for i, param in enumerate(procs):
            processors.append(self.addSubModel(
                            Processor(i, param)))
            self.connectPorts(queue.out_proc[i],
                            processors[i].in_proc)
```

```
                    self.connectPorts(processors[i].out_finished,
                                      queue.in_finish)
                    self.connectPorts(processors[i].out_proc,
                                      collector.in_event)

        # Make it accessible outside of our own scope
        self.collector = collector
```
Listing 6: PythonPDEVS code for the System coupled model.

Now that our DEVS model is completely specified, we can start running simulations on it. Simulation requires an *experiment* file though, which initializes the model with parameters and defines the simulation configuration. An example experiment, again in Python, is shown in Listing 7. The experiment writes out the raw queueing times to a Comma Separated Value (CSV) file. An experiment file often contains some configuration of the simulation tool, which differs for each tool. For PythonPDEVS, the documentation[3] provides an overview of supported options.

```
from pypdevs.simulator import Simulator
import random

# Import the model we experiment with
from system import QueueSystem

# Configuration:
# 1) number of customers to simulate
num = 500
# 2) average time between two customers
time = 30.0
# 3) average size of customer
size = 20.0
# 4) efficiency of processors (products/second)
speed = 0.5
# 5) maximum number of processors used
max_processors = 10
# End of configuration

# Store all results for output to file
values = []
# Loop over different configurations
for i in range(1, max_processors):
    # Make sure each of them simulates exactly the same workload
    random.seed(1)
    # Set up the system
    procs = [speed] * i
    m = QueueSystem(mu=1.0/time, size=size, num=num, procs=procs)

    # PythonPDEVS specific setup and configuration
    sim = Simulator(m)
    sim.setClassicDEVS()
    sim.simulate()

    # Gather information for output
    evt_list = m.collector.state.events
    values.append([e.queueing_time for e in evt_list])
```

---

[3] http://msdl.cs.mcgill.ca/projects/DEVS/PythonPDEVS/documentation/html/index.html

```
# Write data to file
with open('output.csv', 'w') as f:
    for i in range(num):
        f.write("%s" % i)
        for j in range(len(values)):
            f.write(", %5f" % (values[j][i]))
        f.write("\n")
```

Listing 7: PythonPDEVS code for the experiment on the system.

5.5.3 Performance Analysis

After the definition of our DEVS model and experiment, we of course still need to run the simulation. Simply by executing the experiment file, the CSV file is generated, and can be analyzed in a spreadsheet tool or plotting library. Depending on the data stored during simulation, analysis can show the average queueing times, maximal queueing times, number of events, processor utilization, and so on.

Corresponding to our initial goal, we perform the simulation in order to find out the influence of opening multiple processors on the average and maximum queueing time. Figure 5.10 shows the evolution of the waiting time for subsequent clients. Figure 5.11 shows the same results, drawn using boxplots. These results indicate that while two processors are able to handle the load, maximum waiting time is rather high: a median of 200 seconds and a maximum of around 470 seconds. When a single additional processor is added, average waiting time decreases significantly, and the maximum waiting time also becomes tolerable: the mean job is served immediately, with 75% of jobs being handled within 25 seconds. Further adding processors still has a positive effect on queueing times, but the effect might not warrant the increased cost in opening processors: apart from some exceptions, all customers are processed immediately starting from four processors. Ideally, a cost function would be defined to quantize the value (or dissatisfaction) of waiting jobs, and compare this to the cost of adding additional processors. We can then optimize that cost function to find out the ideal balance between paying more for additional processors and losing money due to long job processing times. Of course, this ideal balance depends on several factors, including our model configuration and the cost function used.

Reflective Question What would you have to change in order to use a different queueing discipline or arrival process? How does this compare to the changes needed when using a mathematical model of this same process?

5.6 DEVS Variants

Despite the success of the original DEVS specification, as introduced throughout this chapter, shortcoming were identified when used in some domains. For these reasons, a lot of variants have recently spawned. In this section, we touch upon the three most popular ones, with some remarks on other variants. Note that we make the distinction between variants that further augment the DEVS formalism (*i.e.*, make more constructs available), and those that restrict it (*i.e.*, prevent several cases). Both have their reasons, mostly related to the implementation: augmenting the DEVS formalism makes it easier for modellers to create models in some domains, whereas limiting the DEVS formalism makes some operations, such as analysis, possible or easier to implement.

5.6.1 Parallel DEVS

One of the main problems identified in DEVS is related to performance: when multiple models are imminent, they are processed sequentially. While DEVS does allow for some parallelism, (*e.g.*, between simultaneous external transitions), multiple internal transitions is a common occurrence.

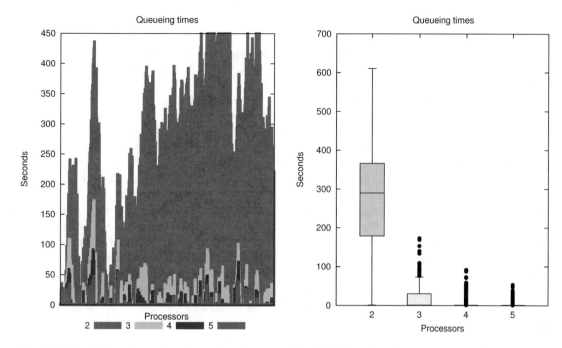

Fig. 5.10: Evolution of queueing times for subsequent customers.

Fig. 5.11: Boxplot of queueing times for varying number of active processors.

Parallel DEVS [72] was introduced as a variant of the DEVS formalism, in which parallel execution of internal transition functions is allowed. This changes the semantics of models though, so it requires changes to the abstract simulator [73]. The proposed changes are therefore not just syntactic sugar: they explicitly modify the semantics of models.

Allowing for parallelism is, however, not a trivial task: several modifications are required, which we briefly mention here. The first logical change is the removal of the *select* function: instead of selecting a model to go first, all imminent models are allowed to transition simultaneously. Whether or not this happens in parallel or not, as it might not necessarily be faster [148], is up to the implementation. This creates some repercussions throughout the remainder of the formalism, as the *select* function was introduced for good reasons.

Since models can now perform their internal transition simultaneously, output functions also happen simultaneously. While this is in itself not a problem, routing might cause the need for events to be merged together, for example when two events get routed to the same model. The abstract simulator was not designed for this, as an external transition was immediately invoked upon the reception of an external event. So in Parallel DEVS, events are always encapsulated in *bags*, which can easily be merged. Bags, or multisets, are a kind of set which can contain items multiple times. This way, multiple bags can be trivially joined, without losing information. Note that order is undefined, as otherwise it would depend on the synchronisation between different output functions: which one is executed before the other. Due to this change in interface, the external transition needs to be altered to operate on a bag of input events, and the output function has to generate a bag of output events.

Problems don't stop there, as internal and external transition might happen simultaneously. Recall that in DEVS, self-loops were not allowed for this exact purpose. In Parallel DEVS, however, two models can perform their internal transition simultaneously, with one outputting an event for the other one. In that case, the model needs to process both its internal transition, and its external transition caused by the other model's transition. Since there is no priority defined between them (that was part of the purpose of the *select* function), they should execute simultaneously. To allow for this, a new kind of transition is defined: the *confluent transition function*. This transition is only performed when both the internal and external transition happen simultaneously. Parallel DEVS leaves open what the semantics of this is, though a sane default is for the internal transition to go first, followed by the external transition.

Thanks to the potential performance gains, many tools favor Parallel DEVS over DEVS in their implementation. Some stick to the elegance of the original DEVS formalism, despite the performance hit.

5.6.2 Dynamic Structure DEVS

Another shortcoming of the DEVS formalism, also present in Parallel DEVS, is the lack of dynamic structure. Some systems inherently require a dynamic structure to create, modify, or delete models during the simulation. While possible in DEVS formalism by defining the superset of all possible configurations and activating only one of them, this has high accidental complexity, and performance suffers. Furthermore, systems might grow extremely big, making it practically impossible to create all possible configurations statically.

To counter these issues, Dynamic Structure DEVS [23] was devised as an extension of DEVS. In Dynamic Structure DEVS, the model configuration is seen as a part of the state, making it modifiable during simulation. Since the coupled model has no state of its own, a *network executive* is added, which manages the structural state of a specific scope. In a separate phase, models can send events to the network executive to request a structural change.

This proposed extension is, however, only a mathematical model as to how it can be made possible. Similar to previous formalisms, an abstract simulator [25] is provided that is structured that way. Real implementations, however, are free to implement this however they want. The network executive might therefore not even exist in the implementation, with all structure changing messages being intercepted in the implementation.

Even though dynamic structure now becomes possible in DEVS models, this formalism is not well suited to handle a huge amount of changes. The work to be done for a change, both for the user and the implementation, is just too time-consuming to execute frequently. But even while highly dynamic models are not ideally suited, infrequent structural changes become very possible.

5.6.3 Cell-DEVS

Another variant of DEVS presented here is the Cell-DEVS formalism. Despite the elegance of the DEVS formalism, it is still difficult to use it in a variety of situations, specifically in the context of cellular models. Cellular Automata [305] are a popular choice in the domain of cellular models, but contrary to the discrete-event nature of DEVS, Cellular Automata is discrete-time based. While discrete-time is a good match with most models in the problem domain of cellular automata, some models would profit from a discrete-event basis. While not frequently a problem, cellular models become restricted to the granularity of the time step, resulting in low performance when the time step is not a good match with the model's transition times.

Cell-DEVS was introduced as a combination of DEVS and Cellular Automata, combining the best of both worlds. Model specification is similar to Cellular Automata models, but the underlying formalism used for simulation is actually DEVS. Due to this change, models gain more control over the simulation time. Furthermore, cellular models can now be coupled to other, not necessarily cellular, DEVS models.

5.6.4 Other Variants

Apart from the formalisms introduced here, many more variants exist that tackle very specific problems in DEVS. We do not have the space here to discuss all of them, though we wish to provide some pointers to some other useful extensions. Examples are other solutions to the dynamic structure problem (DynDEVS [275]), restrictions to make DEVS models analysable (FD-DEVS [152]), and extensions to allow for non-determinism (Fuzzy DEVS [177]). Many of the previously proposed formalisms also have augmented themselves with the changes made to Parallel DEVS, resulting in a parallel version of Dynamic Structure DEVS [24] and Cell-DEVS [273].

Reflective Question What does it mean for there to be extensions to DEVS, while we previously stated that DEVS can be seen as a simulation assembly language?

5.7 Summary

In this chapter, we briefly presented the core ideas behind DEVS, a popular formalism for the modelling of complex dynamic systems using a discrete-event abstraction. DEVS is primarily used for the simulation of queueing networks, of which an example was given, and performance models. It is most applicable for the modelling of discrete event systems with component-based modularity. It can, however, be used much more generally as a simulation assembly language, or as a theoretical foundation for these formalisms.

Future learning directions on DEVS can be found in the Further Reading section, which provides a list of relevant extensions on DEVS, as well as mentions of some of the problems currently being faced in DEVS.

5.8 Literature and Further Reading

Zeigler's book [309] is the default reference for the DEVS formalism and contains further information on other related formalisms. It mostly focusses on the theoretical aspects of modelling and simulation, though some examples are given. A more practical introduction to the formalism of DEVS and examples and how to actually use the formalism and the tool CD++ are given by Wainer [301]. Finally, another text Nutaro [218] develops the design and development of efficient simulators.

The contributions that lead to the the most common variant of DEVS that is in use today have been presented in some landmark papers. Most DEVS research is currently being made on Parallel DEVS [72]. A complete formalisation of simulation formalisms starting from DEVS in [24]. Another paper by Chen [70] presents the symbolic flattening (closure under coupling) that is actually implemented in Parallel Devs. The idea of DEVS as a simulation assembly language. Details are given on why such a language is necessary, and what the implications are is found in [287]. Currently a lot of DEVS tools exist, but there is almost no interoperability in any way between them. Even today, DEVS standardisation is ongoing work [252].

5.9 Self-Assessment

1. Give the formal notation of an atomic DEVS model. Then, give the formal notation of a coupled DEVS model.
2. Give an equivalent atomic DEVS model for a coupled DEVS model.
3. Why does a time advance of 0 need to be possible?
4. Why is the select function necessary?
5. What are the advantages and disadvantages of using simulation in DEVS over mathematical models for performance analysis.

Acknowledgements

This work was partly funded with a PhD fellowship grant from the Research Foundation - Flanders (FWO). Partial support by the Flanders Make strategic research centre for the manufacturing industry is also gratefully acknowledged.

Chapter 6
Statecharts: A Formalism to Model, Simulate and Synthesize Reactive and Autonomous Timed Systems

Simon Van Mierlo and Hans Vangheluwe

Abstract Statecharts, introduced by David Harel in 1987, is a formalism used to specify the behaviour of timed, autonomous, and reactive systems using a discrete-event abstraction. It extends Timed Finite State Automata with depth, orthogonality, broadcast communication, and history. Its visual representation is based on higraphs, which combine graphs and Euler diagrams. Many tools offer visual editing, simulation, and code synthesis support for the Statechart formalism. Examples include STATEMATE, Rhapsody, Yakindu, and Stateflow, each implementing different variants of Harel's original semantics. This tutorial introduces modelling, simulation, and testing with Statecharts. As a running example, the behaviour of a digital watch, a simple yet sufficiently complex timed, autonomous, and reactive system is modelled. We start from the basic concepts of states and transitions and explain the more advanced concepts of Statecharts by extending the example incrementally. We discuss several semantic variants, such as STATEMATE and Rhapsody. We use Yakindu to model the example system.

Learning Objectives

After reading this chapter, we expect you to be able to:

- Specify the behaviour of a autonomous and reactive system using Statecharts
- Understand the semantics of a Statecharts model
- Deploy a Statecharts model onto a software platform (using appropriate tooling)

6.1 Introduction

The systems that we analyse, design, and build today are characterised by an ever-increasing complexity. This complexity stems from a variety of sources, such as the complex interplay of physical components (sensors and actuators) with (real-time) software, the large amounts of data these systems have to process, and the interaction with a (non-deterministic) environment. Almost always, however, complex systems exhibit some for of *reactivity* to the environment: the system *reacts* to stimuli coming from the environment (in the form of *input events*) by changing its internal *state* and can influence the environment through *output events*. Such reactive systems are fundamentally different from traditional software systems, which are *transformational* (i.e., they receive a number of input parameters, perform computations, and return the result as output). Reactive

Simon Van Mierlo
University of Antwerp - Flanders Make, Belgium
e-mail: simon.vanmierlo@uantwerpen.be

Hans Vangheluwe
University of Antwerp - Flanders Make, Belgium; McGill University, Canada
e-mail: hans.vangheluwe@uantwerpen.be

© The Author(s) 2020
P. Carreira et al. (eds.), *Foundations of Multi-Paradigm Modelling for Cyber-Physical Systems*,
https://doi.org/10.1007/978-3-030-43946-0_6

systems run continuously, often have multiple concurrently executing components, are reactive with respect to the environment, and can act proactively by making autonomous decisions. An example is a modern car, whose systems are increasingly controlled by software. Multiple concurrently running software components are interpreting signals coming from the environment (the driver's controls as well as sensors interpreting current driving conditions) and making (autonomous) decisions that generate signals to the car's actuators.

Such timed, reactive, autonomous behaviour needs to be specified in an appropriate language, in order to validate the behaviour with respect to its specification (using verification and validation techniques, such as formal verification, model checking, as well as testing techniques), and to ultimately deploy the software onto the system's hardware components. Traditional programming languages were designed with transformational systems in mind, and are not well-suited for describing timed, autonomous, reactive, and concurrent behaviour. In fact, describing complex systems using threads and semaphores quickly results in unreadable, incomprehensible, and unverifiable program code [181]. This is partly due to the cognitive gap between the abstractions offered by the languages and the complexity of the specification, as well as the sometimes ill-defined semantics of programming languages, which hampers understandability. As an alternative, this book chapter describes the Statechart formalism, introduced by David Harel in 1987 [134]. The syntax and semantics of Statecharts are well-defined and can natively describe a system's timed, autonomous, reactive, and concurrent behaviour. Its basic building blocks are states and transitions between those states. States can be combined hierarchically into composite states, or orthogonally into concurrent regions. Many (visual) modelling tools exist that support the complete life-cycle of modelling a system's behaviour using Statecharts: from design to verification and validation, and ultimately deployment (code generation).

Throughout the sections of this chapter, we introduce the constructs of the Statechart formalism by incrementally building the model of the behaviour of an example system. We explain the syntax as well as the semantics of each construct. The examples are modelled in the Yakindu tool (https://www.itemis.com/en/yakindu/state-machine/), but the techniques can be transferred to any Statecharts modelling and simulation tool with comparable functionality.

Section 6.2 provides background for the rest of the chapter: it explains how we can view a system's behaviour using a discrete-event abstraction, it explains the behaviour of the example digital watch system we are going to model, and it introduces a process model that can be used to build a system (from requirements, through design, down to deployment) using Statecharts. Section 6.3 explains each Statechart construct, starting from the basic state and transition, and progressively introducing more advanced constructs. Section 6.4 explains the semantics of Statecharts (as implemented by STATEMATE) in detail. Section 6.5 explains how Statechart models can be tested, to check whether they satisfy the requirements of the system. Section 6.6 explains how code can be generated from a Statechart model and deployed onto a target platform. Section 6.7 goes into a number of advanced topics. Section 6.8 summarises this chapter.

6.2 Background

This chapter provides background for the rest of the chapter. In Section 8.4, we explain the running example that we will model using Statecharts in the next section. In Section 6.2.2, we look at a system's behaviour using a discrete-event abstraction. Finally, in Section 6.2.3, we present a process model that can be used to model a system using Statecharts, verify that it satisfies the requirements (through simulation and testing), and finally deploy it.

6.2.1 Running Example

The example system we use in this chapter to demonstrate the capabilities of the *Statechart* formalism is a digital watch. A visual representation of the watch is shown in Figure 6.1. A watch is primarily used for keeping time, but has other functions, such as chronometer, an alarm, and a light. A user can interact with the watch by pressing buttons that result in switching between the different modes, editing the current time, turning on the light, . . . It is inherently timed and autonomous, as it changes its internal state without any input from the user.

Fig. 6.1: The running digital watch application.

Some notion of orthogonality is also present, as the system has to ensure time is updated while the chronometer is running, for example.

A full specification of the requirements is given below:

- The time value should be updated every second, even when it is not displayed (as for example, when the chrono is running). However, time is not updated when it is being edited.
- Pressing the top right button turns on the Indiglo light. The light stays on for as long as the button remains pressed. From the moment the button is released, the light stays on for 2 more seconds, after which it is turned off.
- Pressing the top left button alternates between the time display, chrono display, and alarm display modes. The system starts in the time display mode. In this mode, the time (HH:MM:SS) and date (MM/DD/YY) are displayed.
- When in chrono display mode, the elapsed time is displayed MM:SS:FF (with FF hundredths of a second). Initially, the chrono starts at 00:00:00. The bottom right button is used to start the chrono. The running chrono updates in 1/100 second increments. Subsequently pressing the bottom right button will pause/resume the chrono. Pressing the bottom left button resets the chrono to 00:00:00. The chrono will keep running (when in running mode) or keep its value (when in paused mode), even when the watch is in a different display mode (for example, when the time is displayed).
- When in alarm display mode, the time at which the alarm is set is displayed. The default alarm time is 12:00:00.
- When in time or alarm display mode, the watch will go into editing mode when the bottom right button is held pressed for at least 1.5 seconds.
- When in time or alarm display mode, the alarm can be toggled between on or off by pressing the bottom left button.
- The alarm is activated when the alarm time is equal to the time in display mode. When it is activated, the screen will blink for 4 seconds, then the alarm turns off. Blinking means switching to/from highlighted background (Indiglo) twice per second. The alarm can be turned off before the elapsed 4 seconds by a user interrupt (i.e., if any button is pressed). After the alarm is turned off, activity continues exactly where it was left off. Note that after the alarm finishes (flashing ends), the alarm is unset (so it will not go off the next day at the same time).
- When in (either time or alarm) editing mode, briefly pressing the bottom left button will increase the current selection. Note that it is only possible to increase the current selection, there is no way to decrease or reset the current selection. If the bottom left button is held down, the current selection is incremented automatically every 0.3 seconds. Editing mode should be exited if no editing event occurs for 5 seconds. Holding the bottom right button down for 2 seconds will also exit the editing mode. Pressing the bottom right button for less than 2 seconds will move to the next selection (for example, from editing hours to editing minutes).

These requirements now need to be translated to a design of the system. Many options exist: since we are talking about control *software*, a possible choice would be a high-level programming language. But as we have argued

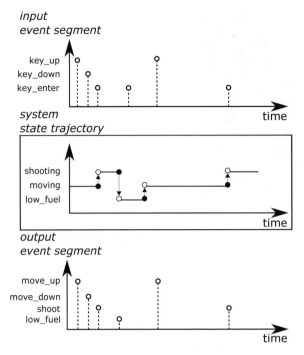

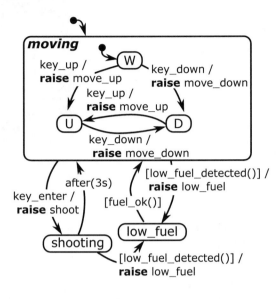

Fig. 6.2: Discrete-event abstraction of an example "tank wars" game.

Fig. 6.3: A possible behavioural model of an example "tank wars" game.

earlier in this chapter, a more appropriate choice (to describe this timed, reactive, autonomous behaviour) is the *Statechart* language.

6.2.2 Discrete-Event Abstraction

Certain system behaviour, in particular the behaviour of control software, can be described using a discrete-event abstraction. In Figure 6.2 a view of the behaviour of an example system is shown – in this case, the system is a "tank wars" game, in which a tank drives around a virtual map by reacting to a player's input through the keyboard. The tank can shoot at the player's command, and it can run out of fuel, at which point it goes into a mode where it can only drive towards a fuelling station (and is no longer able to shoot).

As is clear from this intuitive description, the system reacts to *input* from the environment, and produces *output* to the environment. Such input/output signals can be described by *events*. At the top of Figure 6.2, an input event *segment* is shown. A segment is a finite interval of time, in which a number of events occur. Within such a finite interval, only a finite number of such events can occur (which differentiates discrete-event systems from continuous systems, whose input and output behaviour we can infinitely zoom into, as they are continuous functions). The system *reacts* to the input event segment by producing an output event segment, shown at the bottom of Figure 6.2. The environment (entities interacting with the system) can view the system as a black-box which has an interface (defined by the input events it accepts, as well as the output events it produces). In this case, the player interacts with the system by sending input events corresponding to key strokes: the player controls the tank by pressing the *up*, *down*, and *enter* key. As a result, the system produces output, which describes the reaction of the system to the input it receives. In this case, four output events can be produced: *move_up*, *move_down*, and *shoot*, which signify that the tank starts moving up, starts moving down, or shoots, respectively, and *low_fuel*, which signifies that the tank is low on fuel.

The system has an internal state, which changes over time as a result of input being received, as well as autonomously by the system. A possible system state trajectory for the example system is shown in the middle of Figure 6.2. Three states are defined: *shooting*, *moving*, and *low_fuel*. The first two input events do not cause the state of the system to change, but it does cause two corresponding output events to be raised by the system. The third input event changes the system state from *moving* to *shooting*. And, at some point after that, the

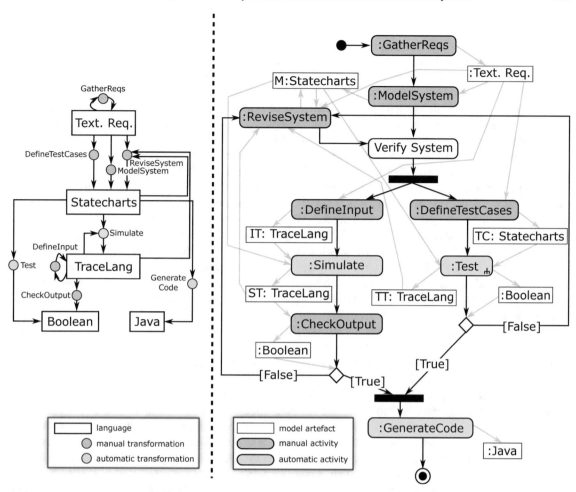

Fig. 6.4: The workflow for modelling, simulating, testing, and ultimately generating code from Statechart models.

system autonomously changes its state to *low_fuel*. From the input, output, and state trace, we can deduce that the system is *timed* (its behaviour includes timeouts), *reactive* (to input from the environment), and *autonomous* (as it can change the system state without any input from the environment).

To describe all possible state trajectories of the system, a state diagram can be used – see Figure 6.3 for a possible model describing the system's behaviour. It shows the different states or *modes* the system can be in: at the highest level, three states (*moving*, *shooting*, and *low_fuel*) are defined (represented by rounded rectangles). The *moving* state has three substates, corresponding to the direction the tank is traveling in. The state of the system can change when a transition (represented by an arrow) *triggers*. A transition triggers due to an (optional) *event* or *timeout*, and an optional *condition* on the total state of the system (including the values of the system's variables). When a transition is triggered, an action is executed, which can change the values of the system's variables, or *raise* an event.

This concludes a high-level description of discrete-event abstractions to describe a system's behaviour, including a possible diagrammatic notation. In the rest of this chapter, the *Statechart* formalism is explained as an example of such a diagrammatic language to describe the timed, reactive, and autonomous behaviour of systems.

6.2.3 Process

As a running example that demonstrates the use of Statecharts, we will develop a Statechart model describing the timed, autonomous, and reactive behaviour of a digital watch. Whenever a system is developed, however, it is important to consider which steps are taken and which artefacts are created in the system development process, and to describe this process in a model. This process, or workflow, will guide us throughout the chapter to design, simulate, test, and ultimately deploy our example system. Figure 6.4 shows a model of this process, in a *Formalism Transformation Graph and Process modelling* (FTG+PM) language [195].

On the right side, a process model (PM) describes the different phases in developing the system. The process consists of several *activities*, either manual or automatic. Manual activities require user input: for example, creating a model starts with a user opening a model editor and ends when the user saves the model and closes the model editor. Automatic activities are programs that are transformational, in the sense that they can be seen as black boxes that take input and produce output. All activities produce *artefacts*, and can receive artefacts as input. Fork and join nodes can split the workflow into parallel branches, where multiple activities are active at the same time. Decision nodes can decide, depending on a boolean value, how the process proceeds.

On the left side, a formalism transformation graph (FTG) is a map of all the *languages* used during system development. Each artefact produced in the process model conforms to a language in the formalism transformation graph. Moreover, it defines the *transformations* between the languages, which can either be manual or automatic. Again, there is a correspondence between activities in the process model and transformations in the formalism transformation graph: the transformations act as an "interface" defining the input and output artefacts, to which the activities in the workflow need to conform.

In our workflow, we will start by defining the requirements of the example system and developing an initial model of the system. This model is subsequently (and in parallel) simulated and tested. A simulation produces an output trace from a given input trace (according to the discrete-event abstraction discussed earlier). This output trace is manually checked and a decision is made whether or not the requirements are satisfied. In the other parallel branch of the workflow, a test case is defined by a generator (which produces input events) and an acceptor, which checks whether the generated output trace is correct. A test runs fully automatically, and again a decision is made whether the requirements are satisfied by the design of the system. If that is not the case, the model of the system is revised until all requirements are satisfied. Once all requirements are satisfied, the system can be deployed by generating appropriate application code. In our case, the system is deployed by generating Python code and coupling it to a visualization of the (simulated) digital watch as shown in Figure 6.1.

6.3 Modelling with Statecharts

The previous section introduces the requirements of our example digital watch example, explains how we can view its behaviour with a discrete-event abstraction, and describes a process for developing the system using the Statechart formalism. In this section, we explain the elements of the Statechart formalism that can be used to model the timed, reactive, and autonomous behaviour of systems. We start with the basic concepts of states and transitions, and gradually introduce the more advanced concepts of depth, orthogonality, broadcast communication, and history. Each new concept's syntax and semantics are explained, and is illustrated by progressively developing the running example's model.

6.3.1 States and Transitions

The basic building blocks of any Statechart model are *states* and *transitions* between those states. They are essential concepts that need to be explained before moving onto more advanced Statechart elements. These basic building blocks have a theoretical underpinning in Finite State Automata [150]. To illustrate the use of states and transitions, two parts of the digital watch behaviour are shown in Figure 6.5 and Figure 6.6, implementing the part of the system that updates the clock value every second, and the part of the system that is responsible for turning on and off the Indiglo light.

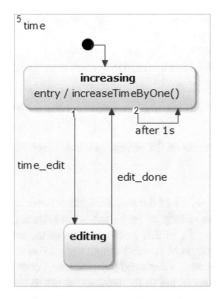

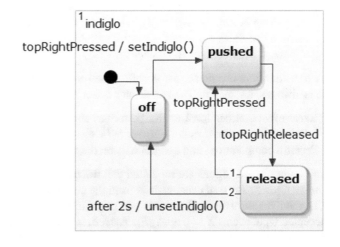

Fig. 6.5: Updating the clock value every second.

Fig. 6.6: Indiglo behaviour.

States model the *mode* a system is in. In the absence of concurrent regions, exactly one state is active at any point in time of the system's execution. A state has a *name*, uniquely identifying it. Exactly one state in the model is the initial state – on system start-up, the state of the system is initialised to that initial state. The visual representation of a state is a rounded rectangle, or roundtangle. To visualize the initial state, a small black circle is drawn, with an arrow pointing to the initial state.

In Figure 6.6, three states are modelled, implementing the behaviour of the Indiglo light: *off* (the default state), *pushed*, and *released*. These names are strategically chosen to reflect the state of the actual ("physical") watch (the display). However, there is no direct link between the name of the state and the state of the display: the names are merely used to quickly convey their meaning to a person looking at the model; the actual behaviour (turning on and off the light) has to be implemented in the link between the Statechart model (or more precisely, its generated code) and the platform code (*i.e.*, the physical digital watch and its embedded software platform, or in our case, a simulated platform in Python). The state *off* represents the light being off; the state *pressed* represents the state where the user is pressing the top right button (turning on the light); the state *released* represents the state where the user has released the top right button.

Besides the state – or mode – the system is in, a system keeps track of a number of *state variables* $\{v_1, v_2, \ldots, v_n\}$. The data type and possible assignments for these variables depend on the data model supported by the specific variant of the Statechart formalism. In case of Yakindu, the implementation-language-independent types *integer*, *real*, *boolean*, *string*, and *void* are defined[1]. These variables with names $\{v_1, v_2, \ldots, v_n\}$ and types $\{t_1, t_2, \ldots, t_n\}$ have values $\{val_1, val_2, \ldots, val_n\}$ where $val_i \in dom(t_i) \ \forall i \in [1, \ldots, n]$

While states describe the current configuration the system is in, transitions model the dynamics of the system and describe how this configuration evolves over time. A transition connects exactly two states: the *source* state and the *target* state. When the system is running, a transition can *trigger* when a number of conditions are satisfied. When the transition triggers, the current state of the system is changed from the source state to the target state. At the same time, the transition's *action* is executed. In general, the signature of a transition is written as follows: $< trigger - event > [< trigger - condition >] / < action >$. The triggering condition of a transition consists of the following elements:

- A triggering *event* (optional), identified by a *name* and a list of *parameters*. In general, an event has the following signature: $< event - name > (< event - params >)$. The event can be an input event (coming from the environment) or it can be internal to the Statechart model. The triggering event can also be a

---

[1]   https://www.itemis.com/en/yakindu/state-machine/documentation/user-guide/sclang_statechart_language_reference#sclang_types

timeout, which is identified by the reserved event name *after* and a parameter denoting the amount of time (using a certain time granularity) that will pass until the timeout triggers.
- A triggering *condition* (optional), which models a boolean condition on the state of the system. The condition can check the values of system variables and check whether a specific state (in another orthogonal component, see Section 6.3.3) is active.

A transition that has no trigger is said to be *spontaneous*. The transition leaving the marker for the initial state is always spontaneous. The transition's action can:

- Raise events, either local to the Statechart model, or to the environment. In general, an event has the following signature: $< event - name > (< event - params >)$.
- Perform computations and assignments on the system's variables.

In Figure 6.6, transitions are modelled that describe the dynamic behaviour of the Indiglo light: it turns on when the *topRightPressed* event arrives from the environment. It stays on as long as the button is not released; if it is, the light stays on for 2 seconds and then turns off again. The *interface* of this model consists of the set of accepted input events $X = \{topRightPressed, topRightReleased\}$, the set of possible output events $Y = \varnothing$, and a set of callback functions $\{setIndiglo, unsetIndiglo\}$. When this model is deployed (by generating code) and placed in an environment, the environment can send input events belonging to the input event set to the running system. It can listen to output events raised by the system and take appropriate actions. Moreover, on system start-up, it has to provide implementations for the call-back functions: these implementations are not known by the Statechart model, as they will change the ("physical") state of the system. In this sense, the same model can be placed in different environments to implement the same behaviour on different "physical" systems; for example, an updated version of the system's hardware might not necessarily entail a change in the behaviour of the system. In this case, we can update the hardware and its associated environment, but leave the Statechart model the same and generate code that is placed in the new environment.

6.3.2 Composite States

A composite state is a collection of substates, which themselves can be basic states or composite states. This allows for nesting states to arbitrary depths. The main purpose of composite states is to group behaviours that logically belong together. Transitions entering a composite state will enter its default state (transitively, to the lowest level). This means that all composite states need to have exactly one default state, as was the case for the Statechart model as well. It is also possible to enter a child state of the composite state directly by targeting it, of course. Transitions going out of a composite state can be thought of as being defined on all of the inner states as well – through a flattening procedure, it is possible to obtain an equivalent Statechart model that only contains basic states and transitions.

For example, in Figure 6.7, the behaviour of the alarm is modelled. A composite state *blinking* is defined: it is entered when the *checkTime* procedure returns true (signifying that the time of the alarm has been reached). Upon entering the *blinking* state, the default *on* state is entered as well (and, if an action is defined on the transition to the default state, it is executed as well). Two outgoing transitions are defined: when the user presses any button, or after 4 seconds, the alarm is turned off. Regardless of the internal active state of *blinking*, when the outer transition is triggered, that active state is exited before the *blinking* state is exited and the *off* state is entered.

One important issue with composite states is that unwanted non-determinism can occur if a substate has an outgoing transition that is triggered on the same event as its composite state's transition. In the flattened version of the Statechart model, this non-determinism will be obvious, since a state will have two outgoing transitions that are triggered on the same event. We can obtain the flattened version of a Statechart model by removing hierarchy: we only retain "leaf" (basic states), by adding the transitions that leave (or enter) their ancestor states on the leaf state. By doing this recursively (starting at the lowest level), we remove hierarchy by progressively adding these "higher-level" transitions to the basic states.

For example, in Figure 6.7, if there was a transition on the *topRightPressed* event from the state *on* to the state *off*, the model is non-deterministic in case the *on* state is active when a *topRightPressed* event is raised by the environment. To resolve such non-determinism (as Statecharts is a deterministic formalism), either the outer-most transition can be chosen – as is the case in STATEMATE [141] – or the inner-most transition can

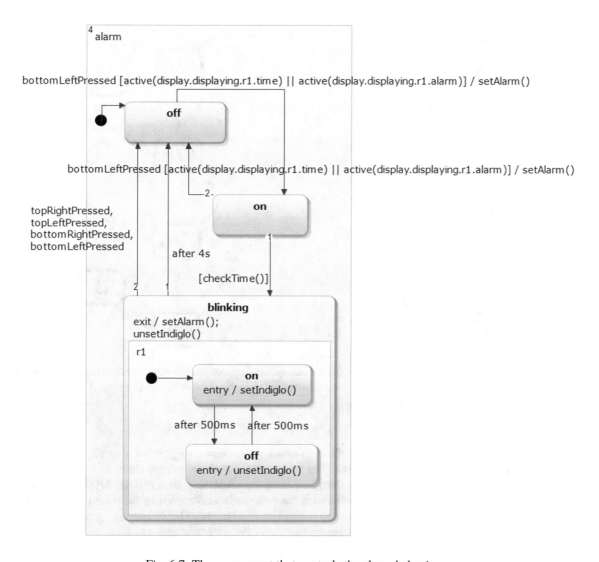

Fig. 6.7: The component that controls the alarm behaviour.

be chosen – as in Rhapsody [139]. These different options are presented in Figure 6.8, which also shows the flattened version of the models. In this tutorial, we assume STATEMATE semantics.

6.3.3 Orthogonal Regions

States can be combined hierarchically in composite states (as explained in the previous subsection), or orthogonally in concurrent regions. While before, exactly one state of the Statechart model was active at the same time, when entering a state which has concurrent regions, all regions execute simultaneously. Each region has a default state, that is entered when the region is entered. Orthogonal regions can react to events concurrently, and communicate with each other. This is done by raising events in one concurrent region that are "sensed" by the other concurrent regions (broadcast communication). A second way in which orthogonal regions can communicate is by querying the state of another orthogonal region: the triggering condition of a particular condition could include a constraint on which state is active in a different orthogonal region.

Figures 6.5, 6.6, and 6.7 all are orthogonal to each other. Indeed: the time needs to be updated, even if the Indiglo light is turned on, or the alarm goes off. Otherwise, the watch would be useless as a timekeeping device. A fourth orthogonal component is shown in Figure 6.9. It controls the display of the digital watch, by switching

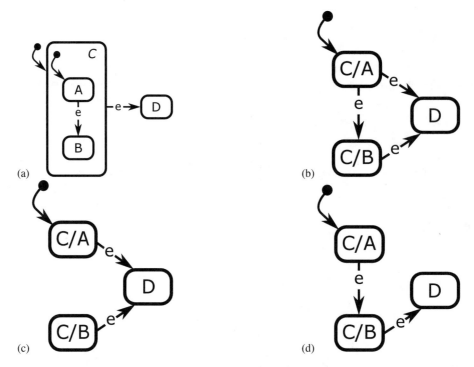

Fig. 6.8: Non-determinism in composite state; (a) an example model containing non-determinism, (b) flattened version: non-determinism in state A, (c) Statemate semantics: outermost transition is prioritized, (d) Rhapsody semantics: innermost transition is prioritized.

(in the *displaying* composite state) between the *displaying_time*, *displaying_chrono*, and *displaying_alarm* states when the user presses the top left button. The user can edit both the current time, as well as the time at which the alarm is set. For this, there is a transition from the *displaying_time* and *displaying_alarm* state to the *editing_time* state, where the user can increase the hours/minutes/seconds by pressing the bottom left button, and change the current selection by pressing the bottom right button. The *display* orthogonal component communicates with the *time* orthogonal component by raising the *time_edit* and *edit_done* events. This will ensure that the time is not updated while the user is editing it.

6.3.4 History

A last element of the Statechart formalism is the *history* state. A history state can be placed in a composite state as a direct child. It remembers the current state the composite state is in when the composite state is exited. Two types of history states exist: *shallow* history states remember the current state at its own level, while *deep* history states remember the current state at its own level and all lower levels in the hierarchy. When a transition has the history state as its target, the state that was remembered is restored (instead of entering the default state of the composite state).

In Figure 6.9, a shallow history state remembers the active direct child of the *displaying* composite state when it is exited (*i.e.*, when a transition to the *editing_time* state is taken). Upon returning from the *editing_time* state, the history state will restore the last active state, to ensure that we end up back in the *displaying_time* or the *displaying_alarm* state, depending on what we were editing. If no history state were present, we would always end up back in the *displaying_time* (default) state.

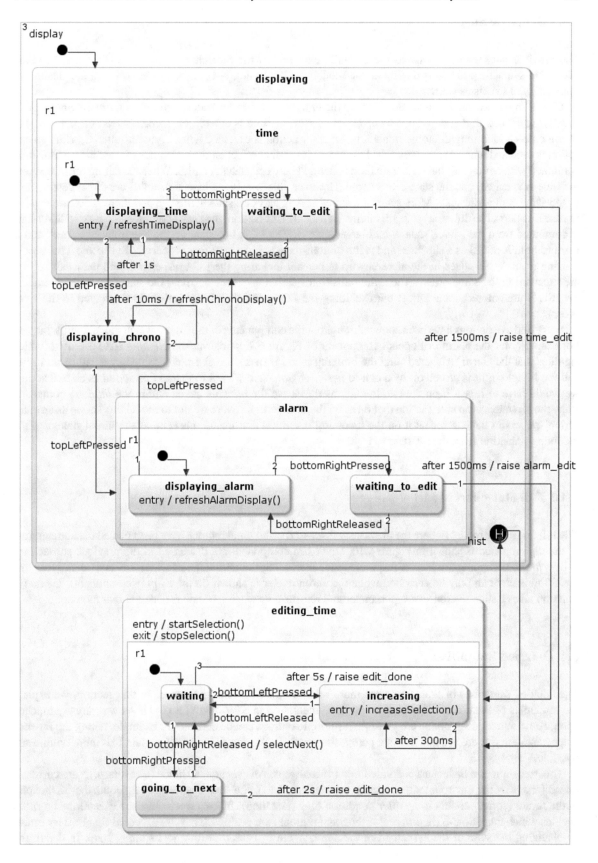

Fig. 6.9: The component that controls the display behaviour.

6.3.5 Syntactic Sugar

The previous subsections discussed the essential elements of the Statechart formalism. There are, however, additional syntactic constructs that make the modeller's life easier, but can be modelled using the "standard" Statechart constructs as well.

One of those "syntactic sugar" additions is the entry/exit action for states, which is a more efficient way of specifying actions that always need to be executed when a state is entered or exited, instead of repeating the action on each incoming/outgoing transition. An entry action is executed when a state is entered, while an exit action is executed when a state is exited. This has an important effect on the semantics of executing a transition combined with composite states. A transition is defined between states *A* and *B*. When executing this transition, the state A is exited, and the state B is entered. However, this is only the case if A and B are direct descendants of the same composite state. More states are exited if A is part of a state hierarchy, and more states are entered in case B is part of a (different) state hierarchy. To execute a transition, the "least common ancestor" (LCA) state is computed from the source state A and target state B. The LCA is a state up the hierarchy of both A and B that has both A and B as a substate (and is the bottom-most state to have that property). To execute a transition, starting from A, the states in the hierarchy up to but not including the LCA are exited (and their exit actions are executed in the same order). Then, the transition's action is executed. Then, the states down the hierarchy towards B are entered, including B but excluding the LCA (and their enter actions are executed in the same order).

Entry and exit actions have been shown throughout the components of the digital watch model in this chapter. For example, for the alarm orthogonal component of Figure 6.7, whenever the *blinking* state is exited (which signifies that the alarm "triggered" and the blinking phase is over), the alarm is disabled by a call to *setAlarm*, and the Indiglo light is turned off by a call to *unsetIndiglo*. Were we not to model this action as an exit action, we would have to repeat it on the two transitions that leave the *blinking* state. Within the *blinking* composite state, the *on* state has an entry action that turns on the Indiglo light. Were we not to model this action as an entry action, we would have to repeat it on the incoming transition that marks this state as the initial state, and the incoming transition from the *off* state.

6.3.6 Full Statechart Model

The full Statechart model (where the previously discussed orthogonal regions have been placed in subdiagrams) of the digital watch is shown in Figure 6.10. One extra component not discussed in the previous subsections is the *chrono* component, which is responsible for increasing the chronometer when it is activated. Note that the chronometer can only be activated when the chronometer is shown (in the *display* component); the *active* function takes a state reference as a parameter and returns *true* when the state is active at that moment.

6.4 Detailed Semantics

The detailed semantics of Statecharts are more intricate than first meets the eye. In this section, we explain the semantics of Statecharts according to its implementation in STATEMATE [141]. As was already touched upon, some semantic variations are possible if multiple options can be chosen (for example, taking the inner or outer transition if there is a conflict in a composite state). In Section 6.7, we will explain that more options are possible.

The execution of a Statechart is divided into (micro)steps. An event queue holds the events that are currently raised (either by the environment, or locally by the Statechart). It also keeps track of timeouts that will expire in the future (corresponding to an *after* condition on a transition). In each microstep, the execution algorithm decides which transition(s) (in different orthogonal regions) are enabled. They are executed (in arbitrary order) by changing the state of the system from the source state of each transition to its target state. In detail, the execution algorithm works as follows:

1. A set of candidate transitions is generated that are enabled by events that were raised locally in a previous step, by timeouts, or by events coming from the environment. The candidate transitions are filtered by

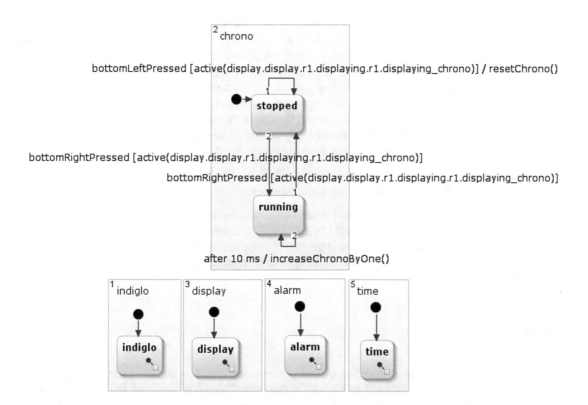

Fig. 6.10: The full Statechart model of the behaviour of the digital watch.

evaluating their condition (which is a boolean function on the total state of the system). The state of the system stays constant until all the transitions execute (at the same time); a function which checks whether a particular state is active does not depend on which transition fires first (as the check is done before any transition fires).

2. The set of candidates is checked for non-determinism (a conflict), of which there are two types:

   - Two candidate transitions can be in conflict if they have the same source state. In the case of STATE-MATE, the execution algorithm cannot give priority to one or the other; as such, a warning is issued to the user and one of the candidates is chosen arbitrarily. More elaborate schemes could assign priority to these transitions based on other criteria; these will be discussed in Section 6.7.1.
   - Two candidate transitions can be in conflict if they are enabled in the same hierarchical state (on any level of the hierarchy). In case of STATEMATE, the top-most transition in the hierarchy is chosen.

3. Once the set of transitions to be executed is known, they are executed by following these steps for each of them (in arbitrary order, since we assume the orthogonal components are independent):

   a. The *Least Common Ancestor* (LCA) of the transition is calculated. This is the state that is lowest in the hierarchy of states that has both the source and the target state as descendant.
   b. The exit set of the transition is calculated. This set contains all the states that need to be exited when the transition is executed. This includes the source state of the transition, as well as all the states in the hierarchy up to (but excluding) the LCA.
   c. For each composite state in the exit set that has a history state as a direct child, the current configuration of the composite state is remembered.
   d. For each state in the exit set (in order, from lowest to highest in the hierarchy), its exit action is executed and the state is removed from the current configuration of the system.
   e. The action of the transition is executed. Actions that are modelled using code (such as value assignments) are immediately executed. Events that are raised are placed in the queue, either to be sent to the

environment as output at the end of the step, or to be sensed by the Statechart in the next iteration of the execution algorithm.

    f. The effective target set of each transition is executed. This is a recursive procedure down to the "leaf" (basic) states. In case the target is a composite state, the target set is extended with the effective target set of its default state. In case the target is a state containing multiple orthogonal regions, the target set is extended with the effective target set of all its orthogonal regions. In case the target is a history state, the target set is extended with the effective target set of the remembered state(s).

    g. All the targets in the effective target set are entered, as long as they are not an ancestor of the source of the transition. The targets are entered from highest in the hierarchy to lowest. Their enter action is executed and they are added to the current configuration of the system. For all the outgoing transitions of the entered states that have a timeout, an event is scheduled in the future.

4. All events that were raised during the step are processed: they are either sent to the environment as output events, or (in case they are local) used in the next microstep.

This process is continued indefinitely – or until an end condition on the execution is satisfied. For example, a particular state could syntactically be designated as the "final" state, or a condition could be defined on the full state of the system. In any case, the execution algorithm is responsible for respecting the timings specified in *after*-events, by pacing the execution in such a way to approach real-time execution (with or without guarantees, depending on the operating system).

6.5 Testing Statecharts

To test a Statechart model, we need to define a trace of input events and an expected trace of output events, as was shown in Figure 6.4. Tests need to be fully automated. Therefore, we need a different tactic from simulating the model and manually providing the input events, while checking the model's reaction. We want to autonomously generate a number of events (timed) in a "generator" and check whether the system raises the correct events in an "acceptor". Basically, an environment interacting with the system is simulated in the form of this generator-acceptor pair.

    To simulate such an environment, either we regard the system as black box and use a mechanism to generate events correctly outside of the model. Alternatively, we can view the system as a white box and model the generator/acceptor pair using Statecharts as well. This has the advantage of instrumenting the model in the same language as it was developed in. Moreover, the Statechart language is appropriate to express the behaviour of the generator and acceptor, as they are timed, autonomous, reactive systems. This is illustrated in Figure 6.11, where we develop a test case for (a part of) the digital watch model. The generator and acceptor are modelled as orthogonal regions alongside the actual system.

    The test case tests the expected behaviour of the digital watch in case the user wants to edit the time. To do this, an input event corresponding to the user pressing the bottom right button is raised after 1 second. The acceptor checks whether this results in the system switching to the *waiting_to_edit* state. Next, after an additional 1.6 seconds, the bottom right button is released, which should result in the system state changing to *editing_time*. Then, 0.4 seconds later, the bottom left button is pressed; this results in the *increasing* state to be activated until the bottom left button is released (after 3 seconds). Then, 5 seconds later, the system should automatically switch to the *displaying_time* state again. If the test ends up in the *passed* state, the *testPassed* callback is called, signifying that the test has passed. Otherwise, the *testFailed* callback is called.

    Due to our white-box approach, we were able to both check the output events produced by the system, as well as its internal state. To be able to check the events raised by the model, we had to change these events to be locally raised, instead of raised to the environment. If we were testing using a black-box approach, the generator and acceptor could be modelled as separate Statechart models, and a communication channel between the generator, the system, and the acceptor could be set up. However, this has the disadvantage of a delay being introduced by the communication channels, which might be difficult to account for in the generator and acceptor. It does allow for testing a system for which we do not have access to the model, but it is outside of the scope of this chapter.

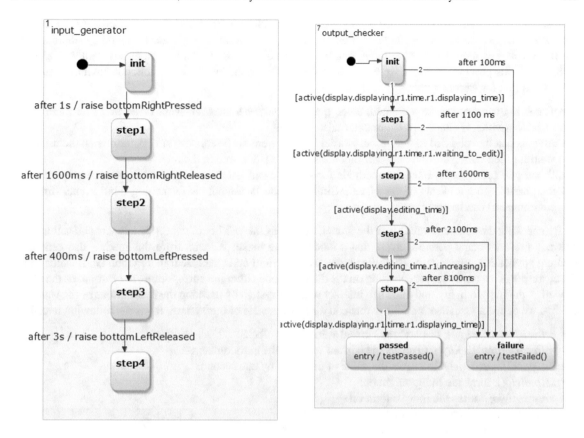

Fig. 6.11: The extra components of the digital watch test.

6.6 Deploying Statecharts

Once we have specified the timed, reactive, autonomous behaviour of a system, it needs to be deployed onto a target platform. The most common method of doing so is by generating code in a high-level language such as C++, Python, or Java. The generated code is then coupled (through its interface of input/output events and callback operations) to a (hand-coded) platform that implements the non-modal (*i.e.*, non-timed, -reactive, or -autonomous) behaviour. Most often, this platform consists of a number of library functions that perform some action, such as displaying information on a screen, sending a signal to an actuator, or querying the value of a sensor. An execution kernel for the platform also has to exist, which is responsible for implementing the behaviour specified by the Statechart – more information on these platforms can be found in Section 6.7.2.

For code generation, we rely on Yakindu. Yakindu is a Statecharts modelling and simulation tool, with the following features:

- A graphical modelling tool for describing systems with the Statechart formalism.
- A neutral action language to use in transition constraints and actions.
- A simulator, to simulate Statechart models to check its behaviour. The simulator allows users to raise events while the simulation is running.
- A code generator interface for generating code to any programming language – pre-defined code generators are provided for Java, C, and C++. The code generator's configurable options include the folder to generate files in, the "execution scheme" (cycle-based or event-driven), whether listeners for external events need to be generated, etc. Yakindu allows for writing custom own code generators, increasing the flexibility of the tool.

Figure 6.3.6 shows the digital watch model, as it was modelled in Yakindu. Central to the figure is the canvas, on which the Statechart model is drawn. The tool is "syntax-directed", which means only syntactically correct models can be constructed. The valid syntactic elements are shown on the right side in a palette. These elements correspond to the ones discussed in the previous sections, along with a few extra syntactic sugar elements, which will not be discussed here. On the left of the figure, an interface for the Statechart model is defined. This interface makes explicit the possible input, output, and locally raised events, as well as any data variables and callbacks. In the previous sections, we have left this definition of the interface implicit, but Yakindu requires to make it explicit for various reasons:

- Transition triggers can be validated, since they can only use an *after*-event or an event declared in the interface (which is either internal or external).
- Actions can be validated to only access variables that were declared, perform operations on them that are valid for their data type, and only call functions that were declared in the interface.
- When generating code, interface methods for output event listeners can be generated, corresponding to the possible output events of the system. Similarly, interface methods for raising input events (from the environment) can be generated.

By checking the syntactic validity of the model, as well as the validity of condition triggers and action code, Yakindu prevents many possible errors that a modeller can make. We can, from this model, also generate a running application. In our case, we will generate the behaviour and visualize it in a Python GUI, which shows the current state of the system (depending on the display mode either the current time, the chrono, or the alarm, as well as the Indiglo light), and allows to interact with the system by pressing the four buttons of the watch. To do this, we define a visualization library that can display the state of our system. It has the following interface:

- *refreshTimeDisplay()*: displays the current value of the clock.
- *refreshChronoDisplay()*: displays the current value of the chronometer.
- *increaseTimeByOne()*: increases the value of the clock by one second.
- *setIndiglo()*: turns the Indiglo light on.
- *unsetIndiglo()*: turns the Indiglo light off.
- *resetChrono()*: resets the chronometer.
- *increaseChronoByOne()*: increases the chronometer by $\frac{1}{10}$ seconds.
- *startSelection()*: starts the selection for editing the time (or alarm).
- *stopSelection()*: stops the selection for editing the time (or alarm).
- *selectNext()*: selects the next element (hours/minutes/seconds) to edit.
- *increaseSelection()*: increases the current selected element (hours/minutes/seconds).
- *setAlarm()*: toggles the alarm.
- *checkTime(): boolean*: returns true when the alarm time is reached.
- *refreshAlarmDisplay()*: shows the set alarm time.
- *addListener(Button, Listener)*: adds a listener for the buttons in the GUI for turning on/off the traffic light, or for the police interrupt.

This library can be instantiated to show (and change) the current state of the digital watch, as is shown in Figure 6.1, where the current time is shown (corresponding to the default display). To connect this GUI to the code generated by Yakindu from the Statechart model, we define appropriate listeners for the buttons in the interface: pressing/releasing the bottom left/top left/bottom right/top right button will send an appropriately named event from the set *{bottomLeftPressed, bottomLeftReleased, topLeftPressed, topLeftReleased, bottomRightPressed, bottomRightReleased, topRightPressed, topRightReleased}*.

The callback functions that are called during the execution of the Statechart are implemented by calling the function with the same name in the interface. This coupling is trivial, but more complex couplings are possible, as long as the callback functions themselves do not introduce waiting behaviour, spawn threads, or take too long to execute.

This development method allows for cleanly separating behaviour (encoded in the model, and generated to executable code by an appropriate code generator) and the presentation (encoded in a visualisation library). More complex control systems benefit from this by separating the control logic from the actuators and sensors, through appropriate interfaces that offer the necessary functionality.

6.7 Advanced Topics

This section explores a number of advanced topics related to Statecharts. We first explore possible semantic variation for Statecharts; we can consider Statecharts not as a single formalism, but a family of formalisms, whose semantics differ slightly. Then, we explain that Statecharts can be executed (deployed) on multiple possible platforms, which has an effect on how the kernel handles the event queue and how time is advanced. Last, we explore ongoing work to introduce dynamic structure to allow for multiple concurrent objects ("agents").

6.7.1 Semantic Variations

In Section 6.4, we've explained the execution semantics of Statecharts with *small steps*: an (unordered) execution of an enabled transition in each orthogonal region. While this is a correct specification, we can ask ourselves several questions:

- When are events that are raised by a transition visible and usable to calculate a new set of enabled transitions?
- How is conflict resolution performed if there are multiple transitions enabled within the same orthogonal region?
- When do we process an external event? Immediately after a small step, or do we wait until the system is in a *quiescent* (*i.e.*, no transitions are enabled) state?
- When are assignments to variables visible? Immediately after a transition is executed, after a small step, or even later?

In their paper, Esmaeilsabzali et al. explore these semantic options for big-step modelling languages (a set of languages that Statecharts belongs to) [94]. These semantic options lead to a *family* of (related) languages: there is no longer *the* Statechart language, but one can configure the language in such a way to suit the application at hand.

At the essence lies the observation that there are three *layers* of steps: small steps, combo steps, and big steps. A big step is a sequence of combo steps, and a combo step is a sequence of small steps. The special property of a big step is that after it ends, environmental input (events sent to the Statechart) and output (events sent to the environment) are processed. One could state that small steps and combo steps are *internal* steps, while big steps lead to *externally visible* changes. A small step behaves the same as defined before, however, the choice can be made to not allow concurrency in a small step: only one transition is executed in that case. Moreover, when concurrency is allowed, a choice can be made whether transitions can *preempt* each other. A combo step groups a sequence of small steps. In the following paragraphs, we explore a number of semantic choices that can be made; these choices configure the Statechart language, resulting in a language "variant".

Big Step Maximality

This semantic choice decides when a big step ends, which always has to be in a "consistent state" (so after at least one small step). One option (*syntactic*) is to designate certain states as "stable" control states; in that case, a big step ends after a series of small steps that end in such a state. A second option is *take one*, which executes exactly one small step. And last, there is the *take many* option, which executes small steps until there are no more enabled transitions.

In the first and last options, a downside is that big-steps might not terminate (either because a designated control state was not entered or because transitions remain enabled). A downside of the second option is the fact that big steps have no clear syntactical scope (while in the other cases the scope of a big step can be deduced from the model).

Combo Step Maximality

This semantic choice is identical to the big step maximality, but then applied to combo steps. The main difference between combo steps and big steps is that after a combo step, no environmental input can be processed.

Small Step Concurrency

While the big step maximality and combo step maximality govern how many small steps can be executed, this semantic option allows for one small step to consist of a *single* transition execution or *many* (concurrent, enabled) transition executions. The advantage of the *single* approach is its simplicity, but it requires that the transitions are ordered non-deterministically. The *many* approach, on the other hand, has the potential for race conditions.

Small Step Preemption

If one transition is an interrupt for another transition (this is the case when both transitions are enabled in orthogonal regions, and one of them has a target state which is not orthogonal to the source states of both transitions, which effectively exits both orthogonal regions), and *many* concurrency is enabled, this semantic option allows to control what happens. Either both transitions are executed (*non-preemptive*, the result being the execution of both the transition's actions, and the target state outside of the orthogonal regions to be active), or only the interrupting transition (*preemptive*) is.

Event Lifeline

The *event lifeline* specifies how long an event "lives" (and can trigger a transition). Both input events and internally generated events have a lifeline (which can be different). There are five options:

1. With *present in whole*, the event is assumed to be present throughout the full duration of a big step. This leads to counterintuitive behaviour (since this might lead to non-causal behaviour, where an event generated by a transition "later" in the big step might cause a transition "before").
2. With *present in remainder*, the event is available in the remainder of the big step. This is an intuitive option, but can lead to unwanted non-determinism: depending on the order in which transitions are executed, a transition can generate an event and immediately enable a transition which would be disabled in a different ordering.
3. With *present in next combo step*, a generated event can only be sensed in the next combo step. This ensures that the ordering of transitions within a combo step cannot affect a transition being enabled or not. However, the ordering is only partial (since combo steps are potentially ordered non-deterministically).
4. With *present in next small step*, a generated event can only be sensed in the next small step. The ordering of transitions within a small step, nor the ordering of combo steps can affect whether a transition is executed in a particular run.

This semantic option has a potentially major impact on the execution of a Statechart model. It also shows the potential benefits of splitting up the executions in big steps, combo steps, and small steps.

Transition Priority

The last semantic option resolves the issue when two or more sets of transitions exist that can be executed in a small step. In that case, a choice has to be made which set is to be executed. A common example is when two transitions in the same state hierarchy are enabled. The *hierarchical* choice lets the priority be determined by the source and destination states of the transitions; for example, in STATEMATE [141], the outermost transition is chosen, while in Rhapsody [139], the innermost transition is chosen. More elaborate schemes are possibly within this choice, however. The *explicit priority* option lets the modeller assign priorities explicitly in the model. Last, the *negation of triggers* requires the modeller to go through all transitions and strengthen the event triggers in such a way that conflicts are avoided. The last two options are tedious to use, but allow for precise control over the priorities.

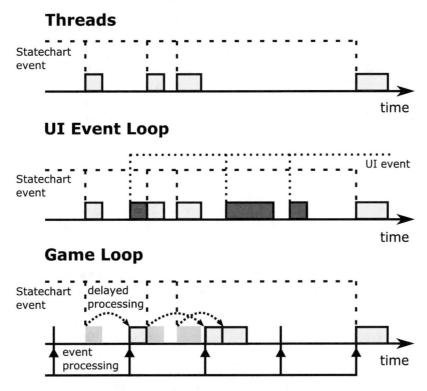

Fig. 6.12: The three runtime platforms.

6.7.2 Execution Platforms

We have discussed deployment of Statechart models in Section 6.6. In particular, we discussed how to provide an implementation for the computation functions that are called in the actions of the Statechart model, and how a "glue" layer has to be constructed that correctly sends input events to the Statechart and interprets output events coming from the Statechart. We did not yet consider *who* the control in the deployed application has; we always consider it is the Statechart. But often, Statechart models are embedded in applications that also require to have control, such as UI systems and game engines. This section explores how the Statechart kernel can work on these platforms by relinquishing certain key control functions.

The semantics of Statecharts (configured according to the semantic options presented in the previous section) are executed on a runtime platform, which provides essential functions used by the runtime kernel, such as the scheduling of (timed) events. The kernel attempts to run the Statechart model in real-time, meaning that the delay on timed transitions is interpreted as an amount of seconds. The raising of events and untimed transitions are executed as fast as possible. It is useful to discuss these platforms, as the Statechart simulators/executors depend on them; moreover, we need different types of platforms for different application scenarios. We distinguish three platforms, each applicable in a different situation. Figure 6.12 presents an overview of the three platforms, and how they handle events.

The most basic platform, available in most programming languages, is based on threads. In its simplest form, the platform runs one thread, which manipulates a global event queue, made thread-safe by locks. Input from the environment is handled by obtaining this lock, which the kernel releases after every (big) step of the execution algorithm. This allows external input to be interleaved with internally raised events. Running an application on this platform, however, can interfere with other scheduling mechanisms (*e.g.*, a UI module), or with code that is not thread-safe.

To overcome this interference problem, the event loop platform reuses the event queue managed by an existing UI platform, such as Tkinter. The UI platform provides functions for managing time-outs (for timed events), as well as pushing and popping items from the queue. This results in a seamless integration of both Statechart events and external UI events, such as user clicks: the UI platform is now responsible for the correct interleaving.

The interleaving is implemented in the "main loop" of the UI platform, instead of (for the threads platform) an infinite *while*-loop.

The game loop platform facilitates integration with game engines (such as the open-source Unity² engine), where game objects are updated only at predefined points in time, decided upon by the game engine. In the "update" function, the kernel is responsible for checking the current time (as some time has passed since the last call to the "update" function), and processing all generated events. This means that events generated in between two of these points are not processed immediately, but queued until the next processing time.

6.7.3 Dynamic Structure

The exclusive use of Statecharts does not scale to the complex (and often dynamic-structure) behaviour of complex software systems, such as (WIMP) graphical user interfaces, games, multi-agent systems, ... Their complexity goes further than the timed, reactive, autonomous behaviour Statecharts is good at representing. In many applications, there is a need for a set of collaborating, concurrent objects, that can appear and disappear at runtime. Object-oriented modelling methodologies address software complexity and allow for the construction of a set of interacting objects, but are not specifically designed for modelling timed, interactive discrete-event systems [137].

A number of approaches, such as OO Statecharts [137] were proposed. More recently, SCCD [281] was presented, combining the structural object-oriented expressiveness of Class Diagrams with the behavioural discrete-event characteristics of Statecharts.

Figure 6.13 presents an overview of the SCCD language, applied to a "bouncing ball" application: the application allows for a user to create a number of balls on the screen. Their autonomous behaviour is simply to move in a certain direction and bounce off the sides of the screen. The user can select and delete balls. As we can see, a number of objects are present: an instance of the *Window* class, and two instances of the *Ball* class are shown. There are a number of links between objects; these links can be used to send or receive events. The object diagram (consisting of objects and links) *conforms* to the class diagram part of the SCCD model. On the other hand, each object has a runtime state, corresponding to one of the states in its Statechart model. A central entity, the object manager, allows for instances to request a new object or link to be created or deleted.

Objects can communicate using events: SCCD specifies a number of event *scopes*. A *narrow cast* allows for an object to send an event over a connection to another specific (set of) object(s). A *broad cast* will send the event to all other objects in the application. An *output event* works similarly to an output events in Statecharts: it is sent to the environment of the application. A special *cd* scope is defined for sending event to the object manager.

In essence, the objects of a Statechart run concurrently and communicate in a more "loose" way than the orthogonal regions of a Statechart. The Statechart model in each object has to finish a big step before events can be sent to other objects or input from other events can be processed. The communication between objects is asynchronous: they are not able to call each other's methods, or change their variable values directly. In this sense, SCCD can be seen as a semantic basis for multi-agent systems, where agents have a well-defined interface through which they can communicate.

6.8 Summary

This chapter demonstrates how to use the Statechart formalism to model the timed, reactive, autonomous behaviour of a system. The chapter is built around a development process that starts from requirements gathering, to build an initial design of the system. The concepts of the Statechart formalism (states, transitions, hierarchy, orthogonality, broadcast communication, and history) are explained by a running example of a digital watch. This design is refined by simulating the model and testing it, by providing an input event segment to the simulation/test and checking the output produced by the model (either manual in a simulation, or automated using an oracle in a test). Ultimately, code is generated from the design model and deployed on a platform.

---

² https://unity3d.com/

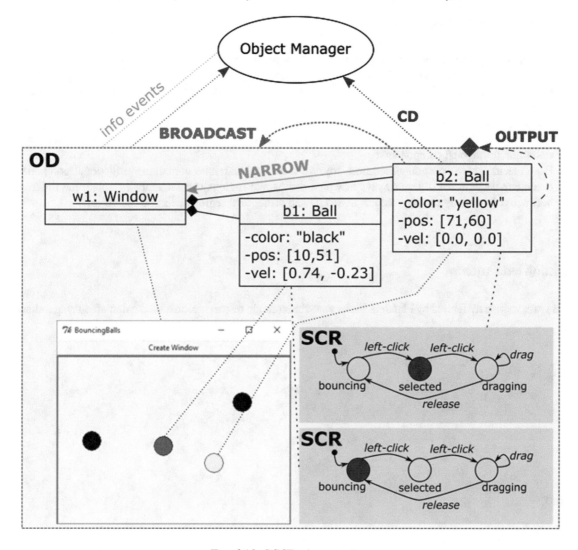

Fig. 6.13: SCCD: An overview.

Coupling between the platform functions and the Statechart output events/callback functions is required to finally arrive at a running system. Last, a number of advanced topics were discussed.

6.9 Literature and Further Reading

A thorough discussion of the origins of the Statecharts visual notation can be found in [136]. The origins of the Statecharts formalism are described in [134]; this work was later complemented with a discussion of its semantics in STATEMATE [141] and Rhapsody [139]. More recently, Esmaeilsabzali et al. have deconstructed the semantics of "big-step modelling languages", a family of languages to which the Statecharts language belongs, and have presented a family of their execution semantics [94]. A standard for the precise execution semantics of UML state machines is being worked on by the OMG [221].

Coordinating concurrently executing Statechart objects has its roots in a 1997 paper by Harel and Gery [137]. SCCD [281] attempt to constrain the possible object operations (creation and deletion of objects, as well as communication between objects) by modelling the allowed object configurations in a class diagram, to which each object diagram (a snapshot in the execution of the model) has to conform.

6.10 Self Assessment

1. Explain the difference between deep history and shallow history.
2. Explain the different ways in which Statechart variants (in particular, Statemate vs. UML 2.0/Rhapsody) resolve non-determinism when outgoing transitions of hierarchically nested states are triggered by the same event.
3. Can a Statechart model ever "block"? Explain why/why not.
4. Draw the flattened version (no hierarchy or orthogonality) of a Statechart model containing hierarchical states and orthogonal components.
5. How are transitions outgoing/incoming from/to a hierarchical state containing orthogonal components executed? Explain the different ways in which orthogonal regions can "interact". Explain the difference between concurrently running objects in SCCD and orthogonal regions within these objects.

Acknowledgements

This work was partly funded by Flanders Make vzw, the strategic research centre for the manufacturing industry.

Chapter 7
Petri Nets: A Formal Language to Specify and Verify Concurrent Non-Deterministic Event Systems

Didier Buchs, Stefan Klikovits, and Alban Linard

Abstract The study of concurrent and parallel systems has been a challenging research domain within cyber-physical systems community. This chapter provides a pragmatic introduction to the creation and analysis of such system models using the popular Petri nets formalism. Petri nets is a formalism that convinces through its simplicity and applicability. We offer an overview of the most important Petri nets concepts, analysis techniques and model checking approaches. Finally, we show the use of so-called High-level Petri nets for the representation of complex data structures and functionality and present a novel research approach that allows the use of Petri nets inside Functional Mock-up Units and cyber-physical system models.

Learning Objectives

After reading this chapter, we expect you to be able to:

- Use common Petri-net patterns to model concurrent processes
- Understand the semantics of Petri-nets in terms of state transitions systems
- Use model checking to systematically check for invariance and reachability properties of Petri-net models

7.1 Introduction

Since the early days of computing the modelling and verification of programs has been an important subject. Nowadays this subject is even more vital as computers are ubiquitous in our current way of life. Computers thrive in all kinds of environments, and some of their applications are life critical. Indeed, more and more lives depend on the reliability of airborne systems, rail signalling applications and medical device software for examples.

Given the importance of the matter, much effort has been invested to ensure the quality of the software. On the organisational side, project management techniques have been devised for the software development process, e.g. the RUP (Rational Unified Process) [174], the Waterfall Model [244], the Spiral Model [40], B-method [189], etc. Most recently, so-called "agile" methodologies, such as SCRUM [261], are taking over the

Didier Buchs
Faculty of Science, Computer Science Department, University of Geneva, Switzerland
e-mail: didier.buchs@unige.ch

Stefan Klikovits
Faculty of Science, Computer Science Department, University of Geneva, Switzerland
e-mail: stefan.klikovits@unige.ch

Alban Linard
Faculty of Science, Computer Science Department, University of Geneva, Switzerland
e-mail: alban.linard@unige.ch

© The Author(s) 2020
P. Carreira et al. (eds.), *Foundations of Multi-Paradigm Modelling for Cyber-Physical Systems*,
https://doi.org/10.1007/978-3-030-43946-0_7

industry [249]. However, software development frameworks can improve software quality only up to a certain point. In fact, they cannot offer complete guarantees for critical systems by themselves as their effectiveness is only based only on empirical evidence [249].

These approaches have the fact in common that all of them require the description of what the system does without prescribing how to do it. That description is called the *specification*. Depending on the development process used, the specification can be informal (e.g. SCRUM), semi-formal (e.g. RUP) or formal (e.g. B-method). Getting the specification right is paramount for the software quality. On the one hand, the specification is used to check if the development team understood the requirements (to answer the question "Are we building the right thing?"). This process is called *validation*. On the other hand, it is used to check if the finished software does what it was meant to do ("Are we building the thing right?"). We call this step *verification*.

A very simple way to do verification is *testing*. In software testing, we use the specification to derive behaviours that we expect from the software. For each expected behaviour we write a *test*. A test is a procedure that exercises the software (or a part of it), and tells if the observed behaviour is as expected or not (according to the specification). Hence, a test can prove that there are errors in the software. However, proving the absence of errors is much more complicated. It implies to write a test for each possible behaviour of the software. The number of behaviours of even simple software is extremely large, meaning that testing is infeasible for proving the absence of errors. Nevertheless, there are some kinds of software that cannot afford to diverge from specification as human life or health depends on it. This need gave birth to a set of verification techniques that can guarantee the absence of errors in a given system: *formal verification*.

Formal verification techniques, a.k.a. *formal methods*, can guarantee the absence of errors in a system up to its modelling. There are several formal methods ranging from *theorem proving* to *model checking*. These techniques aim to build a formal mathematical proof of the program's correctness. This requires of course that the specification is also *formally* described. It further requires that the program itself has a formally specified semantics. In these sections we focus on the modelling phase and the model checking variant of formal methods.

Formally modelling complex systems requires languages that are adapted to the kind of system we are interested in and also must be defined with certain structuring mechanisms. In this chapter, we will mainly describe languages that provide features related to dynamic systems and data types. For structuring mechanisms we propose to consult publications on extensions of algebraic nets such as CO-OPN [32] and LLAMAS [197]. We will not describe them further as they are not absolutely necessary for the understanding of the basic concepts behind formal methods.

7.2 Modelling Concurrency

The modelling of concurrency requires specially adapted formal techniques. Among the numerous existing ones, we observe that all of them use constructs to either explicitly or implicitly describe events, states and synchronisation mechanisms. Moreover, the technique's semantics must exhibit the various behaviours of the modelled system because parallelism and concurrency inherently introduce activity non-determinism into a system. One of the most well-known formalisms is Petri nets, which we will introduce throughout this chapter. However, it is important to understand that most of the explained principles can be translated to other kinds of models such as process algebra [207], state charts [135] and similar.

In order to explain the essence of modelling with Petri nets, we use an example of a car engine throughout this paper. The system describes a very simple combustion engine consisting of foot pedal, engine, carburettor and fuel tank. The system is built from several elementary components, that are similar to automata used in some modelling tools [277]. We will see how all components are combined in a single Petri net that describes the entire system.

Although we could use a model of communicating automata for such problems, the use of a Petri net makes communication between components explicit, whereas it is often implicit and hence unclear when using automata. For instance, the choice of synchronous versus asynchronous communication is a meta-property in automata, rather than an explicit choice.

So, in summary, we will have the possibility to model systems with Petri nets in a condensed way. These systems will be composed of some non-deterministic entities communicating synchronously or asynchronously.

7.2.1 Petri Nets

Petri nets [237] are a graphical, formal modelling language dedicated to the representation of concurrent processes, including communication and synchronisation between them. Petri nets consist of four basic concepts: *places*, *transitions*, *arcs* and *tokens*. Places (represented by circles) model processes and resource containers (such as a fuel tank) of the system, while transitions (represented by rectangles) are used to model system evolution (e.g. the starting of an engine). Places and transitions are connected by arcs. Arcs describe how many tokens (represented as large dots) are taken from a place before and how many tokens are put into a place after a transition is executed. We usually refer to these arcs as the *pre-* and *postconditions* of a transition. Tokens are used to model resources (e.g. the petrol in our example) or control (whether an engine is on or off). Note, that in classic Petri nets only one kind of token is used, meaning that the process control and the resources are both represented equally, which might be confusing for beginners. This feature however offers great flexibility and renders Petri nets a very powerful formalism.

Figure 7.1 shows an example of a Petri net, that represents in a simple way the behaviour of a throttle foot pedal, engine and fuel tank within a car. It represents processes and resources as places: e.g. up, off, fuel or filled. The Petri net also contains transitions such as press, stop or empty. It can be easily observed that the amount of fuel is modelled by the number of tokens within the place fuel, but also that control of the engine is being handled by an individual token which is either in place on or off. An execution of a transition (e.g. start) will *consume* tokens from the precondition states (off) and produce tokens in postcondition states (i.e. on in our example).

The groupings (i.e. the big frames) shown for foot pedal, fuel tank, engine and carburettor have no semantic influence on the system, but merely help system description.

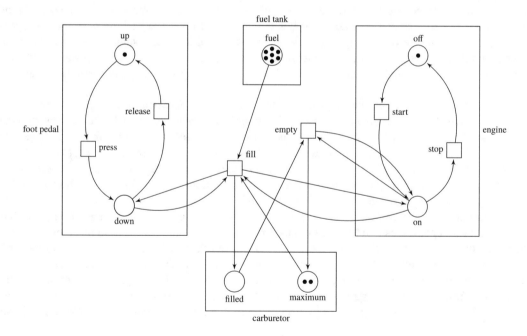

Fig. 7.1: A Petri net that represents a simplified functionality of a car's acceleration

Each places contains a (positive) number of *tokens* (0 by default), that describes how many processes or resources are in this state. We refer to the number of tokens in a place as its *marking*. The combination of markings of all Petri net places is also referred to as the Petri net's *marking* and describes the entire system's state.

Due to their well-studied semantics, they can be simulated, or used for static or dynamic analysis. Their simplicity makes it very easy to transfer them to other forms. For instance, Figure 7.2 shows the marking of the Petri net shown in Figure 7.1 encoded in the Lua programming language[1]. This Petri net can thus be embedded

---
[1] https://lua.org

and used within a software application. We will use the Lua language throughout this article to show how to implement and use Petri nets in practice. The `marking` of the Petri net is a structure that affects to each place its number of tokens. The `initial` marking is the one described initially in the Petri net.

```
function Marking.new (t)
  return {
    up       = t.up       or 0,
    down     = t.down     or 0,
    fuel     = t.fuel     or 0,
    off      = t.off      or 0,
    on       = t.on       or 0,
    filled   = t.filled   or 0,
    maximum  = t.maximum  or 0,
  }
end
```

```
initial = Marking.new {
  up      = 1,
  fuel    = 7,
  off     = 1,
  maximum = 2,
}
```

Fig. 7.2: Declaration of a Petri net marking as Lua code. The `new` function (left) of the `Marking` module is used to create a mapping that represents the Petri net marking. The code on the right shows the usage of this function for the creation of a new marking. Note, that it is only necessary to specify the places that contain tokens, as the other tokens are initialised with `0` by default.

A Petri net state (represented by its marking) is changed by "firing" transitions. In order to more efficiently describe system evolution we will from use markings, and operations thereon to express the behaviour of a Petri net. For convenience, we therefore define the following operators on markings: comparison, addition and subtraction. The comparison operator compares the number of tokens in each place of the two operands for equality or whether one of the operands is smaller/greater. Addition and subtraction operators perform place-wise addition or subtraction of the number of tokens of the operands. These operators allow the concise description of the transition firing semantics. Figure 7.3 provides the operator implementations in Lua.

As stated, the evolution of a Petri net is effected by firing transitions. Such transition firings are dominated by a simple rule: a transition can be fired if it is *enabled*. A transition is called enabled, iff there are enough tokens in all its input places (i.e. the places connected to the transition). When a transition is fired, it removes tokens from its precondition places, and adds new tokens into its postcondition places. It is important to understand that tokens do not "move" from one place to another, but are *consumed* and new (different) tokens are *produced*. The number of tokens consumed and produced are defined using annotations on arcs. By convention arcs annotations stating a weight of 1 token are omitted for legibility reasons.

For an illustration of a transition firing we can look at the `empty` transition from the car engine example above. The transition consumes one token from place `filled`, and another token from place on. Therefore it can be fired if the places `filled` and on contain *at least* one token each. Once fired, the transition produces one token in place `maximum`, and another one in place on. Figure 7.4 shows the Lua code that describes the firing of the transitions `press` and `empty`. It first checks if the input places contain enough tokens. If this condition is met, it performs firing by subtracting the precondition arcs and adding the postcondition arcs to the marking of the Petri net. If the transition is not fireable, the function returns **nil**, the Lua equivalent for NULL or null value.

7.2.2 Common Petri net patterns

Petri nets are better at expressing communication and synchronisation than the traditional automata formalism, as they make these concepts explicit. However, similar to automata, Petri nets have the aforementioned drawback to not distinguish processes and resources. It is up to the modeller to clarify the role of each place, for instance with naming or colour conventions. In order to introduce some common modelling practices we present a few standard patterns which are found in the Petri nets in Figure 7.5:

- Figure 7.5a represents the creation of two processes (q, r) from a process (p), or the release of a resource (r) from a process (p, q).

```
function Marking.__add (1, r)
  return {
    up       = l.up       + r.up,
    down     = l.down     + r.down,
    fuel     = l.fuel     + r.fuel,
    off      = l.off      + r.off,
    on       = l.on       + r.on,
    filled   = l.filled   + r.filled,
    maximum  = l.maximum  + r.maximum,
  }
end
```

Listing 8: Addition operator

```
function Marking.__eq (1, r)
  return l.up        == r.up
    and l.down       == r.down,
    and l.fuel       == r.fuel,
    and l.off        == r.off,
    and l.on         == r.on,
    and l.filled     == r.filled,
    and l.maximum    == r.maximum,
end
```

Listing 9: Equality operator

```
function Marking.__sub (1, r)
  return {
    up       = l.up       - r.up,
    down     = l.down     - r.down,
    fuel     = l.fuel     - r.fuel,
    off      = l.off      - r.off,
    on       = l.on       - r.on,
    filled   = l.filled   - r.filled,
    maximum  = l.maximum  - r.maximum,
  }
end
```

Listing 10: Subtraction operator

```
function Marking.__le (1, r)
  return l.up        <= r.up
    and l.down       <= r.down,
    and l.fuel       <= r.fuel,
    and l.off        <= r.off,
    and l.on         <= r.on,
    and l.filled     <= r.filled,
    and l.maximum    <= r.maximum,
end
```

Listing 11: Less-than-or-equals op.

Fig. 7.3: Addition, subtraction, equality and less-than-or-equals operators used to work on Petri net markings. The evolution of a Petri net can be expressed using these operators.

```
function press (marking)
  if marking >= { up = 1 } then
    return marking
        - { marking.up   = 1 }
        + { marking.down = 1 }
  else
    return nil
  end
end
```

Listing 12: Code of the press transition

```
function empty (marking)
  if marking >= { filled = 1,
                  on     = 1 }
  then
    return marking
        - { marking.filled  = 1,
            marking.on      = 1 }
        + { marking.maximum = 1,
            marking.on      = 1 }
  else
    return nil
  end
end
```

Listing 13: Code of the empty transition

Fig. 7.4: Lua code as examples showing the implementation of transitions

- Figure 7.5b represents the synchronisation of two processes (q, r), or the acquire of a resource (r) from a process (q, p).
- Figure 7.5c represents a choice for process p, that can go either in q branch or in q branch.
- Figure 7.5d represents the collection of processes or resources (q, r) into one (p). When q and r are in mutual exclusion, it can also represent the end of a condition for process p.

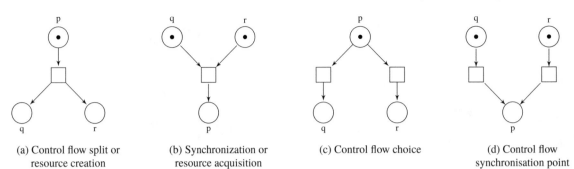

(a) Control flow split or resource creation

(b) Synchronization or resource acquisition

(c) Control flow choice

(d) Control flow synchronisation point

Fig. 7.5: These four common Petri net patterns can be used to express various concepts related to control flow split, merge, choice or synchronisation, but also for the consumption, production, acquisition and release of resources.

7.2.3 Formal syntax and semantics

Since we introduced Petri nets as a formal method, we feel obliged to also provide the syntax and semantics in a more formal setting. We encourage the reader to bear with us through this section, as understanding of these concepts will be necessary for the more complex aspects of Petri nets. In return, we promise to keep ourselves short and only introduce the essential parts. It is also worth mentioning that these formal definitions can be used directly for the creation of tools that manage, simulate and analyse models.

Formally, a Petri net is a *labelled bipartite directed graph*, i.e., a graph whose vertices can be divided into two disjoint finite sets P and T (with $P \cap T = \emptyset$), such that every edge is directed and connects a vertex of P to a vertex of T (preconditions), or a vertex of T to a vertex of P (postconditions). This graph has natural number labels on vertices of P (place markings), no labels on vertices of T (transitions), and positive number labels on edges (number of consumed or produced tokens).

A Petri net is thus described as a tuple $\langle P, T, pre, post, m_0 \rangle$. The set of all possible markings is denoted $M = P \rightarrow \mathbb{N}$. It contains all possible functions that associate tokens to places of the Petri nets. The initial marking m_0 is an element of M that maps all places to their initial number of tokens. The $pre : T \rightarrow M$ and $post : T \rightarrow M$ functions represent arc valuations, and are called the $pre-$ and $post$conditions of the transitions. For every transition, they return a marking that corresponds to the valuations of input (resp. output) arcs of the transition. Note, that their signature is implicitly in curried form, as they can be rewritten as $pre : T \rightarrow (P \rightarrow \mathbb{N})$ and $post : T \rightarrow (P \rightarrow \mathbb{N})$.

The arithmetic and comparison operators $(+, -, =, \leqslant)$ that we introduced in the previous section are formally defined on markings as follows. We require that $\forall m_l, m_r \in M$

$$m_l + m_r = \forall p \in P, (m_l + m_r)(p) \mapsto m_l(p) + m_r(p)$$
$$m_l - m_r = \forall p \in P, (m_l - m_r)(p) \mapsto m_l(p) - m_r(p)$$
$$m_l = m_r \equiv \forall p \in P, m_l(p) = m_r(p)$$
$$m_l \leqslant m_r \equiv \forall p \in P, m_l(p) \leqslant m_r(p)$$

where $a \mapsto x$ signifies that x is the image value of a.

In this chapter we will describe the evolution and behaviour of Petri net using Structured Operational Semantics (SOS) rules such as the example in Equation (7.1). We remind the reader that an SOS rule is composed of a conjunction of premises (above the line) and a conclusion (below the line). The rule states that it can be applied i.e. the conclusion transformation be executed, when the premises are satisfied. Both the premise(s) and the conclusion are expressions (usually called "term predicates").

$$rule : \frac{tpred_1 \wedge tpred_2... \wedge tpred_n}{tpred} \tag{7.1}$$

Each Petri net transition describes possible evolutions of the system. Formally it is a relation between Petri net states, i.e. markings. Thus, the behaviour is described as a transformation of the Petri net marking. The conclusion of SOS rules that describe transition firings are written as $m \xrightarrow{t} m'$, where m and m' are Petri net markings and t is a transition. $m \xrightarrow{t} m'$ states that the system's current marking m leads to m' by firing t.

Using a model checking mentality we can also say that the SOS rule states that, given two markings m and m', firing t is correct if all the premises are correct. The semantics of firing one transition is given by the general rule in Equation (7.2).

$$transition_t : \frac{pre(t) \leqslant m}{m \xrightarrow{t} m - pre(t) + post(t)} \tag{7.2}$$

Example Let us consider the release transition in Figure 7.1. Its semantics is represented by the following rule:

$$release : \frac{(\text{down} \mapsto 1) \leqslant m}{m \xrightarrow{\text{release}} m - (\text{down} \mapsto 1) + (\text{up} \mapsto 1)}$$

Remember that for a marking m to be greater or equal than another one, all individual place markings have to be greater or equal. For this specific premise to be valid, it means that all places in m need to have a greater or equal marking than $(\text{down} \mapsto 1)$ (i.e. all places need to have at least 0 tokens, except for down which needs at least 1 token). If and only if this is the case, then firing the transition will lead to a new marking by removing one token from down and adding one token to up. □

Sequences of transitions also have simple semantics: beginning from a starting state, the transitions within the sequence are applied one by one. Formally, sequences of transitions (noted T^*) are inductively defined as lists. Note, that every transition $t \in T$ can also be seen as a sequence of transitions of length 1. Hence $T \subseteq T^*$. Further, we use a concatenation operator (.) such that $\forall t \in T, \forall s \in T^*, t.s \in T^*$, i.e. every concatenation of transitions is a transition, followed by a sequence of a transitions. Note that a the sequence s can be a single transition. This semantics is given by Equation (7.2) and Equation (7.3).

$$sequence : \frac{m \xrightarrow{t} m', m' \xrightarrow{s} m''}{m \xrightarrow{t.s} m''} \tag{7.3}$$

Figure 7.6 shows the Lua code that performs the computation of a sequence of transition. Each transition is itself a function, such as those we have already defined for press and empty.

```lua
function sequence (marking, transitions)
   for _, transition in ipairs (transitions) do
      marking = transition (marking)
      if not marking then
         return nil
      end
   end
   return marking
end
```

Fig. 7.6: The Lua code that handles the execution of a sequence of transitions on an initial marking. Note that transitions is a list of transition functions as defined above. The ipairs function is an ordered iterator over the elements of transitions.

7.2.4 Deduction Based on Rules

As we have seen above, we can compose new relations from existing rules, as shown for the sequences in Equation (7.3). We refer to this feature as *transitive closure*. However, in order to compute new relations based on existing SOS rules, we need to know exactly how to compute a rule's term predicates, as another rule's premises can be used as a conclusion of another rule.

This process, called deduction, is based on three principles: rule matching, variable instantiation and predicate reduction. These principles require the definition of additional operations on functional terms (the expressions including only application of function) and term predicates (the predicates applied on functional terms).

These principles are related to the use of variables. The operations that manage the values which can be taken by these variables can take must be thoroughly defined.

An assignment (or *substitution*) is defined as a function $\sigma \in \Sigma$, where Σ is the set of partial functions from variables to terms, or $\perp$ if no term is associated to the variable. For instance, a substitution σ can be the replacement of the variable a with the term down $\mapsto 1$, or by $\perp$ if the variable a is undefined. Substitution is extended to apply on terms instead of variables only: for all terms t, $\sigma(t)$ returns the term t where all variables that are defined in σ have been replaced by their associated term in σ. For instance if we take the term $pre(x)$ and apply the substitution $\sigma(x) = $ release onto this term we see that $\sigma(pre(x)) = pre(\text{release})$.

Matching refers to the fact that we would like to identify two terms t_1, t_2 modulo a substitution σ. This means that matching checks if there exists a substitution σ such that $\sigma(t_1) = \sigma(t_2)$. For instance, the substitution $\sigma = \{x \mapsto \text{release}, y \mapsto \text{release}\}$ can be used to match $pre(x) = pre(y)$.

We would like to introduce a simple example defining the comparison predicate over natural numbers. According to the rules given in Equation (7.4): $a, b \in \mathbb{N}$ are variables over natural numbers.

$$R1 : \frac{a \leqslant b}{a \leqslant b + 1} \qquad R2 : \frac{a \leqslant b}{a + 1 \leqslant b + 1} \qquad R3 : \frac{}{0 \leqslant 0} \qquad (7.4)$$

We assume that the reader is familiar with the operations based on addition of natural numbers. Their very definitions can be given as an exercise to the reader. Deductions are computed using a special *Infer* function. $Infer(R)$ is defined as the deduction process for a set of rules R. In our specific case Infer will be defined as $Infer(\{R1, R2, R3\})$ and we will use Infer to compute the possible deduction on the predicate $\leqslant$.

In such deductions rules are applied on top of each other, the correspondence being the possible matching of one premises with the conclusion of another rule placed above. Correct deductions have to be finite and also must rely on a base case, i.e. a rule without premises (in our case R3).

Example The following examples show possible deductions, based on the rules R1, R2 and R3. By applying a rule whose conclusion that matches the provided term, we can find a premise, which we then use to match against another rule. This means we generally work bottom-up and try to find matching rules which will lead us closer to R3.

On the left, we try to deduce that $1 \leqslant 2$. We observe that $1 \leqslant 2$ can be the conclusion of the rule R1, leading us to infer that the premise of this rule must be $1 \leqslant 1$. Using rule R2 and the $1 \leqslant 1$ as conclusion, we see that the premise for this rule is $0 \leqslant 0$. Now we can apply R3, which does not have any premise and we successfully deduced that $1 \leqslant 2$.

On the other hand we can look at the example to the right. Our goal here is to deduce that $4 \leqslant 2$ is true. A first application of rule R2 will get us to $3 \leqslant 1$ and another application of R2 to the premise $2 \leqslant 0$. At this point, none of the rules can be applied any more This means that, according to the rules we defined, we cannot reach a base case and that $4 \leqslant 2$ must be incorrect.

Note, that in both cases we could have applied different rules than the ones that we actually used. On the left hand we could have swapped the rules and first applied R2 before using R1. On the right side we could have also used R1 instead of R2. In either case we would have reached the same conclusions. In general, inferences can be performed in any order by using any rule that is applicable. The trick is to use rules that quickly lead to an end (either a clash or a base-case).

$$R1\frac{R2\frac{R3\frac{}{0 \leqslant 0}}{1 \leqslant 1}}{1 \leqslant 2}$$

$$R2\frac{R2\frac{R?\frac{???}{2 \leqslant 0}}{3 \leqslant 1}}{4 \leqslant 2}$$

According to the principles we gave, we can say that $Infer(R)$ is the least set that includes the results of all substitutions possible on all conclusions of all rules that do not have premises:

$$\sigma(post(r)) \in Infer(R) \text{ if } pre(r) = \emptyset, \forall r \in R, \forall \sigma \in \Sigma$$

and all rules' substituted conclusions where a substitution exists so that a rule premise is in the set:

$$\sigma(\phi(post(r))) \in Infer(R) \text{ if } \phi(pre(r)) = \phi(d), \forall r \in R, \sigma \in \Sigma \exists \phi \in \Sigma, \forall d \in Infer(R)$$

where $pre(r)$ are is the premise and $post(r)$ the conclusion of the rule

This last rule is called modus ponens, Example 7.2.3 shows such a deduction for a sequence of transitions. $Infer(R)$ is then obtained by fixpoint application of the modus ponens rule. For rules with positive numerator (no negation, only conjunction) this set always exists according to fixpoint theorem of Knaster-Tarski [264]. Nevertheless, the set can be infinite.

Example Let us consider the sequence press.release in Figure 7.1. The semantics of this sequence are represented by the following simple deduction using the SOS rules of press and release. What we see is that the conclusion of the rule (the firing of press.release and creation of a new marking) has two premises: 1. the conclusion of the press rule, and 2. the conclusion of the release rule. Each of these is part of an SOS rule with it's own premises, which has to be satisfied for this rule to be applicable.

$$\cfrac{\cfrac{(\text{up} \mapsto 1) \leqslant m}{m \xrightarrow{\text{press}} m' = m + (\text{up} \mapsto 1) - (\text{down} \mapsto 1)} \qquad \cfrac{(\text{down} \mapsto 1) \leqslant m'}{m' \xrightarrow{\text{release}} m'' = m' + (\text{down} \mapsto 1) - (\text{up} \mapsto 1)}}{m \xrightarrow{\text{press.release}} m''}$$

7.2.5 Reachability Graph

We saw that applying a transition to a Petri net marking generates a new marking. In order to study system evolutions in a more efficient manner, we can try to represent transitions and the thereby created markings as an automaton. The automaton's states are the markings and the automaton's transitions correspond to the Petri net's transitions. We refer to such an automaton as the *reachability graph* of a Petri net. This is due to the fact that it shows, starting from an initial state, the possible states of the system that can be reached.. To fully understand the definition of the automaton, it is noteworthy that the automaton's input alphabet is the Petri net's set of transitions.

Note, that depending on the initial marking, a Petri net has different reachability graphs, as different transitions might be enabled. To visualise this, we present the reachability graphs of the Petri net of Figure 7.1 in Figure 7.15 (at the end of this chapter). The two graphs correspond to the possible system evolutions with 0, 1, 2 and 3 initial tokens in fuel in Figures 7.15a – 7.15d, respectively. We can clearly see the repeated groups of markings that correspond to matching behaviour.

These groups of reachability graph nodes differ by the number of tokens available in the fuel place. To aid the legibility, we use different shades of gray to distinguish markings with different token numbers in place fuel, where the lightest is 0 and the darkest is 3. The last group at the bottom of the figures (almost white) shows the behaviour when fuel is empty. We can easily observe that the states of 0 tokens in Figure 7.15a is repeated in Figure 7.15b, but this time with 1 token (signalled by the darker shade of grey). In the same way, we can find the group of Figure 7.15b in Figure 7.15c, where however, every state is one shade darker, as there is one more token in state fuel. The reachability graph of the marking shown in Figure 7.1 with seven tokens in fuel has 84 states. We can observe from the graphs in Figure 7.15 that a Petri net with 0 tokens has 4 states and 12 states with 1 token. The number of states increases by 12 for each additional token added initially to the fuel place.

The advantage of Petri nets is that, due to their better representation of concurrent processes and resources, they can be *orders of magnitude* smaller than their equivalent automata. For instance, [300] managed to simulate the state of several multi-threaded programs to perform intrusion detection. Simulation was performed on-the-fly during the program execution by wrapping system calls, and thus required an efficient representation and computation time. This goal was achieved using Petri nets, and it would not have been possible using the equivalent automata on the available resources.

Formally, the reachability graph of a Petri net is defined by a function $s : M \to T \to M \cup \{\bot\}$ that returns the successor for any marking and transition, or $\bot$ if none exists. This function is derived from the following basic rules:

$$\forall m \in M, t \in T, \begin{cases} s(m)(t) = \bot & \Leftrightarrow \nexists m' \in M, m \xrightarrow{t} m' \in Infer(\{transition_{t_i} | t_i \in T\}) \\ s(m)(t) = m' & \Leftrightarrow m \xrightarrow{t} m' \in Infer(\{transition_{t_i} | t_i \in T\}) \end{cases} \quad (7.5)$$

The relation can be restricted to the subset of states that are reachable from the initial state m_0 of the Petri net. This subset is easily defined using the sequences of transitions of Equation (7.3).

We define that the state space (SS) is the set of reachable markings starting from an initial marking m_0 for a given set of transitions T

$$SS(T, m_0) = \{m \in M \mid \exists s \in T^*, m_0 \xrightarrow{s} m \in Infer(\{transition_{t_i} | t_i \in T\} \cup \{sequence\})\} \quad (7.6)$$

By convention, for a given Petri net it is natural to implicitly define $SS(m_0) = SS(T, m_0)$.

The (naive) reachability graph generation algorithm is given in Figure 7.7. This function iterates over the reachable markings, initially only initial, and tries to apply all transitions. The function returns the set of reachable markings, each one annotated with the transitions that can be fired, and the corresponding successor marking. The explored set contains the already explored markings, whereas the encountered set contains markings that have not been explored, or that have been encountered again since their exploration. The code shows a simple implementation that iterates over all previously seen markings to ensure uniqueness. More efficient implementations may use a hash table.

7.2.6 Monotony

Petri nets exhibit an interesting property: monotony. Monotony means that if a transition can be fired for one marking, then it is also fireable for any marking greater than the evaluated one. Extended to the reachability graph, monotony means that all sequences of transitions that exist in the reachability graph of a Petri net also exist in reachability graphs of the same Petri net with greater initial markings. Trivially expressed we could say that adding tokens to a Petri net can only add new behaviours, but never inhibit existing ones.

Formally, monotony is defined by Equation (7.7), which can be derived from the rules in Equation (7.2) and Equation (7.3). This rule defines monotony on both, a transition and a sequence of transitions.

$$monotony : \frac{m \xrightarrow{t} m'}{m + \delta \xrightarrow{t} m' + \delta} \quad (7.7)$$

In our example monotony means that any sequence of actions in our engine can also be performed if we add tokens to any state of the Petri net (such as adding more fuel). We can observe the inclusion of reachability graphs in Figure 7.15 for initial markings fuel = $\{0, 1, 2, 3\}$. The sequence stop.release.start.press is observed several times, but with different initial markings, leading to a varying number of tokens in the fuel place for each one.

```lua
local markings = {}

function unique (marking)
  for _, m in ipairs (markings) do
    if marking == m then
      return m
    end
  end
  markings [#markings+1] = marking
  return marking
end
```

```lua
function reachable ()
  local explored    = {}
  local encountered = {
    [unique (initial)] = true
  }
  while next (encountered) do
    local marking = next (encountered)
    encountered [marking] = nil
    if not explored [marking] then
      explored [marking] = true
      for _, transition in ipairs (transitions) do
        local successor = transition (marking)
        if successor then
          successor = unique (successor)
          marking      [transition] = successor
          encountered [successor ] = true
        end
      end
    end
  end
  return explored
end
```

Fig. 7.7: Source code function that enumerates reachable markings (right). The code iterates over the already encountered markings and tries to find all possible successors by attempting to fire all individual transitions on that marking and adding the successful attempts into the list of encountered markings.

The unique function ensures that each marking corresponds to only one entry in the Lua memory. It is required by the explored [marking] and encountered [successor] lookups.

To summarise, we can see that the reachability graph of a Petri net is always included in the reachability graph of the same Petri net with a greater initial making. While the markings of the sequences' states may differ for each one, the transition sequences are preserved.

In fact, since monotony is a property of all Petri nets, we can observe the following lemma. It states that the state space (and thus the reachability graph) inferred from just the transitions is equal to the state space inferred by the transitions and the monotony rule that we can define as:

$$SSM(T, m_0) = \{m \in M \mid \exists s \in T^*, m_0 \xrightarrow{s} m \in Infer(\{transition_{t_i} | t_i \in T\} \cup \{sequence, monotony\})\} \quad (7.8)$$

So that the following statement holds:

$$SS(T, m_0) = SSM(T, m_0) \quad (7.9)$$

We also know that behaviours are preserved when extending markings:

$$m \leq m' \Rightarrow SS(T, m) \leq SS(T, m') \quad (7.10)$$

Where $\leq$ on markings compare markings one by one.

7.3 Properties of Petri Nets

In the previous section we introduced Petri nets and the rules for the exploration of the reachable state space of an initial marking. In this section we investigate the properties of Petri nets using representatives of state properties and transition properties.

7.3.1 Marking Properties

The state properties or in Petri net-lingo *marking properties* of a Petri net can be used to analyse one configuration of a Petri net. One of the most common representatives for these kinds of properties is the *bound*. A place's bound refers to the minimum (lower bound) and the maximum (upper bound) number of tokens that can be reached by any marking. The minimum and maximum number of tokens of all places are commonly referred to as the *bound* of a Petri net. This means that the Petri net bound is composed of

- the greatest marking that is contained by any reachable marking (lower bound),
- and the smallest marking that contains any reachable marking (upper bound).

Formally, we can define the lower and upper bound as follows:

$$lower = \max\left(\{m \in M \mid \forall m' \in SS(T, m_0), m' \geqslant m\}\right) \tag{7.11}$$

$$upper = \min\left(\{m \in M \mid \forall m' \in SS(T, m_0), m' \leqslant m\}\right) \tag{7.12}$$

Figure 7.8 shows the code that computes the bound of a Petri net, using its reachable markings. The code works by iterating over all markings and computing the minimum and maximum number of tokens for each place. There exist algorithms to compute the bound without the use of the set of reachable markings. These algorithms use structural properties of the Petri net instead, such as the one shown in [188].

The importance of bounds is paramount as they are often used to detect dead parts in the model, i.e. places that can never contain any tokens. Bounds computation can further detect places that can have an infinite number of tokens. However, as an infinite number of tokens would (theoretically) require an infinite reachability graph, such a bound cannot be detected using algorithms such as the one presented below. In order to detect an infinite bound, we have to use a *coverability graph*, an adaptation of the reachability graph for which a detection of repetitive sequences is added. This is explained for example in [238].

In the Petri nets community the term *k-boundedness* is often used, where $k \in \mathbb{N}$ is a natural number. This notion of bound refers to the upper bound at the place level and helps describing a Petri net. We say that a place p is k-bounded iff it contains at most k tokens in any reachable marking i.e. $\forall m \in SS(T, m_0), m(p) \leqslant k$. We can further refer to Petri net as being k-bounded iff all its places are k-bounded, i.e. $\forall p \in P, p$ is k-bounded.

```
function bounds (marking)
  local markings = reachable (marking)
  local bound    = {
    up       = { minimum = math.huge, maximum = 0 },
    down     = { minimum = math.huge, maximum = 0 },
    fuel     = { minimum = math.huge, maximum = 0 },
    off      = { minimum = math.huge, maximum = 0 },
    on       = { minimum = math.huge, maximum = 0 },
    filled   = { minimum = math.huge, maximum = 0 },
    maximum  = { minimum = math.huge, maximum = 0 },
  }
  for marking in pairs (markings) do
    bound.up.minimum = math.min (bound.up.minimum, marking.up)
    bound.up.maximum = math.max (bound.up.maximum, marking.up)
    ...
  end
  return bound
end
```

Fig. 7.8: Source code stub of the function that calculates the Petri net's bounds

7.3.2 Sequence Properties

While marking properties provide useful information about the state of a system, they do not give any information about *how* the system evolves. In contrast, sequence properties express the possibility to fire transitions in the reachability graph. One such property describes the liveness of a transitions. This property assigns a liveness level ($l_i \in \{l_0, l_1, l_2, l_3, l_4\}$) to each transition of the Petri net, so that it provides information about whether a transition is fireable, can/will be fireable or will never be fireable. The individual levels are defined as follows:

l_0: the transition is *dead*, i.e. can never fire;
l_1: the transition can fire at least once in at least one path of the reachability graph;
l_2: the transition is *quasi-live*, i.e. can fire infinitely often in at least one path of the reachability graph;
l_3: the transition is *live*, i.e. can fire infinitely often in all paths of the reachability graph;
l_4: the transition can always fire.

Note that the liveness constraints are ordered and that if a transition is l_k-live, it is also l_{k-1} live, except for l_0 (dead). The liveness level of transitions is a useful information to debug models, as this level should in theory match the intended behaviour of the process using by the transition. In practice, l_0-live transitions should almost never occur in models: finding a l_0-live transition is usually a clear indication of a bug within the model as it should not appear at all if the model was correct. Transitions that are l_1-live can be fired at least once, and have no really interesting meaning. They can appear in any behaviour, such as the representation of scripts or programs that terminate. l_2-live transitions trigger infinitely often, and are thus found in the behaviour of programs that do not terminate, for instance web servers. They are distinguished from l_3-live transitions, because the latter can fire infinitely often in all execution paths. Thus, l_2-live transitions are used usually in looping processes that can terminate, whereas l_3-live transitions are used usually in looping processes that cannot terminate. l_4-live transitions are in practice very rare, as they represent an action that can always happen. They can be found in user interfaces, where one process records the inputs and has thus transitions that can always be fired.

In order to find the liveness level of transitions, we can employ the algorithm provided in Figure 7.9. The computation works on the set of reachable markings and further uses the information about a reachability graph's strongly connected components (calculated using *Tarjan's strongly connected components algorithm* [263]). The algorithm works in several stages, treating each transition individually as follows:

1. the transition is initially set as both, dead (l_0) and always fireable (l_4); these flags are unset as soon as the transition can respectively be fired or cannot be fired in the reachability graph;
2. by iterating over the reachable markings, the transition is set as l_1 live if and only if there is a marking in which it can be fired;
3. l_2 flag is set using the strongly connected components of the reachability graph: a transition is l_2 live if a component contains a marking in which the transition can be fired; and that stays within the component;
4. l_3 flag is set similarly using the strongly connected components: a transition is set as l_3 live if and only if it is l_2 live, and can be fired within a component that has no outgoing transitions (that is called a final component).

7.3.3 Invariants

The actual aim of modelling is often to reason over certain problems. Invariants pose a convenient way of reasoning over Petri nets independent from their actual marking.

Invariant reasoning is part of the structural analysis of Petri nets. Generally, invariants are distinguished between place (P-)invariants and transition (T-)invariants. While the former is useful for deadlock detection [214], T-invariants serve the modelling of logic programs [215] and Horn clauses [187] as pointed out in [199].

The goal of invariant creation is the establishing of proofs based thereon. In general, proofs based on invariants have an algorithmic part that consists of computing invariants and a more analytic part, which is written by a human and describes the reasoning necessary for building the proofs. There is a large number proofs we can perform with invariants, ranging from state based proofs such as mutual exclusion to liveness properties. Such invariant-based proofs are interesting alternatives to the ones based on temporal logic properties. Although they tend to be less powerful, they can nevertheless provide parametric proofs more easily. This comes however at the cost of human contribution while temporal logic proofs are automateable.

```
function liveness (marking)
  local markings   = reachable (from)
  local components = tarjan (markings)
  for _, transition in ipairs (transitions) do
    transition.liveness = {
      l0 = true,
      l1 = false,
      l2 = false,
      l3 = false,
      l4 = true,
    }
    for marking in pairs (markings) do
      if marking [transition] then
        transition.liveness.l0 = false
        transition.liveness.l1 = true
      else
        transition.liveness.l4 = false
      end
    end
    for _, component in ipairs (components) do
      for marking in pairs (component) do
        if  marking [transition]
        and marking [transition].tarjan.component == component then
          transition.liveness.l2 = true
        end
      end
      transition.liveness.l3 = component.is_final and transition.liveness.l2
    end
  end
  return transitions
end
```

Fig. 7.9: Liveness level detection algorithm.

7.3.4 Formal Definition of Invariants

Each state $m \in SS(T, m_0)$ can be associated with an observation value. The observation is a summary of the state that contains only information relevant to the properties that are verified. Given a set of observations O, which observations are obtained through functions $\omega : M \to O$ that return an observation for any possible marking in M, ω is an invariant if and only if it returns the same observation for all the reachable states of the system. For instance, in Figure 7.1, the foot pedal is either up or down. In the marking, this means that there is one token in one of the places and none in the other at all times. Therefore $\omega : m \mapsto (m(\text{up}) + m(\text{down}))$ is a function that always returns the value 1 for all reachable states and is thus an invariant. More formally, ω is an invariant iff $\exists o \in O, \forall m \in SS(R, m_0), \omega(m) = o$.

The definition above for invariants is very general. It can be used if the users define the invariants themselves, but cannot help in deducing invariants from the structure of the Petri net. It is interesting to consider instead only functions that are linear combinations of elementary observation of places.

Usually, elementary observations are functions that for any place and marking return the marking of the given place. The elementary observations is thus defined as the function $o_e : P \to (M \to \mathbb{N})$ such that $\forall m \in M, \forall p \in P, o_e(p)(m) = m(p)$ These elementary observations can be instantiated using any place of the Petri net, for instance $o_e(\text{up})$ is an observation. The set of all elementary observations is included into the observations. For convenience, we consider that writing the name of a place is equivalent to the elementary observation for this place. Hence we will use up to also refer to the function that returns the marking of this place: $m \mapsto m(\text{up})$.

Observations can be composed as linear combinations of observations. For any observations $o_1, \ldots o_n \in O$, and constants $c_1, \ldots c_n \in \mathbb{N}$, an observation can be built to compute $c_1 * o_1 + \ldots c_n * o_n$, for instance $1 * o_e(\text{up}) + 1 * o_e(\text{down})$ (the places with coefficient 0 are not shown). The set of observations is thus defined by the linear combinations of elementary observations:

$$O = \{\Sigma_{p \in P}(c_p * o_e(p)) \mid \forall p \in P, c_p \in \mathbb{N}\}$$

Following this definition, if ω_1 and ω_2 are invariants, then $\omega_1 + \omega_2$ is also invariant. The addition of two invariants $\omega_1 + \omega_2 : M \rightarrow O$ is defined as: $\forall m \in M, (\omega_1 + \omega_2)(m) = \omega_1(m) + \omega_2(m)$. Also, if ω is an invariant then $\forall k \in \mathbb{N}, k * \omega$ is an invariant. The multiplication of an invariant with a constant is defined as: $\forall m \in M, (k * \omega)(m) = k * \omega(m)$. To compute invariants it is necessary to find the appropriate linear combination of elementary observations. The well-known Farkas algorithm can be employed for the finding of invariants, as explained in the next section.

7.3.5 Computing P-invariants

Discovering the P-invariants can be done using the Farkas algorithm [97]. The Farkas algorithm is an iteration process operating on a structure which corresponds to a Petri net's incidence matrix. For spatial reasons we do not provide the implementation of the algorithm but describe the (basic) functionality in textual form.

Its principle is to represent and manipulate a matrix where columns represent the transitions of the Petri net, and lines represent place markings or linear combinations of place markings. The algorithm works iteratively, performing the following actions:

1. add new lines to the matrix by building linear combinations of already existing lines, the aim of this process being to nullify columns (i.e. create columns that only contain zeros);
2. remove the lines that were used based on the summation;
3. remove the columns that are nullified;
4. remove all lines that are already expressed through other lines.

The process terminates when all columns are nullified. The solutions are linear combination of place markings that have a constant observation, and are thus invariants.

Example Below is an example of how to compute invariants following the given algorithm, on a very simple Petri net shown in Figure 7.10.

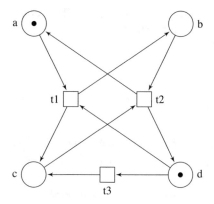

Fig. 7.10: Example of Petri net used to compute invariants

The successive F_i structures show the advance in Farkas algorithm. Each F_i is the result of one step, computed from the F_{i-1} structure. The first one (F_0) is the incidence matrix of the Petri net, that represents for each place and transition the difference between the weight on post arcs that rely the place and the transition, and the weight on pre arcs that rely them also.

$$F_0 = \begin{array}{c} a \\ b \\ c \\ d \end{array} \begin{bmatrix} -1 & 1 & 0 \\ 1 & -1 & 0 \\ 1 & -1 & 1 \\ -1 & 1 & -1 \end{bmatrix}$$

The first iteration is to eliminate t_1. To do it, we have to find linear combinations of rows such that the column for t_1 is always zero. Obviously, both $a + b$, $c + d$, $a + c$ and $b + d$ are correct linear combinations. There is no need to represent greater coefficients, such as $2 * a + 2 * b$, thus only the minimal combinations are kept. Because all rows a, b, c, d have been used to compute the new rows, the rows a, b, c, d are removed from the matrix in F_1'. All columns that contain only zeros are also removed in F_1'', as they are useless for the remaining of the iterations.

$$
F_1 = \begin{array}{c} a \\ b \\ c \\ d \\ a+b \\ c+d \\ a+c \\ b+d \end{array}
\begin{array}{ccc} t_1 & t_2 & t_3 \\ \left[\begin{array}{ccc} -1 & 1 & 0 \\ 1 & -1 & 0 \\ 1 & -1 & 1 \\ -1 & 1 & -1 \\ 0 & 0 & 0 \\ 0 & 0 & 0 \\ 0 & 0 & 1 \\ 0 & 0 & -1 \end{array}\right] \end{array}
\qquad
F_1' = \begin{array}{c} a+b \\ c+d \\ a+c \\ b+d \end{array}
\begin{array}{ccc} t_1 & t_2 & t_3 \\ \left[\begin{array}{ccc} 0 & 0 & 0 \\ 0 & 0 & 0 \\ 0 & 0 & 1 \\ 0 & 0 & -1 \end{array}\right] \end{array}
\qquad
F_1'' = \begin{array}{c} a+b \\ c+d \\ a+c \\ b+d \end{array}
\begin{array}{c} t_3 \\ \left[\begin{array}{c} 0 \\ 0 \\ 1 \\ -1 \end{array}\right] \end{array}
$$

As there is only one transition t_3 remaining, the second iteration is to remove it. Only one linear combination can do it: $a + b + c + d$, obtained by two ways: by adding the lines $a + b$ and $c + d$, and by adding the lines $a + c$ and $b + d$. But the Farkas algorithm requires to use only the rows that are not zero, thus $a + b$ and $c + d$ are not taken into account. As previously, the rows that have been used to compute the new ones are removed ($a + c$ and $b + d$) in F_2", and the nullified column is also removed in F_2''.

$$
F_2 = \begin{array}{c} a+b \\ c+d \\ a+c \\ b+d \\ a+b+c+d \end{array}
\begin{array}{c} t_3 \\ \left[\begin{array}{c} 0 \\ 0 \\ 1 \\ -1 \\ 0 \end{array}\right] \end{array}
\qquad
F_2' = \begin{array}{c} a+b \\ c+d \\ a+b+c+d \end{array}
\begin{array}{c} t_3 \\ \left[\begin{array}{c} 0 \\ 0 \\ 0 \end{array}\right] \end{array}
\qquad
F_2'' = \begin{array}{c} a+b \\ c+d \end{array}
\left[\;\;\right]
$$

After this iteration, there are no more columns to remove. The remaining rows represent invariants of the Petri net: $a + b$ and $c + d$, that are named for later use:

$$
i_1 = a + b
$$
$$
i_2 = c + d
$$

As previously discussed, invariants are formulas that do not change applied on reachable markings. This means that for any sequence s and markings m, m', such that $m \xrightarrow{s} m'$, the invariant i applies at both ends of the sequence, and even at each state within the sequence: $i * m = i * m'$. In our example i is either i_1 or i_2, stating that the observation of $a + b$ and $c + d$ remains constant, independently of which transitions are fired.

7.3.6 Using Invariants for Proving Properties

Proving properties on models can be done by exploration of the reachability graph. Nevertheless, in most practical applications this method is difficult to use due to the state space explosion problem. The previous section defines invariants on Petri nets, that can be computed without the need to create the full reachability graph. In fact, invariants can be computed without even having to specify an initial marking. This section outlines the power of invariants by showing how to use them to prove properties on models. The usual way of performing these proofs is to establish invariants, and follow a case by case analysis of the possible state of the Petri net. This is done using in the reasoning the fact that invariants are respected in the whole reachability graph and also for the initial marking.

Proofs can usually be expressed as case analysis of the reachability graph of the Petri net. The proof process is then based on the following steps:

1. compute the set of invariants
2. apply invariants to the initial marking to obtain the corresponding observations
3. define an arithmetic reasoning based on the observations, invariants and case analysis, where the cases are in general markings of the state space that respect the invariants.

Using the invariants from Example 7.3.1, we can prove the principle of mutual exclusion of a and b in the Petri net above. The reasoning steps for this proof are as follows:

1. Application of the Farkas algorithm onto the incidence matrix provides us with the Petri net's invariants: $i_1 = a + b$ and $i_2 = c + d$.
2. Next, we calculate the invariant values using the initial marking $m_0 = \{a = 1, b = 0, c = 0, d = 1\}$. This leads to the inference that the value of $i_1(m_0) = m_0(a) + m_0(b) = 1$.
3. Since i_1 is an invariant, this observation needs to hold on any marking in the reachability graph, i.e. $\forall m \in M, m(a) + m(b) = m_0(a) + m_0(b) = 1$
4. Equipped with the knowledge that a place-marking cannot be negative ($\forall p \in P, \forall m \in M, m(p) \geqslant 0$) we can deduce that $m(a) \in \{0, 1\}$ and $m(b) \in \{0, 1\}$
5. Lastly, we can easily observe that the mutual exclusion has to hold since if $m(a) = 1$ and $m(a) + m(b) = 1$, then $m(b)$ has to have the value 0. Similarly, $m(a) = 0 \wedge m(a) + m(b) = 1 \Rightarrow m(b) = 0$

While this trivial proof can easily be computed by hand, the same principle can be used to create more complex, semi-manual, structural proofs. In practice such proofs are supported by powerful proof checkers such as Coq [30] to verify that hand written proofs are correct. The detailed description of this technique surpasses the scope of this book, but we encourage the reader to consult publications dedicated to this topic.

7.4 Techniques for Model Checking

Invariants can often be computed using less time and memory than a Petri net's the entire reachability graph. In some cases, however, such as during the computation of causality properties (for instance "if an event A occurs, then an event B must occur later"), the need to perform the generation of the state space is still required. Although the computation is straightforward (see the algorithm in Figure 7.7), this technique does not scale well since many models have an exponentially growing number of states with respect to to their size. It is not uncommon for a model's reachability graph to have more nodes than the number of atoms in the universe (10^{82}).

The size of the reachability graph of a Petri net usually depends on two factors: the number of places and transitions within the Petri net and the number of tokens in each place. For instance, the reachability graph of our example in Figure 7.1 increases with the number of tokens in the fuel place. It could also increase if we add more places, for instance to represent several positions of the foot pedal (up, middle, down for instance).

In order to still be able to calculate the state space, research led to the development of various approaches to overcome the increasing size of the state space. Below, we will briefly outline a few examples.

- *Symmetry reductions* [74] allow us to perform model checking on a smaller system, by analysing one representative component instead of several identical components. For instance, instead of analysing the behaviour of several databases, identified by their names, symmetry reductions focus on analysing the behaviour of one anonymised database.
- *Partial order reductions* [125, 163, 278] are based on the commutativity of concurrent actions. When performing model checking of asynchronous systems, action interleaving requires an arbitrary order between the events. To treat all possible cases, all permutations of the order must be considered, resulting in an exponential explosion of the number of traces and states. Partial order reductions allow to check only a subset of the behaviours, by removing executions that differ only by the order of independent transitions. This technique can only be applied when the property we want to check does not depend on the removed executions.
- *Büchi Automata* [296]: Some temporal logic formulæ (namely linear temporal logic ones) are checked against the set of infinite executions of the model. A way to perform this verification is to transform the formulæ into automata that accept the valid executions. Such automata are called Büchi automata. For

verification, the model and the formula are both encoded as Büchi automata which accept the languages that represent the executions of the model and the valid executions with respect to the property to verify. Model checking then consists of checking that the language of the executions is included in the language of the property.

- *Bounded Model Checking (BMC)* [33]: This approach is a kind of "degraded mode" of standard model checking, as the formula is checked only for executions of a maximum length k (where k is the sequence length from the initial marking). If no problem is detected, then k is increased until the formula does not hold any longer, or k reaches an upper bound (called the *completeness threshold of the design*). Note that there is no guarantee for executions longer than this upper bound. Since these kinds of problems can be reduced to propositional satisfiability problems, they can use very efficient SAT or SMT solvers.

- *Distributed Model Checking* [121, 22, 21]: One of the main problems when performing model checking is memory exhaustion. The idea of distributed model checking is to distribute the states over several computers in order to increase the overall available memory. This technique has a drawback: the transmission of data between computers poses as a bottleneck. Various algorithms have been designed to perform a "good" distribution of states so that most transitions are local to the same computer. They have reached their limits however due to the rise of modern, highly efficient CPUs. Distributed model checking is in practice most efficient when the time required to compute successor states is much longer than the time required to transmit states.

- *Parallel Model Checking* [257, 20]: With multi-core architectures, the trend is nowadays to perform parallel model checking instead of distributed model checking. In this approach, all the states are represented and computed on the same computer, but several execution processes are used to speed up the computation. This approach still grows more difficult with increasing parallelism, because of possible memory contention within storage for computed states. The same approach is used to perform parallel model checking on GPUs.

- *Symbolic Model Checking (SMC)* [58, 59, 77]: Instead of explicitly representing states and transition relations, this approach only represents and manipulates sets of states or sets of transitions using Decision Diagrams [55]. This representation allows at the same time to share some common parts of the states and reduce computations when applying the transition relation, in the best cases logarithmic with respect to the size of the model.

Some techniques have been developed that combine several approaches [92]. For instance, symmetry and partial order reductions are also often coupled with static reductions techniques that reduce the size of the specification with respect to the property to check, or with state compression as in [96], where some states are represented by the difference with their predecessor. Similarly, parallel model checking is easily used in combination with the other approaches discussed above. One exception is symbolic model checking because Decision Diagrams require a unicity table and computation caches, that need to be locked and are thus bottlenecks.

7.5 Data manipulation in Petri nets

The previous sections present the use of simple Petri nets models. While this type of Petri nets are sufficient to represent simple systems, they lack several features that are useful when dealing with complex cyber-physical systems, such as modularity, time constraints, and data manipulation. Such simple models encode data as a number of tokens. For instance, the fuel level is represented by a certain amount of tokens within the fuel place.

The basic Petri nets we introduced so far are commonly referred to as *Place/Transition* (P/T) nets. As we have seen, they are well-suited for the modelling of process control, synchronisation and resource flow within a system. However, in more complex situations the information used might not be easily representable as simple, black tokens. We can easily imagine cyber-physical systems where the transmission of more complex information is required. A P/T net modelling a very simple drone controller may have transitions up, down, turn left, turn right, move forward and move backward, which all modify the state. However, if we now add possibilities to perform multiple actions at the same time, we would have to add more transitions such as move up & forwards. The number of transitions within this relatively simple system would grow exponentially.

It would be more efficient to represent such information directly in the tokens. In order to do so, we need a token for vertical movement that could express to either move up, move down or stay, another one for changing directions (left, right, stay) and a third one for the forward/backward movement. We can then send these three tokens into a transition that would perform the action depending on the token values.

This simple scenario shows the necessity for *High-level Petri nets* (HLPN). HLPN were developed with the goal to attach a values to tokens, such as described before. Transitions and arcs are extended with *guards*, which evaluate the token values and assert a certain configuration.

7.5.1 Drone Controller

Our new High-level Petri net drone controller is presented in Figure 7.11 On the first view we can see three distinct groupings for vertical movement, horizontal movement and orientation. Each one of these three groups contains one place with an initial token with value stay. The behaviour of the groups follows the same scheme. From the initial place we can use transitions to create tokens with different values. The transitions' preconditions are guarded. This means that if the token's value is stay, we can only fire transitions whose guards require a stay-token (there are two in each group). Once we fire the transition, the stay-token is consumed and a new one is produced. For example from the place altitude we can either fire a transition move up, which produces an up-token, or move down, that creates a down-token. This behaviour matches the pushing of a joystick on the drone's physical remote to either direction. There are two more transitions within this group that are guarded by up and down, respectively. Firing these transitions will consume the token respective token and produce a stay token. In other words, we can use them to "reset" the token (i.e. to stop movement into that direction). On an actual drone remote, this behaviour matches the releasing of the drone's altitude joystick to neutral position. The groups that express the horizontal movement and orientation behave correspondingly.

Next to these three groups we can see the central move transition. This transition is responsible for the actual displacement of the drone Move consumes the tokens from vertical, horizontal and orientation. Interestingly, the arcs for these tokens are not guarded by concrete token values, but by variables (a, d and m). These variables are assigned when move fired and take the tokens' values. The transition further uses a token from place drone and a token from place battery. Note, that both drone and battery have different token types (i.e. can hold different valuations): While battery contains multiple, classical black tokens, the drone-place only holds one data-record token whose value stores information about the drone's current *position* (as $\langle x, y, z \rangle$-coordinates) and *angle* (stored as anti-clockwise deviation angle from north[2]).

Firing move will consume the three movement indicator tokens, one battery-token and the drone-token (assigned to variable s). It then produces equivalent tokens to the consumed ones in each one of the movement places, plus a new token in the drone-place, whose value is updated to match the new state of the drone[3]. Producing equivalent movement tokens to the one that has been consumed, has the effect that the behaviour of the drone will remain continuous until it is actively altered. In the example, move's consumption of an up token will produce another up in the altitude-place. Hence, when firing move again, the drone will continue rising, unless the token has been modified (using the transitions within the altitude group)

Note, that since there is no way to produce tokens in the battery-place, this is the resource that limits the number of times we can trigger the central transition and hence move the drone.

Our drone controller further has a safety-mechanism integrated. This mechanism is shown as a guard on the central transition and expresses that we cannot fire this transition if the drone's position z-position is lower than 50 (centimetres) and the altitude token is specifies a downward movement (i.e. has value down).

7.5.2 Formalising High-level Petri nets

As we saw in the example, HLPN use various additional concepts such as variables and expressions. In this section we will focus on the adaptation of our existing Petri net formalism to integrate these concepts.

Variables and expressions To formally express the concepts that were used in the example above, we need to define matching and filtering of tokens. For this reason we introduce the notions of variables and expressions: We define a set of variables V (e.g. s,a,d,m), and a set of expressions over data and variables (noted E).

---

[2] I.e. the value 90 describes West, 180 South, and 270 and −90 stand for East

[3] In the figure the update is simply expressed as $s + f(a, d, m)$, where we assume the existence of a function f which can provide the $\langle x, y, z \rangle$-coordinate and orientation angle change produced depending on the group tokens.

Since variables are expressions or parts thereof, V is a subset of E: $V \subseteq E$. We further define a function $variables : E \rightarrow \mathcal{P}(V)$, which returns the set of variables that are used by an expression.

If an expression does not contain any variables, we call it a *ground expression*. The set of ground expressions $E^{\varnothing}$ is defined as $E^{\varnothing} = \{e \mid e \in E \wedge variables(e) = \varnothing\}$. In order to express concrete values (such as stay or up) this set has to be non-empty.

Binding Equipped with these tools we can now express the binding of variables (such as binding of the altitude token to the variable a). A binding $\sigma \in \Sigma$ is a partial function from variables to ground expressions: $\sigma : V \rightarrow E^{\varnothing} \cup \{\bot\}$. We remind ourselves that a substitution, as introduced in Section 7.2.4, is denoted noted $\sigma(e)$. This means that it is the application of a binding σ on an expression $e \in E$ and replaces variables with their value in the binding. Note, that a substitution does not have to replace all variables within an expression. Variables that do not appear in the binding remain in the expression. Formally:

$$\forall e \in E, \exists e' \in E^{\varnothing}, e' = \sigma(e) \wedge variables(e') = variables(e) \setminus \{v \mid v \in V \wedge \sigma(v) \neq \bot\}.$$

Transition guards In high-level Petri nets transition guards are used to prevent transitions to fire with unwanted token values and thereby stop unwanted behaviour. In the drone controller we use the guard $\neg(a = down \wedge s.position.x < 50)$ to prevent the drone from moving down when its altitude is below a certain threshold. Formally, transition guards are functions that evaluate to Boolean values ($\mathbb{B} = \{\top, \bot\}$). Transitions can only be fired iff the guard evaluates to $\top$.

Integration of the concepts Using the above definitions, we modify the definition of a Petri net to incorporate the new concepts. Specifically we change the following:

1. Add ground expressions to tokens within markings as a means to hold data
2. Add (variable) expressions to arcs in order to filter tokens or bind variables
3. Add Boolean guards over expressions to allow filtering of tokens and disabling transitions for certain token-values.

Formally, HLPN are described as tuples based on the structure of P/T nets (as defined before), but with added and modified fields: $\langle V, E, variables, P, T, pre, post, guard, m_0 \rangle$, where:

- $V, E, variables$ are the sets of variables and expressions, and the $variables$ function as introduced above;
- $M = P \rightarrow \mathcal{M}(E)$ is the set of all possible multisets over expressions. It replaces the former marking definition ($P \rightarrow \mathbb{N}$);
- $guard : T \times \Sigma \rightarrow \mathbb{B}$ defines the allowed bindings for each transition;
- $m_0 : P \rightarrow \mathcal{M}(E^{\varnothing})$ associates to each place a multiset of ground expressions as the initial marking.

This definition differs resembles the P/T definition from before, except for the introduction of variables and expressions and the new definition of marking, where a marking consists of a multiset of expressions rather than a natural number.

In fact, we can look at a classical P/T net is a special kind of high-level Petri net where the set of variables is empty ($V = \varnothing$) and only one expression ($E = \{\bullet\}$) exists. Note that since transition guards are usually defined as Boolean expressions over expression comparisons, a P/T net's $guard$ function always returns true: $\forall t \in T, \forall \sigma \in \Sigma, guard(t, \sigma) = \top$.

The semantics of a transition t is given in Equation (7.13). It differs from the previous transition semantics given in Equation (7.2) by adding the substitution of a binding within the pre and $post$ functions.

$$transition_t : \frac{\exists \sigma, \sigma(pre(t)) \leqslant m, guard(t, \sigma) = \top}{m \xrightarrow{t} m - \sigma(pre(t)) + \sigma(post(t))} \tag{7.13}$$

The use of expressions in HLPN requires the extension of the substitution to include all expressions in the function image. Formally, $\forall m \in M, \forall p \in P, \sigma(m)(p) = \sigma(m(p))$. This substitution also has to be extended to multisets, since HLPN markings are defined as such. Applying a substitution on a multiset is performed by applying the substitution to each element: $\forall es \in \mathcal{M}(E), \sigma(es) = [\sigma(e) \mid e \in es]$, where [ ] denotes a multiset by intention.

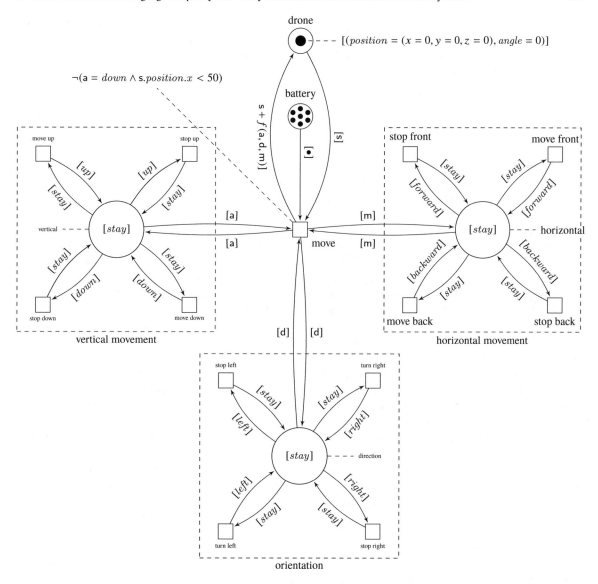

Fig. 7.11: Petri net representing the control of a drone, using data as tokens and arc valuations. The control is composed of three modules, that represent the state of the controller joysticks: one to move upwards or downwards, one to rotate left or right, one to move forwards or backwards. The drone moves when the *move* transition is fired. It captures the state of the controller to move the drone. Places are labelled with multisets of tokens, shown as $[token_1, \ldots token_n]$. Each token is a structured data, that can be a simple token (•), an atom (*stay*) or a record or subdata with named fields, such as $(position = (x = 0, y = 0, z = 0), angle = 0)$. The *position* is a position relative to the initial position of the drone, and the *angle* is an angle on the horizontal plane and signifies the deviation from north. Arcs are labelled by multisets of tokens, that can use variables, for instance x. Transitions have implicitly a guard that always returns ⊤, except the *move* transition that has an explicit guard $\neg(a = down \wedge s.position.x < 50)$.

7.5.3 Other High-level nets

In the last couple of decades, researchers hae come up with numerous extensions and modifications of the P/T net formalism to ease the modelling of complex systems. This section introduces four representatives of such complex variants of Petri nets.

Coloured Petri nets (CPN) [158] CPN are an extension of Petri nets where each token can be of a certain *colour*. Each colour is a set of values, comparable to a type definition. The set of colours has to be clearly specified for a CPN. Arcs are modified to allow the specification of the colour to use in pre- and postconditions.

The drone example in Figure 7.11 is a CPN with the following colours:

- $CVertical = \{up, down, stay\}$
- $CHorizontal = \{forward, backward, stay\}$
- $CDirection = \{left, right, stay\}$
- $CFuel = \{\bullet\}$

Unfortunately, not all data types can be easily represented as colours, for instance, the drone state is a dictionary or tuple. The corresponding colour contains all the possible combinations of positions and angles. Since this set is infinite (x, y, z and *angle* are rational), this solution is not usable in practice, and the domains have to be made discrete and bounded.

Symmetric nets [71] Symmetric nets, formerly known as "well-formed Petri nets", are a special kind of Colored Petri nets that use only simple data types: tuples of constants in finite and ordered domains. The data structure is thus very limited, and cannot easily represent data of varying size, as well as data in *a priori* unbound domains. Operations on data are also very restricted: it is only possible to obtain the successor or predecessor of a value, and test equality between two values. There are no operations allowed on the tuples themselves.

Despite all these limitations, symmetric nets are successfully used in practical problems because they offer a formalism that is convenient for efficient structural analysis and model checking. The limited expressiveness of these types increases the number of properties that can be verified, in particular for the structural analysis of models, such as computation of bounds or invariants.

Algebraic Petri nets (APN) [297] Algebraic Petri nets are a special form of Colored Petri nets, that allow the use of abstract algebraic data types [87] as colours. Such algebraic data types consist of a signature and axiomatisation and hence allow the user to represent custom data types. The advantage of APNs is that every data type used has a precise axiomatisation and consequently proofs can be done by theorem proving without the usual limitation of finiteness of model checking. Using APNs, parametric and under-specified systems can be modelled and also be verified in a more systematic way.

Timed Petri nets [231] The concepts of time and of Petri nets are quite opposite: while time determines the occurrences of events in a system, Petri nets consider only their causal relationships. Several variants of Petri nets have been defined with the notion of time. The three most common are: Time Petri nets, Timed Petri nets, and Petri nets with Time Windows. In Time Petri nets, transitions are labelled with time intervals, that define the time at which the transition can be fired, after it has been enabled (has all its preconditions met). In Timed Petri nets, time is also put within tokens, that have thus an age, and transitions are labelled with time intervals that define the age at which tokens can be consumed. In Petri nets with Time Windows, transition are given time intervals, this means that transition can fire (not mandatory) only in this time interval.

In this chapter we do not consider time with Petri nets for simplicity, and only focus on the causal relations associated to Petri nets.

7.6 Combining Model Semantics and Simulation

The recent rise of computers in everyday life is especially of importance when their purpose is to react to and act upon environment changes. We refer to such systems as cyber-physical systems (CPS). Such systems consist of a software part (e.g. a controller program) and a hardware side that usually consists of sensors and actuators.

While smaller systems, such as heating/light systems that measure presence have been installed and used for a long time, the trend towards Internet-of-Things applications, "smart systems" (such as new-generation cars and trains), and general large-scale systems that include hundreds, sometimes even thousands of components drives the need for means to verify and validate such systems.

The problem of these highly heterogeneous systems lies in finding the right means to model each part of the system. While former approaches to find *the one* modelling language or tool failed, nowadays the trend is reversed. Modern research is looking to model every part of the system with its most appropriate modelling formalism. Subsequently the individual components are combined and simulated together. This approach, called co-simulation [126] has shown promising research results, but comes with one important question: How should we combine models that were developed using different syntaxes and semantics?

One of the approaches that aims to answer this question are *Functional Mock-up Units* (FMUs) [36]. The FMU formalism provides a homogeneous interface, the Functional Mock-up Interface (FMI). Each component is wrapped inside its own FMU and is henceforth executed using the FMI. The individual FMUs' inputs and outputs are connected to one another in order to allow information to be transmitted within a system. The semantics of such a composition are dominated by a so-called *master algorithm* that is responsible for passing control to individual components, relaying signals and choosing appropriate time step sizes which suit all units in the system.

In this section we introduce Petri Net Functional Mock-up Units (PNFMUs) [180]. This new type of FMU wraps a Petri net within the FMU in order to provide access to the efficient evaluation and calculations we introduced in the previous sections. Using Petri nets it is possible to detect deadlocks and possible system evolutions.

The FMI standard strictly defines the qualities of a valid FMU. In order to comply with this standard however, it is necessary that PNFMUs overcome three major challenges:

1. Time evolution Similarly to a Mealy or Moore machine, the Petri net formalism doesn't explicitly define a time concept. Often, the firing of a transition is accepted as a time unit. To overcome this limitation, the periodic or aperiodic wrappers presented in [272] can be used for this purpose. In addition, these wrappers has also been discussed in [83] when considering the semantic adaptation, giving the possibility to automatically generate FMUs from a domain specific language that solve this problem.

2. Inputs/Outputs adaptation Obviously, the inputs and outputs type and nature of a given formalism can differ from the simple value affectation of a variable defined by the FMI standard. Therefore, an adaptation must be performed as well for inputs and outputs of a FMU. The authors of [83] provided solutions with their domain specific language to overcome this limitation too.

3. Non determinism In the standard, FMI API functions are mathematically modelled as total functions [272]. This means that calling an FMU's API with the same parameters should always yield the same result. An initial, yet naive, approach to represent a Petri net state would be to consider it as a single marking. However, in situations where a Petri net has more than one fireable transitions at a given state, the evolution function of the FMI API (*doStep*) cannot yield a deterministic result. This particularity is clearly shown by the Petri net and its reachability graph in Figure 7.12. In this example, either t_1 or t_2 can be fired from its initial marking, leading to different system evolutions. This issue must be addressed to be able to represent a Petri net within an FMU.

Since the two first subjects of interest have been extensively discussed in the referenced publications, the non determinism of the formalism is considered here. Now that the problems related to the consideration of formalisms as FMUs has been addressed, let's discuss now of the PNFMU formalisation.

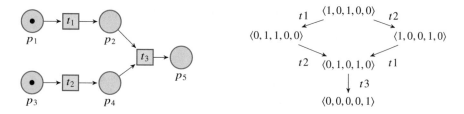

Fig. 7.12: A simple Petri net and its reachability graph. The markings are encoded as $\langle p_1, p_2, \ldots, p_n \rangle$, where p_i are the number of tokens within the places, which the ordered by their index. $\langle 1, 0, 1, 0, 0 \rangle$ encodes the initial marking on the left, stating that there is one token in the first and third place each (i.e. p_1 and p_3).

7.6.1 PNFMU Formalisation

The need for Petri nets within an FMI system is closely tied to the need to analyse system evolutions and reachability of possible states. In order to do so, we need to define the formal basis of an PNFMU, which we base on the FMU formalisations presented in [272] and [51].

A standard FMU is defined as the structure $F = \langle S, U, Y, D, s_0, set, get, doStep \rangle$. Given a Petri net PN = $\langle P, T, pre, post, m_0 \rangle$, we define that a PNFMU is a tuple $PNFMU = \langle S, U, Y, D, s_0, set_\mathbb{N}, get_\mathbb{N}, doStep_\mathbb{N} \rangle$.

A PNFMU's internal states S are all possible markings of the Petri net, not just the reachable ones (i.e. $reach(s_0) \subseteq S$). U and Y are the Petri net's places which are writeable and readable, respectively. D remains the dependency of outputs on inputs and is used to avoid circular dependencies when composing FMUs. s_0 is set to the initial marking m_0.

The biggest difference between FMU and PNFMU is that the former only operates on one state. On the contrary, PNFMU are designed to explore and operate on a Petri net's reachability graph. This change is reflected in the three functions set, get and $doStep$, which are adapted to perform operate on sets of states as follows:

$set : \mathcal{P}(S) \times U \times \mathcal{P}(\mathbb{N}) \to \mathcal{P}(S)$ operates on a set of states (markings), modifies the marking of a certain place in each marking and returns a set of new states. Intuitively, the function iterates over the values to set and modifies returns a new state for each state that is modified. In total the function returns $n \times m$ markings, where n is the count of states and m the number of values entered into this function. Using the reachability graph from above, the call $set(\{\langle 0, 1, 1, 0, 0 \rangle, \langle 1, 0, 0, 1, 0 \rangle\}, p_5, \{2, 4\})$ returns the set $\{\langle 0, 1, 1, 0, 2 \rangle, \langle 1, 0, 0, 1, 2 \rangle\}, \{\langle 0, 1, 1, 0, 4 \rangle, \langle 1, 0, 0, 1, 4 \rangle\}$. Note, that none of the four returned markings is reachable from s_0 by transitions.

$get : \mathcal{P}(S) \times Y \to \mathcal{P}(\mathbb{N})$ recuperates the set of place-markings of a set of states.

For example, $get(\{\langle 0, 1, 1, 0, 0 \rangle, \langle 1, 0, 0, 1, 0 \rangle\}, p_3)$ returns $\{1, 0\}$, while $get(\{\langle 0, 1, 1, 0, 0 \rangle, \langle 1, 0, 0, 1, 0 \rangle\}, p_5) = \{0\}$.

$doStep : \mathcal{P}(S) \times \mathbb{N} \to \mathcal{P}(S) \times \mathbb{N}$ executes system evolutions. Given a set of states and a time step $h \in \mathbb{N}$, $doStep$ returns the length of the longest sequence h' (with $0 \leq h' \leq h$) that can be executed and the states that are reached. Note, that $doStep$ only operates states that are reachable from s_0, any states $s \notin S$ are not considered. Furthermore, contrary to standard FMUs, PNFMU's $doStep$ is defined over natural numbers and fails if $h \notin \mathbb{N}$. $doStep$ is defined as follows:

$$doStep(s, 0) = (s, 0) \tag{7.14}$$

$$doStep(s, 1) = \begin{cases} (succ, 1) & \text{s.t.} \quad succ = \left\{ s' \mid \exists t \in T, \exists s_i \in \left(s \cap SS(T, s_0) \right), s_i \xrightarrow{t} s' \right\} \wedge succ \neq \emptyset \\ (s, 0) & \text{otherwise.} \end{cases} \tag{7.15}$$

For $h \geq 2$: $\tag{7.16}$

$$doStep(s, h) = \begin{cases} (s, 0) \text{ st. } (s, 0) = doStep(s, 1) \\ (s'', h'' + 1) \text{ st. } (s', 1) = doStep(s, 1) \wedge (s'', h'') = doStep(s', h - 1) \end{cases} \tag{7.17}$$

The $doStep$ function for the example Petri net net above is given by the equations above. To show its execution, here is the first four steps of $doStep$ for pn':

- $doStep(\{\langle 1, 0, 1, 0, 0 \rangle\}, 1) = (\{\langle 0, 1, 1, 0, 0 \rangle \langle 1, 0, 0, 1, 0 \rangle\}, 1)$;
- $doStep(\{\langle 1, 0, 1, 0, 0 \rangle\}, 2) = (\{\langle 0, 1, 0, 1, 0 \rangle\}, 2)$;
- $doStep(\{\langle 1, 0, 1, 0, 0 \rangle\}, 4) = (\{\langle 0, 0, 0, 0, 1 \rangle\}, 3)$.

Table 7.1 compares the definitions of the individual components of both, the FMU and PNFMU.

7.6.2 PNFMU Example

Figure 7.13 shows some possible evolutions of a PNFMU that wraps the Petri net of the above example. Out of the infinite sequences of actions possible, we choose five traces that are being presented as an evolution tree.

- First, the main branch (center) shows the $doStep$ evolution of the system. Note that after the third $doStep$ the returned state remains unchanged and the returned h' is 0. This indicates a deadlock, as no further evolution is possible.

Component	FMU	PNFMU
S	a set of internal states of F	the set of all possible markings; S
U	a set of input variables over values $\mathbb{V}$	a set of input places; $U \subseteq P$, $\mathbb{V} = \mathbb{N}$
Y	a set of output variables over values $\mathbb{V}$	a set of output places; $Y \subseteq P$, $\mathbb{V} = \mathbb{N}$
$D \subseteq U \times Y$	a set of input-output dependencies specifying which outputs depend on which inputs	
$s_0 \in S$	the initial state of F	the initial marking of PN; $s_0 = m_0$
set	sets the value of an input variable, returns the new state; $set : S \times U \times \mathbb{V} \to S$	sets the value of an input place in the given states, returns the new states; $set : \mathcal{P}(S) \times U \times \mathcal{P}(\mathbb{N}) \to \mathcal{P}(S)$
get	returns the value of an output variable; $get : S \times Y \to \mathbb{V}$	returns the an output place's values in all given markings; $get : \mathcal{P}(S) \times Y \to \mathcal{P}(\mathbb{N})$
doStep	attempts simulation step, returns actual step size and new state; $doStep : S \times \mathbb{R}_{\geq 0} \to S \times \mathbb{R}_{>0}$	attempts simulation step, returns actual step size and new states; $doStep : \mathcal{P}(S) \times \mathbb{N} \to \mathcal{P}(S) \times \mathbb{N}$

Table 7.1: Comparison of the structures of FMU and PNFMU.

- We observe another trace using only $doStep$ actions. The $doStep(s, 4)$ attempts to find the states reachable with a sequence lenght of 3. However, the reachability graph dictates that only two steps are possible. Hence the action returns the end of the longest sequence that has been reached (in this case 1) and
- On the left branch, we see the creation of two markings of which one is not in the reachability graph of m_0. The subsequent $doStep$ therefore ignores this state when performing the calculation.
- The right branch describes the evolution after a set action on the initial state. This set creates one single, non-reachable marking. Therefore, the $doStep$ action on this state returns the empty set of states.
- Executing another set creates a reachable marking. This means that the $doStep$ is possible and succeeds.

We can easily see that, using PNFMUs, it is possible to study the possible evolutions of a system and analyse whether the execution of certain sequences leads to a reachable marking (i.e. a "good state"). We can also find deadlocks situations, i.e. states where a $doStep$) returns 0.

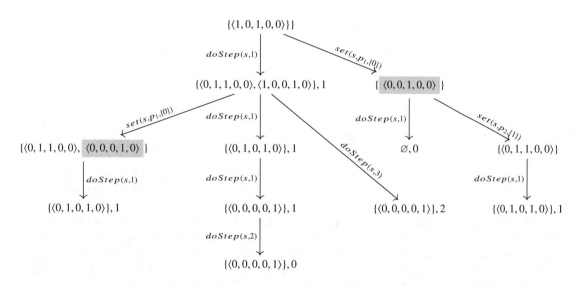

Fig. 7.13: A tree displaying possible system evolutions. Nodes are the (sets of) states and arcs are annotated with the action performed (where s is the variable that stores the previous states). Markings that are not in the reachability graph of s_0 are gray .

7.6.3 PNFMU Composition

The FMI standard defines FMU composition in a very simple manner. An FMU's outputs are directly connected to another FMU's inputs. This is usually performed by the *master algorithm* using the *set* method. Generally, the master algorithm performs such updates in two steps: 1. update all FMUs' input values; 2. perform system evolution by calling *doStep* on each FMU.

The composition of PNFMUs is slightly more complex, as it is necessary to handle sets of states. In general, three main scenarios can be distinguished:

One-to-One/One-to-Many Connecting a PNFMU that is currently in one, single state to another PNFMU is most trivial of the possibilities. Upon update, the marking(s) of the second PNFMU's inputs are set to the first PNFMU's output places' values. If the second PNFMU has multiple states, all states are updated.

As represented in Figure 7.14a the place p_5 of $PNFMU_1$ is linked to the place p_0 of $PNFMU_2$, drawn in bold. The execution scenario is quite trivial. In fact, for every marking of $PNFMU_2$, the number of tokens of place p_0 is set to the number of tokens of the place p_5 from the $PNFMU_1$.

Many-to-One The case represented Figure 7.14b is slightly more complicated. $PNFMU_1$, which has multiple markings, is connected to $PNFMU_2$ that has one single marking. In other words, there is two distinct numbers of tokens in the place p_5 in $PNFMU_1$, respectively 0 and 1. In consequence, the variables update step should create one marking for each values of $PNFMU_1$'s output variable. In the current case, $PNFMU_2$ should have two markings after the update, one for $M(p_1) = 0$ and another for $M(p_1) = 1$.

Many-to-Many After a step evolution, it is possible that both PNFMUs contain more than one marking. In fact, the *Many-to-Many* relationship is a generalisation of the *Many-to-One* case. This case requires to write all output values of one PNFMU to all inputs of the other. Effectively, the Cartesian product of possible states is created, where all combinations of $PNFMU_1$'s values are applied to all input places of $PNFMU_2$.

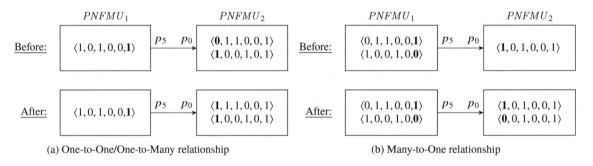

(a) One-to-One/One-to-Many relationship   (b) Many-to-One relationship

Fig. 7.14: FMU relationships before and after the update

7.6.4 Advanced Composition Mechanisms

Modularity is a mandatory principle to apply Petri nets to real world-sized systems. Modular extensions of Petri nets allow to create complex models by combining smaller entities. They facilitate the modelling and verification of large systems by applying a divide and conquer approach and promoting reuse. Modularity includes a wide range of notions such as encapsulation, hierarchy and instantiation. Over the years, Petri nets have been extended to include these mechanisms in many different ways. The heterogeneity of such extensions and their definitions makes it difficult to reason about their common features at a general level. An approach has been proposed to standardise the semantics of modular Petri nets formalisms, with the objective of gathering even the most complex modular features from the literature. This is achieved with a new Petri nets formalism, called the *LLAMAS* Language for Advanced Modular Algebraic Nets (LLAMAS)[197]. This framework can be envisioned to extend current work on $PNFMU$ and for the abstract description of the master algorithm.

7.7 Tooling

The vast amount of research invested in the Petri nets domain not only led to many important, theoretical findings but also saw the development of tools and software that can be used to model, analyse and simulate Petri nets. In the following section we will describe some tools that can be used to develop models based on Petri nets and to perform model checking thereon. Some of the tools described below are currently unavailable or are not able to perform model checking for Petri nets directly. Nevertheless, we will introduce them since they contribute interesting concepts and serve as potential candidates for extension to the formalism. Note, that since the domain is subject of intensive research, new ideas are often presented in in academic tools which oftentimes remain proof-of-concept implementations and not advanced into mature, reliable software.

7.7.1 Tools for Petri net Modelling and Verification

Since the Petri net formalism is based on a graphical syntax, matching model editors, composers and visualisers are an important important part of the Petri nets tool chain. Over the years, a lot of tools have been implemented to perform these tasks, but only a few have survived long enough to be well-recognised by the Petri nets community, or to be used in industry.

CPN Tools [206] (http://cpntools.org) is a tool for editing, simulating, and analysing high-level Petri nets. It supports basic Petri nets, timed Petri nets and Coloured Petri nets. It features a graphical editor, a simulator and includes a state space analysis component.

CPN-AMI [132] (http://move.lip6.fr/software/CPNAMI) is a formal modelling platform based on Petri nets, more specifically targeting Symmetric nets. It provides a graphical user interface to create and edit models, run tools built upon the platform and obtain their results. The CPN-AMI platform uses tools developed by several research teams to perform structural and behavioural analysis of models.

The platform features software to graphically create Petri nets modularly or using a scripting interface. Other tools allow to compute structural properties, such as bounds, invariants, syphons, traps, and liveness. The platform is also able to generate the state space of a Petri net, and perform CTL and LTL verification using various model checkers.

ITS-Tools ITS-TOOLS [267] (http://ddd.lip6.fr/) is a successor of CPN-AMI. It allows its users to create models using a textual language or a graphical editor for Petri nets. This tool is able to perform behavioural analysis (safety, CTL and LTL model checking) on models expressed in Place/Transition, Symmetric and Time Petri nets, as well as some other formalisms.

AlPiNa & StrataGEM: Algebraic Data Types and Term Rewriting tools In recent years, the University of Geneva's SMV group has produced two tools for model checking and model editing: the Algebraic Petri Nets Analyzer (AlpiNa) [56, 57, 151] and StrataGEM [4] [190, 39]. AlpiNa is a model checker dedicated to Algebraic Petri Nets. StrataGEM marries the concepts of Term Rewriting to the efficiency of Decision Diagrams, in order to perform efficient model checking. While AlPiNA is adapted to high-level specifications using Petri nets, StrataGEM focuses on the low-level ones.

These two tools heavily rely on algebraic data types and term rewriting techniques to represent systems and their semantics. They are able to compute large state spaces, and to evaluate reachability properties, as proven in the Model Checking Contest [171].

SNAKES [230] (https://snakes.ibisc.univ-evry.fr) is a Python library that provides a framework to define and execute many sorts of Coloured Petri nets. A key feature of SNAKES is the ability to use arbitrary Python objects as tokens and standard Python expressions in many points, for instance in transitions guards or arcs.

Renew: Renew [62] (http://www.renew.de) is a tool that supports the development and execution of object-oriented Petri nets, a specific kind of Coloured Petri nets. Its main feature is its integration with the Java programming language: Petri nets can be labelled by Java code, and thus call Java methods in transition guards or on arcs.

Petri net kernel (ePNK): [164] (http://www.imm.dtu.dk/~ekki/projects/ePNK/) is a platform for developing Petri net tools based on the PNML [147] transfer format. Its main idea is to support the definition of

---

[4] https://github.com/mundacho/stratagem

Petri net types, which can be easily integrated into the tool, and to provide a simple, generic graphical editor, which can be used for graphically editing nets of any plugged in type.

Tina [29] (http://www.laas.fr/tina) The TIme Petri Net Analyser is a toolbox for the editing and analysis of Petri nets, and Time Petri nets. It features a graphical editor, and a set of tools to perform structural analysis (e.g. invariants), behavioural analysis (e.g. reachability and coverability graphs) and LTL model checking.

TAPAAL [157] (http://www.tapaal.net/) is a tool for the modelling, simulation and verification of Timed Petri nets. It offers a graphical editor for drawing Petri nets, a simulator for experimenting with the designed nets, and is able to check the bound of the model, and to verify properties expressed in a subset of CTL. TAPAAL can translate its models to the format of the UPPAAL tool. In the same domain, Open-Kronos [298] uses the Büchi automata approach to verify real-time systems.

The Petri nets repository [146] (http://pnrepository.lip6.fr) is a recently created repository of Petri net models. It includes models imported from several sources, such as the Model Checking Contest [170, 169], the former PetriWeb repository [127], and the Very Large Petri nets benchmark suite (http://cadp.inria.fr/resources/vlpn/). Its main feature is that this repository provides access to models and corresponding, computed properties through both, a web interface and an API.

Many other tools exist to perform model checking on systems specified with formalisms that are not Petri nets, such as for instance automata or communicating processes. Among them, the following tools are worth mentioning:

SPIN [149] (http://spinroot.com) is a software model checker that verifies specifications written in PROMELA (PROcess MEta LAnguage), adapted to the representation of asynchronous distributed systems. The tool uses the Büchi automata approach to verify LTL properties on the models.

SPOT [192] (https://spot.lrde.epita.fr) is a Büchi automata library rather than a full model checker. It is intended to be coupled with an engine able to compute the transition relation of the system, in order to build a LTL model-checker. This design allows it to be used with any kind of formalism. This library implements great number of useful algorithms related to LTL model checking: LTL parsers, LTL formulae syntactical simplification, translation to several flavours of Büchi automata, automata simplification, and of course emptiness check algorithms for them. It is considered one of the best candidates for operational model checking [245].

UPPAAL [27] (http://www.uppaal.com) UPPAAL is an integrated tool environment for modelling, simulation and, verification of real-time embedded systems. Typical application areas of UPPAAL includes real-time controllers and communication protocols in particular, those where timing aspects are critical.

SCADE [265] (http://www.esterel-technologies.com) is an industrial-grade environment for the development of critical embedded systems. It is coupled with a model checker and used in industry in the domain of synchronous systems.

Note that this list is non-exhaustive and the provision of a complete list is out of the scope of this chapter. Its purpose is to provide a short overview over some of the more popular tools to help the decision of an appropriate one. The choice of a tool should however depend on various criteria, such as the Petri net class (P/T net, Coloured Petri nets, Timed Petri net), the type of properties that should be tested by the tool, the required efficiency of the tool and the execution environment. We encourage the reader to consult other resources[5] to find additional guidance towards a better informed decision process.

7.7.2 Evaluation of Model Checking Techniques

The Petri nets community is eagerly hosting and participating in various model checking contests in order to evaluate the various model checking techniques and tools and to discover approaches that might be particularly well-suited for certain model classes and types. Some of the more prominent ones are following, but note that this list is non-exhaustive.

The Hardware Model Checking Contest [34] was first held in 2007, and is now associated with the CAV (Computer Aided Verification) and FLOC (Federated Logic Conference) conferences. It focuses on circuit verification by means of model checking based on SAT-solvers. This event ranks the three best tools according to a selected benchmark.

---

[5] such as https://en.wikipedia.org/wiki/Model_checking

The Verified Software Competition [165] takes place within the Verified Software: Theories, Tools and Experiments (VSTTE) conference. This competition is a forum where researchers can demonstrate the strengths of their tools through the resolution of five problems. The main objective of this event is to evaluate the efficiency of theorem proving tools against SAT-solving. Started in 2010, it has now become a yearly event.

The Competition on Software Verification [4] is an event associated with the conference on Tools and Algorithms for the Construction and Analysis of Software (TACAS). Aimed at the verification of safety properties of C programs, it has been held yearly since 2012.

The Satisfiability Modulo Theories Competition [54] takes place within the context of the CAV conference. Held yearly since 2005, its objective is to evaluate the decision procedures for checking the satisfiability of logical formulas.

The SAT Competitions [159] proposes to evaluate the performance of SAT solvers. Initiated in 2002, it is held every two years since 2007, and identifies new challenging benchmarks at each edition.

The Model Checking Contest [169] at the Petri Nets conference puts emphasis on the specification of parallel and distributed systems, and their qualitative analysis.

The main problem of these contests is the difficult choice of comparison metrics. It is seemingly easy to compare tools for a set of models. Drawing conclusions from the achieved tool results in order to compare of model checking *techniques* is a much more complex endeavour, as each tool merges several techniques to be efficient. Moreover, the efficiency of model checking techniques highly depends on each individual model's characteristics. For instance, partial order reductions are showing good results for highly parallel systems with few synchronisations, but are less efficient when the degree of synchronisation increases. However, since verification techniques are often combined with other ones, it is difficult to clearly attribute good performance to an individual technique.

7.8 Summary

In this chapter we have shown a way to develop cyber-physical systems that include concurrent dimension. Our proposal is to consider to model first our system as many entities cooperating. Each entities can be modelled using the most convenient formalism to the problem that is considered. We propose for concurrent systems the use of Petri nets. We explain the interest and capabilities of Petri nets for modelling complex systems in an abstract way, i.e. by masking details that are not useful. Such modelling techniques hide concerns related to distribution, naming, communication mechanism and representation of data structures. We provide also their semantics through deduction systems in order to explain precisely all notions we are using such as states, marking and state space. It must be noted that the key aspect of the semantics of Petri net is the potential non-deterministic behaviours that we can exhibit from a model. The impact of this dimension is the potentially very large number of states of a system. We then develop some formal ideas to analyse Petri nets such as marking exploration and the use of invariants. While there is a complexity barrier for marking exploration, the use of invariants need human intervention.

In the rest of the chapter we have been interested into the idea of managing complexity by applying the *divide to conquer* principle. It means to consider separately the modelling and verification into small units. We also promote by this model decomposition the possible integration of heterogeneous modelling which is very important for the intrinsically heterogeneous cyber-physical systems.

The need for simulation of heterogeneous systems is also covered by our approach. It is based on the construction of functional units (FMU) and their cooperation into a larger system by specific master algorithms. Petri nets being a formalism often used to represent concurrent executions, defining its state as a set of markings was discussed and formally presented, avoiding the violation of the FMI API determinism. Furthermore, the API standard was extended with functions to add more observability and functionalities on the considered PNFMU. This study showed that deadlocks can be detected when simulating a PNFMU with the standard API, yet without being able to know the actual markings in which the Petri net is in a deadlock. Then, the composition of PNFMUs was introduced, yet partially leaving out of the study time evolution and outputs/inputs adaptation. Moreover, since the state of a PNFMU is now defined as a set of markings, a multi states relationship of the composition between FMUs was presented.

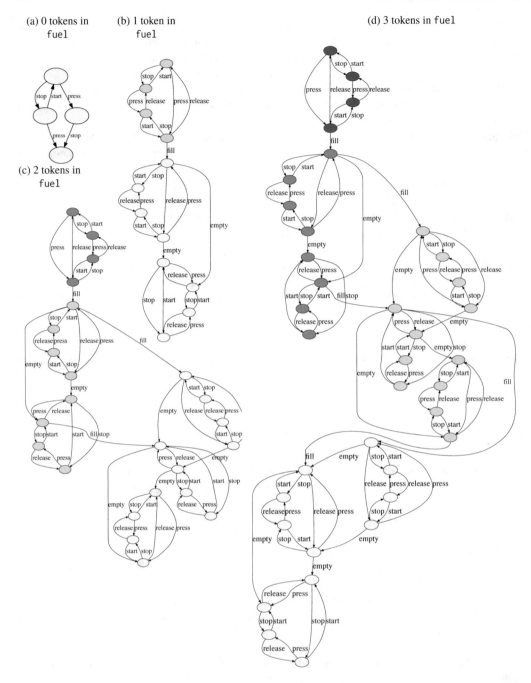

(a) 0 tokens in fuel

(b) 1 token in fuel

(d) 3 tokens in fuel

(c) 2 tokens in fuel

Fig. 7.15: The reachability graphs of the Petri net in Figure 7.1 with different initial markings. Note how the smaller graphs can be found as subgraphs inside the larger ones. This is due to the monotony of Petri nets as discussed above. Moreover, the node colour is representative of the number of tokens in place fuel, ranging from 0 (lightest) to 3 (darkest).

7.9 Literature and Further Reading

For a deeper understanding of concurrency modelling and Petri nets, we refer the interested reader to the following works. An excellent book by Reisig [239] is explaining fundamental analysis techniques for Petri nets. Through well-chosen examples, it also shows how to model several well-known systems that can be useful for beginners and newcomers to the domain of discrete dynamic system modelling and analysis. M. Diaz [85] also provides a look at basic techniques for modelling with Petri nets. Several extensions of place transition nets are explained and their practical use for concrete applications detailed. This book will probably be of more interest for the curious Petri-netter who wants to dive into various modelling options such as stochastic Petri nets, Timed Petri nets and their corresponding formal techniques.

Finally, [26] is a reference for people who are interested in exciting and practicable directions related to performance modelling and cost estimations of system through quantitative methods. The authors presented useful methods based on Markov chains and develop these techniques for the purpose of stochastic Petri net analysis.

7.10 Self Assessment

1. What is a Petri net marking? What is the initial marking?
2. What is the difference between a reachability and a coverability graph?
3. Can you list some interesting properties of Petri nets related to states? And related to transitions?
4. What does monotony express in the Petri net context?
5. What are Petri net invariants? What purpose do invariants serve? Can you name the two types of invariants and state their difference?
6. Can you name different types of High-level Petri nets and state their purpose? In these Petri nets, can you say which of the transitions are live or not?
7. For these two Petri nets, can you say which transitions are live and which ones are not?

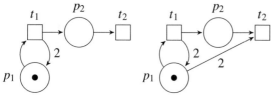

8. In these Petri nets, what are the bounded places. What are the place bounds of this Petri net?

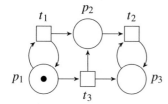

9. Can you identify which of these transitions are live, and which ones are not?

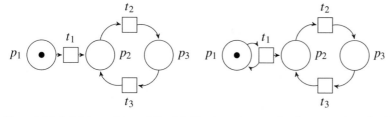

10. For which initial marking of P_1 this Petri net has no dead state or has dead state?

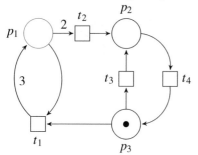

Chapter 8
AADL: A Language to Specify the Architecture of Cyber-Physical Systems

Dominique Blouin and Etienne Borde

Abstract This chapter is devoted to formalisms for describing system architectures, and in particular to the Architecture Analysis and Design Language (AADL). AADL is an Architecture Description Language (ADL) well suited for the modelling of embedded and cyber-physical systems. The architecture is central in Multi-Paradigm modelling for Cyber-Physical Systems as it provides a description of the overall system and the environment into which it will operate. From such description, other models of other languages and formalisms such as those described in this book can be generated and augmented to study other aspects of the system, which is essential for its validation and verification. After a brief introduction to ADLs and their role in MPM4CPS, the AADL will be presented and its use illustrated with the modelling, analysis and code generation for a simple Lego Mindstorm robot for carrying objects in a warehouse. A simple top-down architecture-centric design process will be followed starting from the capture of stakeholder goals and system requirements followed by system design, design analysis and verification and finally automated code generation.

8.1 Learning Objectives

After reading this chapter, the reader should have sufficient knowledge of the family of AADL languages and modelling approaches for:

- Modelling a simple cyber-physical system at the different levels of abstraction of requirements, functional architecture, physical architecture, software architecture and deployment.
- Modelling reusable component libraries and product families.
- Understand basic timing and scheduling concepts and AADL properties for performing scheduling and latency analysis from the aforementioned models.
- Understand basic code generation concepts and AADL properties for performing automatic code generation with the RAMSES tool [235] from the aforementioned models.

8.2 Introduction

This chapter introduces Architecture Description Languages (ADLs) and in particular the SAE Architecture Analysis & Design Language (AADL). We start by a short introduction on the problems of building nowadays complex cyber-physical systems and explain the overall approach to solve these problems as promoted by the AADL community, which is based on architecture-centric virtual integration.

We then move on to the introduction of AADL by a short overview of the language followed by a detailed introduction of the language constructs and semantics. We have chosen to introduce AADL by presenting a

Dominique Blouin and Etienne Borde
Telecom ParisTech, Paris, France
e-mail: {dominique.blouin,etienne.borde}@telecom-paristech.fr

© The Author(s) 2020
P. Carreira et al. (eds.), *Foundations of Multi-Paradigm Modelling for Cyber-Physical Systems*,
https://doi.org/10.1007/978-3-030-43946-0_8

typical development process so that it better illustrates how the language can be used. Therefore this process is first introduced followed by the Lego Mindstorm line follower robot example that is used to illustrate AADL modeling. We then present the detail modeling of the example for each step of the development process. Note that we have chosen to introduce the language constructs in an incremental way, by only introducing the constructs that are necessary for the modeling corresponding to the current step of the development process. In this manner, the reader can immediately see how the constructs are used in the context of the development process.

We illustrate the modeling process by presenting successive refinement steps of the models down to automatic code generation. Traceability links established between each step are also illustrated as well as basic analysis capabilities of AADL on the models of the deployed system. In the end, C code is automatically generated that can be compiled and deployed on the Lego robot for building a real working system.

8.2.1 Increasing Systems Complexity and Unaffordable Development Costs

The complexity of CPSs is constantly increasing due to the increasing number of functions that these systems are required to perform, which often must include more and more intelligence and must be more and more interconnected. In addition, these systems must satisfy an increasing number of constraints due to their operating environments, which are often hostile and limited in resources. Therefore, these systems are becoming more and more difficult to develop at affordable costs. This is particularly true for the avionics domain, for which a measure of complexity can be obtained from the number of Software Lines of Code (SLOC) embedded in aircrafts. This increase in complexity is illustrated in figure 8.1 where a plot of the number of onboard SLOCs as a function of years for the most common aircrafts built by Airbus and Boeing since the 70's is shown. For each constructor, the slope of the curve indicates that the number of SLOCs has roughly doubled every four years resulting in a non-linear increase in systems complexity.

The development effort required to develop these systems has been shown to increase exponentially with the number of SLOC. For example, while the F35 military aircraft has approximately 175 times the number of SLOC of the F16, it is estimated that it required 300 times the development effort of the F16 [253]. The result of this, as shown in figure 8.1, is that we are no longer able to develop more complex aircrafts with traditional development methods, since their development costs is not affordable. This is illustrated on the figure by the dark blue line whose slope is pegged after roughly year 2010.

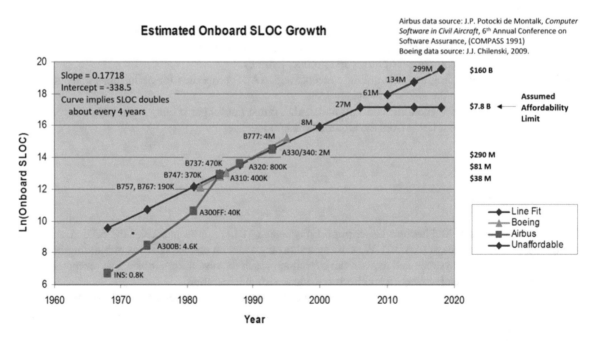

Fig. 8.1: The evolution of the number of SLOCs with time in commercial aircrafts (reproduced from [133])

In order to understand how this limit is reached, let us consider the traditional V-cycle model (figure 8.2), which is the most commonly used engineering process for safety-critical embedded systems. The numbers in blue on the figure indicate the percentage of errors introduced at the various phases of the cycle using the traditional development methods. As can be seen, a large majority of these errors (70%) are introduced at the *early* phases (requirements engineering and design) of the cycle, while the majority of these errors are only discovered much later at system integration and operation time. As a result, the cost of fixing these errors is dramatically high, since they often require the upfront design to be modified and parts of the system to be re-implemented. As a matter of fact, studies have shown for large projects that on one hand, rework due to introduced errors may account for 60% to 80% of the total software development costs [133]. On the other hand, the cost of software development, which could be as high as 70% in 2010 keeps increasing and could reach up to 88% of the total system-development cost by 2024 [133]. These figures show the potential for achieving high cost reductions by the discovery of flaws as early as possible during development.

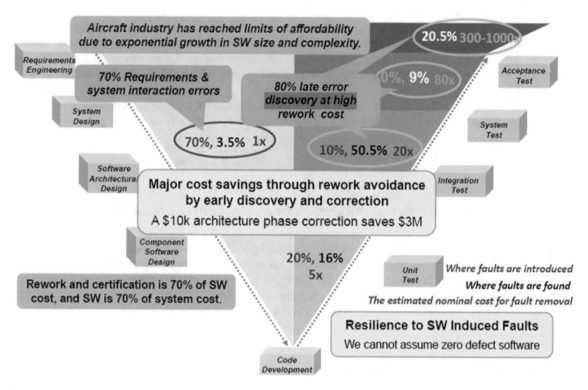

Fig. 8.2: V-cycle development process model annotated with figures for introduced errors and costs to removed them (reproduced from [101])

8.2.2 Mismatched Assumptions in Collaborative Engineering

But how are these errors leading to these rework costs are introduced? In order to understand this, let us consider a number of well-known errors that led to important damages and costs.

Ariane 5 Rocket

In June 1996, the Ariane 5 rocket exploded during its inaugural flight less than 40 seconds after departure. This was due to an integer overflow in the software used to control the side velocity of the rocket. That error propagated to an error in the navigation system, which triggered self-destruction of the rocket to prevent ground damages. This software bug cost 370 million dollars and is known to be the most expensive bug in history [10].

The integer overflow occurred because the software of Ariane 4, which had much smaller velocities, was reused for Ariane 5. It was forgotten that this software was working under the *assumption* of a maximum velocity that could be represented in software. Such assumption was no longer matched by Ariane 5, which had much greater velocities.

iTunes

When dual core processors were first used in computers, the iTunes software was crashing randomly when ripping music CDs [156]. The software was multi-threaded with one thread determining the sound level of tracks while a second one was converting the audio data. A single core processor would always sequentially execute the first task and then the second one. On a dual core, the two concurrently executing threads were attempting to update the same music data without proper synchronization.

In this case, the software was implicitly assuming sequential execution of tasks, which assumption was no longer matched by dual core processors.

Airbus 380 Cables

When the first Airbus 380 aircraft was assembled in Toulouse in France, the wires and their harnesses turned out to be a few centimeters too short for the cabin, which led to several billions of extra costs [5].

The problem was traced to the fact that different design teams had used different versions of a Computer Aided Design (CAD) tool to create the drawings. Version 5 of the tool was a rewrite of version 4 and the calculations of bend radii for wires as they follow the air frame were inconsistent across the two versions. It was eventually realized that the issue was pervasive throughout the design. Again, this issue can be traced to different assumptions on the way calculations of wire bending are performed between different teams using different tools.

These few examples show that an important source of errors is due to assumptions a system or component makes for proper operation not being matched by the environment in which it is used. This points out the need for improved system integration performed early for ensuring consistency of the overall system is preserved during the constantly evolving collaborative designs.

8.2.3 System Architecture Virtual Integration: Integrate, Analyze then Build

A solution to this mismatch assumptions problem is promoted by System Architecture-centric Virtual Integration (SAVI), which makes use of models of the systems that can be virtually integrated and analyzed for early fault discovery before the system is physically built [253]. Such process is illustrated in figure 8.3 where the left side of the well-known V-cycle development process is augmented with parallel left side validation activities making use of models for the domains covered by the system at the appropriate levels of abstraction at each phase.

8.2.4 Architecture-Centric Authoritative Source of Truth

The aforementioned virtual integration process is supported by an architecture model playing the role of Authoritative Source of Truth (ASoT) [253]. The model provides a global view of the system and its environment and is a central place into which other models detailing the different parts of the system can be integrated. It forms a reference model capturing all properties relevant for determining if the system meets its requirements.

Such ASoT model is depicted in figure 8.4. Other models for the various analyses required by the virtual integration process may be generated from the architecture model. Properties such as safety, security, real-time performances, resources consumption, etc. estimated from the analysis activities supporting virtual integration are then used to determine if the system meets its requirements.

This architecture-centric approach therefore makes architecture models first class citizens in the development of CPSs in order to support virtual integration activities, thus reinforcing the importance of ADLs, which is the topic of this chapter. Such architecture-centric virtual integration process can actually lead to substantial system development costs reduction as shown by a study on the Return of Investment (ROI) for the SAVI Initiative [133]. According to this study, the cost reduction for a 27-MSLOC system can be as high as $2 billion out of an estimated $9 billion total cost, which represents a cost saving of about 26% due to early correction of requirements and design faults.

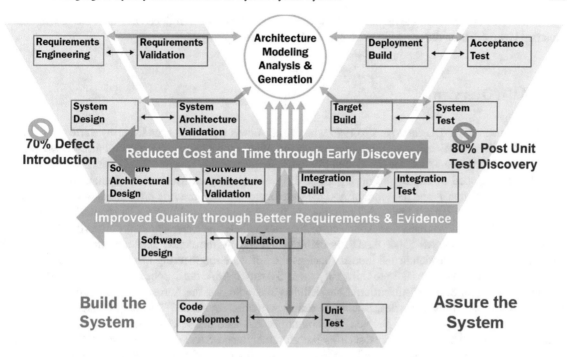

Fig. 8.3: Traditional V-cycle development process model augmented with architecture-centric virtual integration validation activities (reproduced from [205])

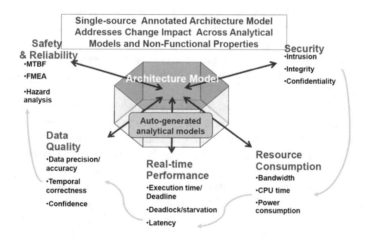

Fig. 8.4: Architecture-centric single source of truth modelling approach (reproduced from [98])

8.2.5 Organisation of the Chapter

The previous sections of this chapter illustrated the central role of ADLs for architecture-centric system architecture virtual integration in order to address the challenges of developing complex CPSs. Therefore, the rest of this chapter introduces ADLs in greater details, and in particular AADL, which is an ADL well suited for the modelling of both the cyber and physical parts of CPSs, and the deployment of the cyber part over the physical execution platform part.

AADL will be introduced in a tutorial-like fashion by presenting the modelling of a simple robot CPS whose purpose is to carry objects in a warehouse. This example is introduced in section 8.4. Even if the AADL language itself does not impose any development process in particular, in order to better illustrate the use of AADL, the aforementioned top down design V-cycle process model augmented with SAVI will be followed (figure 8.3), starting from requirements modelling down to automated code generation from the software architecture model.

For each step of this process, the required AADL constructs will be first introduced and their usage will then be illustrated by modelling the corresponding part of the robot example system.

8.3 AADL Overview

The development of AADL dates back to 1999 when Bruce Lewis working for the US army started a committee to make AADL an SAE standard. Initially standing for Avionics Architecture Description Language, AADL was first developed as an ADL for the avionics domain. It was strongly inspired from another language called MetaH and developed by Honeywell during a DARPA research project. Much of the syntax and the strongly typed property of MetaH (and therefore of AADL) were borrowed from the Ada programming language, which is dedicated to safety-critical embedded systems. As it was soon realized that AADL could be used for any embedded system, it was renamed to Architecture Analysis and Design Language thus preserving the acronym.

As an SAE standard AS-5506, AADL is being developed by the AS2C subcommittee, which includes participants from both academia and industry such as the Software Engineering Institute (SEI) of the Carnegie Mellon university, ISAE, Kansas State University, the U.S Army, NASA, the European Space Agency, INRIA, the Russian Academy of Science, Adventium Labs, Ellidiss Technologies, the Aerospace Corporation, Honeywell, Rockwell-Collins, Airbus industries, Boeing, Dassault Aviation, Toyota and Telecom ParisTech, which is the institution of the authors of this chapter. The committee has also active collaborations with other research initiatives and standardization bodies such as the aforementioned SAVI initiative, the ARINC653 working group and The Open Group Real-time. Despite that the development of AADL was started long time ago, its development is still an ongoing work, for which there has already been 2 major releases while a third one (AADL 3) is in preparation at the time of writing.

A comparison of the capabilities of AADL with other well-known ADLs such as SysML, MARTE and UML is depicted in figure 8.5, according to the intended use and domain covered by the languages. Both SysML, which covers the domain of the physical world and UML, which covers the domain of software were developed mostly for modelling with limited analysis capabilities, due to the weak semantics inherited from their high level of genericity. Conversely, AADL and MARTE were given a stronger semantics due to their more specific covered domain of embedded systems, with AADL performing better on the analysis domain due to its longer history and better maturity[1].

A few approaches have made use of UML and SysML for embedded systems modelling such as respectively Papyrus UML-RT [227] and TTool [274]. However, the only way to make these languages useful for embedded systems is to specialize them using the UML profile extension mechanism. We note however that while the advantage of using UML profiles is that existing tools can be reused with the extended language, it has the drawback of less flexibility in tailoring the language for the domain, including the introduction of accidental complexity due to unnecessary features inherited from the generic language also visible from the extended language. Such is also the case of MARTE, which while showing characteristics similar to those of AADL in terms of intended use and domain suffers from its implementation as a UML profile. The fact that AADL was directly implemented as a DSML avoids this accidental complexity and provides constructs better fitted for the intended domain and use.

While the intended domain of AADL was originally avionics (it was initially named Avionics Architecture Description Language) and soon extended to embedded systems, its version 2 is an adequate language to also model CPSs. Version 2 added abstract components that can be used to model at the system level of abstraction, with similar level of abstraction than SysML. It also introduced annexes for behavior and fault tolerance modelling. In addition, the hardware modelling capability initially developed for embedded systems in combination with the system and abstract components can be reused for modelling the physical parts of CPSs (plant model), with a more appropriate coverage of the domain than what could be done with the too generic SysML block concept. Finally, the software modelling and deployment capabilities of AADL make it an ideal language for addressing the challenges due to the rapidly growing cyber part of CPSs [101].

There are 3 main tools for editing and analysing AADL models. The Eclipse-based Open Source AADL Tool Environment (OSATE) [224], which is developed by the SEI at Carnegie Mellon University, is a reference implementation as it is developed by the leading language architecture team. There is also another Eclipse-based

---

[1] Note that this comparison taken from [3] is only illustrative and not very precise as one could argue that the UML should somehow overlap with MARTE and AADL regarding the covered domains since all three languages cover parts of the software domain.

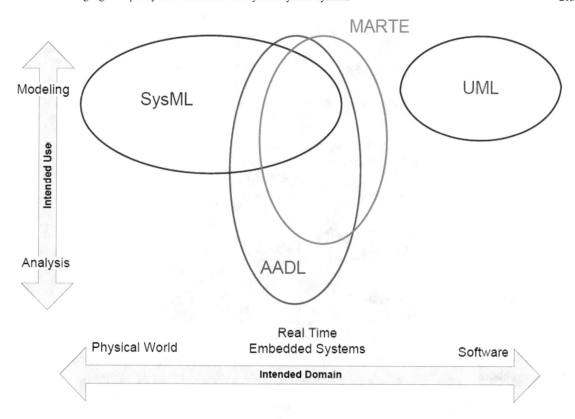

Fig. 8.5: Comparison of well-known ADLs in terms of intended use and domains (reproduced from [3])

tool named MASIW [200], which is developed by the Russian academy of science and finally, commercial tools such as STOOD and AADL Inspector developed by Ellidiss Technologies are also available [7].

8.4 CPS Running Example

The running example that will be used throughout the rest of this chapter consists of a robot system whose purpose is to carry objects in a warehouse from one point to another. The robot shall follow a path indicated by a line drawn on the floor of the warehouse. Several robots may be used in the warehouse and their paths may cross each other. Therefore, the robot should be able to stop momentarily when an obstacle is detected on the line and resume its trajectory once the obstacle is gone. The robot system should be able to carry objects in a minimal amount of time, and the cost of manufacturing such robot should be as low as possible so that the product has advantages over its competitors on the market. For illustration purposes, we will also add a few performance requirements to illustrate the need to perform performances analyses such as timing and latency.

Due to its low cost thus making it easily available, the Lego Mindstorm robot [182] in its NXT version will be used in a configuration as shown on figure 8.6, where two wheel assemblies are provided ❶ with a light sensor for line following ❷ and a sonar for obstacle detection ❸.

The robot will be running a simple line follower application described in [220] for following the edge of a thick black line as illustrated in Figure 8.7. Since the light sensor has a given field of view, the observed light intensity is inversely proportional to the part of the field of view of the sensor occupied by the black line. This light intensity will be used to compute the turn angle of the robot.

The turn angle will be computed using a PID (Proportional Integral Derivative) control algorithm. Such algorithm has the advantage of providing a much smoother line following behaviour compared to the two simpler algorithms depicted in Figure 8.8, since the computed turn angle variable has a much finer grained set of states (theoretically continuous) compared to the left (bang bang) and middle approaches where the turn angle variable has respectively 2 (left or right) and 3 (left, straight or right) states.

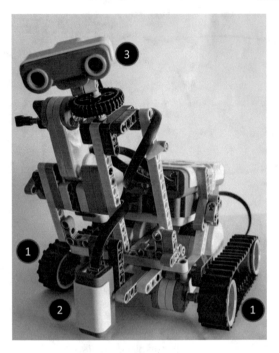

Fig. 8.6: The Lego NXT Mindstorm robot configured to execute a simple line following application with obstacle avoidance

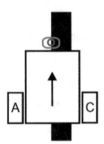

Fig. 8.7: Robot and light sensor following the edge of a line (reproduced from [208])

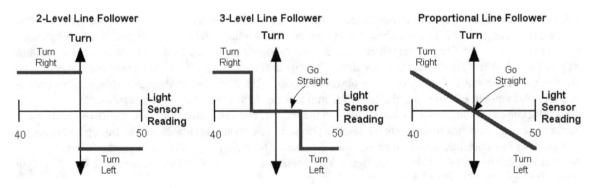

Fig. 8.8: Three ways of controlling the robot with the adopted PID method on the RHS of the figure (reproduced from [208])

8.5 The Development Process

While AADL does not impose any specific development process, it is typical for safety-critical systems to follow a V-cycle development process model augmented with architecture-centric system architecture virtual integration as discussed earlier and illustrated in Figure 8.8. The modelling of the line follower robot CPS will follow such process. Introducing AADL by following a development process allows to better illustrate how AADL can be used and what can be achieved with it.

The architecture-centric virtual integration process is partially supported by the ALISA (Architecture-Led Incremental System Assurance) set of notations and workbench. ALISA is an add-on to the AADL language supported by the OSATE tool that will be used for this tutorial. It originated from the Requirements Definition and Analysis Language [37, 236], which was originally planned to become a requirements annex for AADL.

ALISA augments AADL with a set of languages for the modelling of stakeholders, stakeholder goals, system and software requirements, verification plans and verification methods, and assurance cases for the incremental development of high-assurance systems. Similar to SysML, requirements can be modelled and allocated to architecture elements responsible for verifying them. Requirements can be decomposed and refined incrementally as the architectural design is refined and decisions are made. Complete modelling of assurance cases with arguments and claims linked to requirements is supported making ALISA ideal for the certification of safety-critical systems. Although a thorough description of ALISA is beyond the scope of this chapter, its basic requirements modelling capability will be demonstrated with the running example of this chapter. For instance, a few performance requirements will be captured and linked to the scheduling and latency analyses performed on the running example.

RDAL, and consequently ALISA, were strongly inspired by another work that is worth mentioning in this chapter. It consists of a set of 11 best practices described in the Requirement Engineering Management Handbook (REMH) [184]. The REMH was commanded to Rockwell Collins by the Federal Aviation Administration (FAA) to survey requirements engineering practices in industry [183]. Following this survey, the authors recommended 11 practices for the requirements management of safety-critical systems based on lessons learned from the Requirements Engineering (RE) research and results from the industry survey. The proposed practices can be adopted piecemeal, with minimal disruption of an organizations development process to address the slow or non-existent industry adoption of RE.

For this chapter, the left part of the V-cycle process model decomposed into steps inspired by the best practices of the REMH will be used to illustrate the use of AADL. Such decomposition is depicted in figure 8.9, where the boxes of the central column depicts the steps of the process, with their input and output models shown as boxes respectively located on the left and right sides of the figure. The language used for the models is specified following the ":" symbol. The dashed borders indicate steps that are not covered in this chapter.

The process starts by capturing an overview of the system to be developed creating models of the stakeholders, goals and operational contexts for the system. Such models are then used to derive use cases scenarios for the system, which together with the contexts and goals allow creating a functional architecture for the system. Then, a physical plant model for executing the system functions is specified. Next, the software model is derived from the functional model and deployed on the physical plant. From the deployed system model, analyses such as scheduling and latency can be performed. The models may then be modified to meet performance requirements and optimize non functional properties. Then, an operating system platform is selected to execute the software. The models are then refined for the given operating system using the RAMSES tool [235]. This refined AADL model can then be analysed again providing more accurate results due to specifics of the execution platform taken into account. Depending on the analyses results, the input models at any step of the process may be modified in an iterative manner until the design meets the requirements. Then, code can be automatically generated using RAMSES.

8.6 Modelling the Line Follower Robot with AADL

This section presents the modelling of the running example introduced in section 8.4 by following the development process introduced in section 8.5. Note that for didactic purposes, we will not present the complete set of models but only the parts relevant for the language concepts being taught. For more information, the reader can view the complete set of models made available from [63].

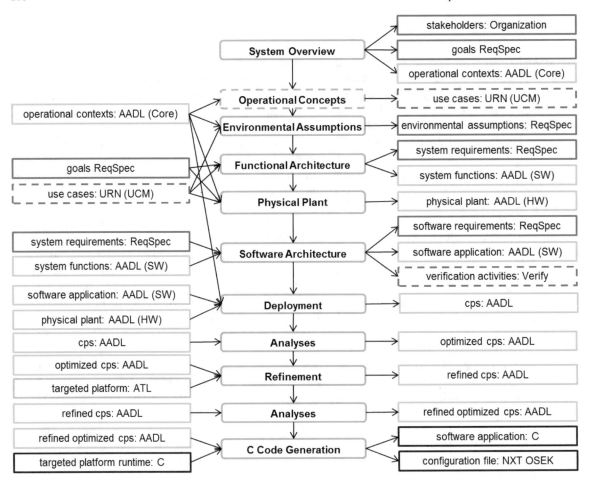

Fig. 8.9: The process for designing the CPS system (left side of the V-cycle of Figure 8.3) and the employed notations

8.6.1 System Overview

The first step of the process consists of specifying the overview of the system to be developed. It includes specifying why the system is needed by stating its purpose and goals, as well as how he system will interact with its environment. Specifying the system and its environment provides a sound understanding of the *boundary* of the system. Such boundary is defined by the set of monitored and controlled variables representing the precise interaction of the system with its environment. Identifying a correct system boundary is difficult since one needs to ensure that all interactions of the system with its environment on which proper system operation depends have been considered, for all the operation modes of the system.

Stakeholders and their Organizations For the running example, let us assume that company *Tartempion Warehouse Equipment* is developing our robot system introduced in section 8.4. Many stakeholders, each of which having their own concerns about the system are involved in the development of the system. Example stakeholders are the customers who will use the system, the designers of the system, the marketing in charge of selling the system, etc.

Stakeholders and their development organisation can be captured with the *organisation* notation of ALISA. An *organisation* has a name and can declare stakeholders. A *stakeholder* also has a name (that must be unique within the containing organisation) and optional description, role, email, phone and supervisor stakeholder. An organisation is declared in a separate file with the extension *org*.

Running Example Specification

The specification of Tartempion Warehouse Equipment Ltd, which develops the line follower objects carrier robot is shown in Listing 14.

```
organization Tartempion_Warehouse_Equipments_Ltd

stakeholder Customer [
    description "The customer of the line follower robot system"
]

stakeholder Marketing [
    description "The people in charge of marketing the line follower
        robot system"
]
```

Listing 14: Stakeholders and their organizations

Stakeholder Goals

As stated in section 8.4, the purpose of the robot is to carry objects in the warehouse by following a predefined trajectory. In addition, it must also be able to stop momentarily upon obstacles detection since other robots working in the same warehouse may be crossing the followed line. Furthermore, the time taken to carry objects should be as low as possible as well as the cost of producing the robot.

Constructs

Goals and requirements modelling is supported by the *ReqSpec* notation of ALISA. A stakeholder goal must have a name and can have several optional attributes such as a title, a description, a category, a rationale, etc. Providing rationale is of particular importance especially for non-trivial goals as it allows to quickly understand why the goal exists. Furthermore, providing rationale avoids questioning the goal over and over again when new people are introduced to the project.

Another important characteristic for goals is traceability to the system that should achieve them. This can be represented in ReqSpec through the *for* construct, which must refer to an AADL classifier representing the system to be built. AADL classifiers will be introduced in section 8.6.1.

Running Example Specification

A specification for the goals of our running example is shown in Listing 15. Goal *G_Behav_1* is first defined, which consists of carrying objects from one point to another by following a trajectory marked on the floor. *G_Behav_1* has *Customer* as stakeholder and has a *Behaviour* category since it relates to the functions of the system we want to build. Note the rationale on *G_Behav_2* that allows understanding why it exists. As will be seen in section 8.6.4, these goals will be transformed into requirements verifiable by the system.

```
stakeholder goals Line_Follower_Robot_Behavior for
    Line_Follower_Functions::Cary_Object [

    goal G_Behav_1 : "Objects_Transportation" [
        description
                "The robot should be able to carry an object between two
                    specified points by following a predefined trajectory
                    in the warehouse."
        stakeholder Tartempion_Warehouse_Equipments_Ltd.Customer
        rationale "This fulfills the main need of customers."
        category Quality.Behavior
    ]
```

```
goal G_Behav_2 : "Obstacle_Avoidance" [
    description "The robot should be able to avoid obstacles along
        the path."
    stakeholder Tartempion_Warehouse_Equipments_Ltd.Customer
    rationale
            "There may be several robots working on the warehouses
                and therefore, it is important to avoid damaging the
                robots and the carried object"
    category Quality.Behavior
    ]
]
```

Listing 15: Behavioural stakeholders goals

Listing 16 shows some performance goals for the system. As opposed to behaviour goals, performance goals can be set with a level of achievement that can be useful when performing design optimisation. Such goals can also be transformed into verifiable requirements that will set bounds on the level of achievement of the goals.

```
stakeholder goals Line_Follower_Robot_Perf for Line_Follower_Robot_Cps::
    Line_Follower_Robot_Cps [

    goal G_Perf_1 : "Minimal Cost" [
        description "The cost of producing the robot should be minimal."
        stakeholder Tartempion_Warehouse_Equipments_Ltd.Customer
            Tartempion_Warehouse_Equipments_Ltd.Marketing
        rationale "The robot should be cheap so that it is easy to market
            ."
        category Quality.Cost
    ]

    goal G_Perf_2 : "Minimal_Transportation_Time" [
        description "The time taken to carry objects should be minimal."
        stakeholder Tartempion_Warehouse_Equipments_Ltd.Customer
            Tartempion_Warehouse_Equipments_Ltd.Customer
        rationale "The robot ."
        category Quality.Performance
    ]
]
```

Listing 16: Quality stakeholders goals

The *for* elements of these goals refer the AADL classifier of the system that is being developed to meet the goal. This is introduced in the following section that presents the modelling of the system, its environment and its contexts of use with AADL.

System, Environment and Contexts of Use

Modelling not only the system but also its environment and their interactions is of primary importance. From these interactions, a set of monitored and controlled variables can be identified. The purpose of the behavioural requirements then consists of specifying the precise relationship between the monitored and controlled variables for all contexts of use of the system and all possible values of the monitored variables.

It is rarely the case that there is only one context of use of a system. For instance, for our running example, the robot carrying an object following a line can be thought of as the normal context of use of the system. But other contexts may exist such as when the robot is under maintenance. Ideally, all contexts of operation of the system should be identified and modelled but for our robot example, we will only present the normal context.

Constructs

The system and its environment can be modelled with AADL using component *types* and their *features*, component *implementations* and their *subcomponents*, component *packages* and component *properties*. Together these constructs form the core of the AADL language.

Component Types and Features (Component Interfaces)

In AADL, a component type declaration is used to provide an interface specifying how a component can interact with other components, without the need to provide details of the component's internal composition. This is achieved by declaring *features*, which are connecting points for connecting the component to other ones. A feature can be typed by a component type declaration. Such typing is used to restrict the connection to other features. A feature must have a direction that can be in, out or in and out. Feature directions also restrict how components can be connected to each other.

Features can be of several kinds, but for modelling at the system level, we only introduce abstract and feature group kinds of features for now. Other kinds of features specific to hardware and software components will be introduced later as they are needed.

An *abstract feature* is a placeholder to be refined to a concrete feature and is to be used for incomplete component type declarations.

A *feature group* is a special kind of feature used to group component features or other feature groups. It is therefore a modelling facility to ease the connection of components with many features, since the connection of a feature group represents the connections of all contained features. Feature group features are declared in a feature group type used for compatibility verification.

Component Implementations (Internal Component Composition)

The internal composition of a component in terms of its subcomponents and their inter connections is specified in AADL with *component implementation* declarations. A component implementation must have an associated component type specifying the component's interaction features as explained above. The advantage of separating component type and implementation declarations is that several implementations can reuse the same type, similar to interfaces of the Java programming language.

Component types and implementations are grouped under the name of component *classifiers*. Such classifiers are used for typing subcomponents in a component implementation. Connections can be declared inside component implementations to connect subcomponents between each other (or to the containing component), according to the features declared in the subcomponent's type.

Component Categories

Components classifiers are divided into 13 categories providing semantics for the domain of embedded systems. These categories are divided into 3 root categories; *composite*, *software* and *execution platform* (hardware).

The composite category is subdivided into two sub-categories:

- *Abstract*: Abstract components are used to represent incomplete components for modelling at a high level of abstraction, similar to SysML blocks, when no specific information is yet known about the kind of component that is modelled. An abstract component implementation can contain subcomponents of any category and can be contained by components of any category. Abstract components can be refined into any of the other AADL concrete component categories, which is useful as more information is known about the system under design.
- *System*: A system represents an assembly of interacting components. A system can have several modes, each representing a possibly different configuration of components and their connectivity contained in the system.

Each component category has its own rule specifying which categories of subcomponents can be contained in a component implementation. This will be detailed along with the *software* and *execution platform* component categories later as we refine the design of the line follower robot example system.

Component Extension and Refinement

Similar to object oriented programming languages, AADL component types and implementations (classifiers) can be extended thus inheriting the features / subcomponents and connections of the extended component. New features / subcomponents and connections can be added to the extending classifier to provide more details. In addition, the type of the inherited features and subcomponents can be *refined*, according to a compatibility rule chosen by the modeller among a set of predefined compatibility rules.

Packages (Organisation of Component Declarations)

Component declarations are contained in *packages*, which can have public and private sections. Package can use declarations from other packages provided that they are imported into the using package.

Properties

Finally, AADL provides a rich sublanguage for modelling *properties*. A specificity of AADL is that property types are defined at the model level and not in the AADL metamodel. This allows users to define their own properties. The AADL standard however defines a set of predefined properties for most common analyses such as timing, scheduling and resource consumption. This avoids everyone defining its own set of properties and contributes to interoperability of the specifications.

A property must have a type (integer, real, range of integer or real values, enumeration, reference, etc.) and optionally a unit type, which can also be defined by users. A property also has an applicability clause restricting the component category or classifier to which property values can be set.

Properties can be set at various places in an AADL specification (e.g.: on a component classifier, a feature, a subcomponent, a connection, etc.). Because component classifiers can be extended, a complex search algorithm must be decided to determine property values. Such algorithm is illustrated in figure 8.10. For example, to determine the value of a property of a subcomponent, the algorithm first search if a value is set on the subcomponent itself (#1). If not, it then searches for a value on its implementation (#2). If no value is found, the extended implementation if it exists is searched (#3). If no value is found, then the component type of the subcomponent is searched for (#4), followed by its extended component if any (#5). If still no value has been found, a value may be searched on the containing component itself (#6), provided that the property type is declared as "inherit", meaning its value can be inherited by the contained subcomponents. Finally, a property type may be set with a default value, which will then be used if no value has been found by the previous searches.

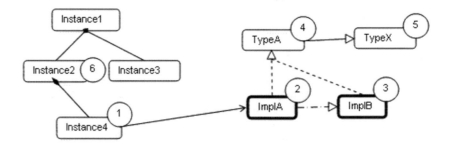

Fig. 8.10: Property value determination (reproduced from [247])

Properties and units are declared in *property sets*, which similar to component packages can be imported to be used by other declarations.

Running Example Specification

The previously introduced core AADL concepts will be understood better by using them to model our system overview, which consists of the system, its environment and its contexts of operation. This will complete the modelling of the system overview step as depicted in the process model of figure 8.9.

We propose a modelling approach for which separate AADL packages stored in different files will be created for the warehouse, the robot physical system, the system functions, the software application and finally the integrated line follower robot CPS. In doing so, AADL component libraries are created independently and become reusable over several projects.

Warehouse

The AADL package declaration for the warehouse is given in Listing 17. The package declares a public section into which two abstract component types are created for the floor and the warehouse. These components of the environment are modelled with the abstract component category, since at this modelling step, it is not meant to know more about these components other than their interaction points. One could also think of using the system category instead of abstract. However, the abstract category is preferred since the AADL system category has the more precise semantics of embedded systems.

An abstract component implementation is declared for the warehouse (line 16), whose name is prefixed (before the dot) by the associated component type (*Warehouse*). It declares a subcomponent for the floor typed by the previously defined *Floor* component type. Note that no component implementation is provided for the floor since its internal composition is of no interest for this modelling. Only the interaction points, which are provided by component types as features are needed in order to model how the system interacts with its environment.

```
package Warehouse
public
    with Physics;

    abstract Floor extends Physics::Reflecting_Object
        features
            applied_force: in feature Physics::Force;
            reacted_force: out feature Physics::Force;
    end Floor;

    abstract Warehouse
        features
            light_source: feature Physics::Light;
    end Warehouse;

    abstract implementation Warehouse.basic
        subcomponents
            floor: abstract Floor;
    end Warehouse.basic;

end Warehouse;
```

Listing 17: Warehouse AADL package

The component type for the warehouse declares a *light_source* abstract feature (line 13 of Listing 17) representing the light emitted by the warehouse lightening system. The feature is typed by a *Light* abstract component type representing the light physical electromagnetic radiation. This component type is declared in a reusable *Physics* package shown in Listing 18. The Physics package is made available to the Warehouse package by adding a *with* clause for the package (line 3 of Listing 17) and by prefixing the imported types by the name of the package that declares them followed by the :: delimiter symbol (line 7 of Listing 17).

```
package Physics
public

    abstract Reflecting_Object
        features
            light_in: in feature Light;
            light_reflected: out feature Light;
```

```
            sound_in: in feature Sound;
            sound_reflected: out feature Sound;
    end Reflecting_Object;

    abstract Light
    end Light;

    abstract Sound
    end Sound;

    abstract Force
    end Force;

    abstract Power
    end Power;

    abstract Power_Consuming_Object
        features
            power_in: requires bus access Power_Bus;
        properties
            Classifier_Substitution_Rule => Type_Extension;
    end Power_Consuming_Object;

    bus Power_Bus
    end Power_Bus;

end Physics;
```

Listing 18: Physics AADL package

Both the line and the floor component types extend a *Reflecting_Object* component type provided by the Physics package of Listing 18. Since the line follower robot is going to observe the reflected light to follow the trace, the component type provides abstract features for modelling this interaction of the robot with both the floor and the line components. Since a sonar device will also be used by the robot to detect obstacles, features for the reflected sound are also added to the *Reflecting_Object* abstract component type. A *Sound* abstract component type is declared to model the sound physical vibration and used to type these features. Similarly, the floor component type declares features of the *Force* physics type in order to model its interaction with the robot for propulsion.

Robot CPS

Now that classifiers have been declared for the environment of the system (warehouse and the floor), component types are required for the line to follow and for the robot CPS itself. A new package is created for this as shown in Listing 19. Note that here the *system* category is intentionally used for the CPS (line 8) and not the *abstract* one, which was used for components of the environment previously modelled. This is because we want to use the semantics of the AADL system category, since our robot is made of software and hardware components.

```
package Line_Follower_Robot_Cps
public
    with Line_Follower_Software, Robots_Library, Warehouse, Physics,
        Physics_Properties;

    abstract Line extends Physics::Reflecting_Object
    end Line;

    system Line_Follower_Robot_Cps
        features
```

```
        light_sensor_in: in feature Physics::Light;
        sonar_in: in feature Physics::Sound;
        sonar_out: out feature Physics::Sound;
        force_left_wheel: out feature Physics::Force;
        force_right_wheel: out feature Physics::Force;
        force_gripper: out feature Physics::Force;
    end Line_Follower_Robot_Cps;
```

Listing 19: Libe follower CPS AADL package

The *Line_Follower_Robot_Cps* system type declares features for its interaction with its environment such as the *light_sensor_in*, the *sonar_in* and *sonar_out*, the *force_left_wheel* and the *force_right_wheel* and the *force_gripper*, which represents the force on the object held by the gripper of the robot. Note that a single bi-directional feature could have been used to represent the pair of *sonar_in* and *sonar_out* features. However using separate features allows better distinguishing the monitored and controlled variables of the system.

System Context of Operation

After having provided component classifiers for the system and for its environment, those can be instantiated as subcomponents of an enclosing component implementation for the Warehouse. This is for modelling the system and its environment. This is shown in Listing 20 where a component type is created for the warehouse of the robot extending the generic warehouse (line 1).

```
abstract Warehouse_Robots extends Warehouse::Warehouse
    properties
        Physics_Properties::Illuminance => 150.0 lx applies to
            light_source;
end Warehouse_Robots;

abstract implementation Warehouse_Robots.normal extends Warehouse::
    Warehouse.basic
    subcomponents
        line: abstract Line;
        line_follower_robot: system Line_Follower_Robot_Cps;
        obstacle: abstract Physics::Reflecting_Object;
    connections
        floor_robot_light_sensor_in: feature floor.light_reflected ->
            line_follower_robot.light_sensor_in;
        line_robot_light_sensor_in: feature line.light_reflected ->
            line_follower_robot.light_sensor_in;
        obstacle_robot_sonar_in: feature obstacle.sound_reflected ->
            line_follower_robot.sonar_in;
        robot_sonar_out_obstacle: feature line_follower_robot.sonar_out
            -> obstacle.sound_in;
        force_left_wheel_floor: feature line_follower_robot.
            force_left_wheel -> floor.applied_force;
        force_right_wheel_floor: feature line_follower_robot.
            force_right_wheel -> floor.applied_force;
        light_source_line: feature light_source -> line.light_in;
        light_source_floor: feature light_source -> floor.light_in;
        Warehouse_Robots_normal_new_connection: feature light_source ->
            obstacle.light_in;
    properties
        Physics_Properties::Curvature_Radius => 99.0 mm applies to line;
end Warehouse_Robots.normal;
```

Listing 20: Normal context of operation of the robot CPS

A property value is set to characterise the *illuminance* of the light source in *lx* units (line 3). The property is declared to apply to the *light_source* feature inherited from the extended *Warehouse* classifier. An abstract component implementation (line 6) is created for modelling the normal context of use of the robot CPS, that is when it is carrying an object. It extends the generic warehouse component containing the floor subcomponent by adding other subcomponents for the line to follow, the robot CPS and an obstacle object.

One problem that is faced with the textual representation of Listing 20 is the difficulty of perceiving how subcomponents interact with each other. One advantage of AADL is that it proposes two notations; a textual and a graphical one. Therefore, users can choose the notation that is most appropriate depending on what is being viewed. Only the textual notation has been used so far, which was appropriate to display component types. However for component implementations containing connected subcomponents, the graphical notation is more appropriate. Therefore, a diagram of the AADL graphical notation is shown in figure 8.11 for the normal operation of the robot CPS. This diagram is obviously much easier to read than the code of Listing 20. However, the diagram does not show all information. For instance, the illuminance property of the warehouse is not visible on the diagram. The diagram is just a view while the textual always contain all information of the model.

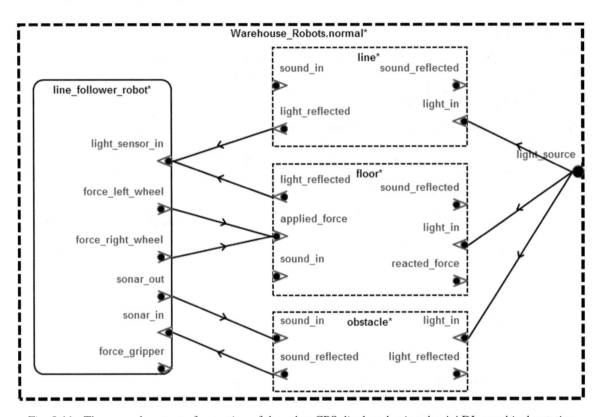

Fig. 8.11: The normal context of operation of the robot CPS displayed using the AADL graphical notation

In order to better interpret the diagram, figure 8.12 summarises the AADL graphical notation for component categories. Systems are represented as rounded box while abstract components as dashed rectangles. Abstract features are represented by the symbol ">" with a dot inside and connections as lines with arrows indicating the direction of the represented flow.

The diagram of figure 8.11 can be called a context diagram as it describes the context of operation of the system and clearly distinguishes it from its environment. This distinction is achieved by the different graphical representations of system and abstract components, which further justifies the choice of modelling the system to be built with the system category as opposed to abstract for the other entities of the environment.

The diagram also shows how the robot system contained in the warehouse interacts with its environment, represented by abstract features and connections for sensing the obstacle through the sonar, and the floor and line to follow through the light sensor. The interactions of the wheels with the floor for propulsion are also represented. From this context diagram, the system boundary is precisely identified for the given context as being the set of monitored (in) and controlled (out) features of the robot system that are *connected* to entities of the environment. Also note that for this context of operation, the *force_gripper* feature is not connected as it is assumed that the carried object is contained by the robot system and is therefore not part of the environment. For a different context for which the robot is picking the object, that object would then be part of the environment as it would not yet have been picked by the robot but would be interacting with the robot's gripper actuator. In this context, the *force_gripper* feature would then be connected to the object to be carried.

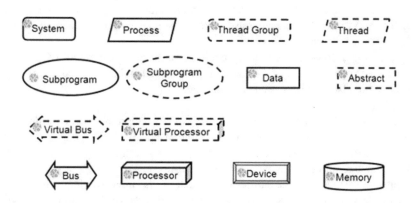

Fig. 8.12: The AADL graphical notation for component categories

When developing a system, it is often challenging to be aware of all the interactions the system may have with its environment. One must not forget interactions that must be taken into account for safe operation of the system in the given context and decide which ones can be neglected. A context diagram can help identifying the neglected interactions by searching for features of the system and components of its environment of the same type that are not connected, thus indicating a neglected interaction. For instance, such is the case of the *light_reflected* and *light_sensor_in* features of the obstacle and the robot CPS, which remain unconnected. Although the obstacle actually does reflect the light from the warehouse, its interaction with the light sensor of the robot is neglected, as the light sensor will be close enough to the line and not be influenced by this light. Such neglecting can also be modelled as a verifiable environmental assumption constraining the environment of the system. Environmental assumptions modelling will be presented in section 8.6.3.

Note that only the line following normal context of operation is presented here. Another context could be modeled for when the robot is under maintenance. This context would be modeled as another abstract component implementation for the same warehouse component type, where only the interacting subcomponents and connections for the context would be instantiated.

8.6.2 Operational Concepts

The next step in the modelling process of figure 8.9 consists of developing scenarios that describe how the system will be used in the contexts provided by the system overview. Use cases are a good way to do this. Several languages exist for use case modelling. The Use Case Maps sublanguage of the User Requirements Notation [155] and its Eclipse-based tool jUCMNav [1] are well suited for this, since the modeled use cases and scenarios can be simulated for their validation. This turned out to be extremely useful in the modelling of the isolette thermostat example provided by the REMH [184] as several errors were discovered in the natural language use cases. However, presenting use case modelling is beyond the scope of this chapter and the reader can refer to [38, 37] for more information.

8.6.3 Environmental Assumptions

The introduction of this chapter on AADL pointed out how well known design faults leading to catastrophic errors such as the Ariane 5 bug were due to mismatched assumptions between a system and its environment.

Therefore, identifying the environmental assumptions on which a system depends for correct operation is essential for reusing the system in different contexts and avoiding misuse of the system.

Assumptions can specify conditions that must be met by one or several entities of the environment of the system for proper operation. They can also constrain the types, ranges, and units of the monitored and controlled variables of the system.

Let us assume that the robot CPS is using a simple passive light sensor, which measures the light reflected by the line and the floor in its field of view. An obvious environmental assumption is that a minimum quantity of light is emitted by the light source of the warehouse. This is essential for proper operation of the line following robot otherwise it cannot see the line to follow.

Another example assumption, would be to impose a limit on the curvature radius of the line to follow, since there is a limit on the steering capability of the robot given its speed and the carried mass. A limit on the mass of the carried object would also be a reasonable assumption.

Constructs

Environmental assumptions are modelled using the *ReqSpec* notation as *system requirements*, but however assigned to elements of the environment and not to the system itself. Similar to goals, system requirements can have a *description*, a *rationale*, a *category*, etc. Furthermore, a requirement can also be traced to a goal that it transforms into a verifiable requirement using the keyword *see*.

Requirements must be verifiable by the entities they constrain. In order to achieve this, they can be set with a predicate expression or a verification activity that can be registered in the workbench and referred by the requirement. A predicate expression can for instance compare the value of properties on the architecture model with bounds set for those values and captured in the requirements set.

Running Example Specification

Listing 21 shows the modelling of environmental assumptions for the robot carrier CPS with the ReqSpec notation. A system requirements specification named *Line_Follower_Robot_Env_Assumptions* is created and assigned via the *for* keyword to the *Warehouse_Robot.normal* system implementation representing the normal context of operation of the robot (see figure 8.11). Both assumptions are captured as system requirements but constraining the *light_source* and the *line* subcomponents of the warehouse as defined by the *for* keyword (lines 3 and 11). Such assumptions are set with the *Kind.Assumption* category to further distinguish them from the system's requirements.

```
system requirements Line_Follower_Robot_Env_Assumptions for
   Line_Follower_Robot_Cps::Warehouse_Robots.normal [

   requirement EA_1: "Minimum Warehouse Luminosity" for light_source [
      description "The power of the light source shall not be less than
         the Minimum Illuminance value"
      rationale "Otherwise the light sensor of the robot will not be
         able to give proper readings given its sensitivity and its
         calibration that was performed under these conditions of
         minimum lightening. Study my_study has shown that the value of
         100 lux is a lower bound for warehouses illuminance"
      category Kind.Assumption
      val Minimum_Illuminance = 100.0 lx
      value predicate #Physics_Properties::Illuminance >=
         Minimum_Illuminance
   ]

   requirement EA_2: "Minimum Curvature Radius" for line [
      description "The curvature radius of the line to be followed by
         the robot shall not be lower than the specified
         Minimum_Curvature_Radius value"
      rationale "Otherwise the robot given its speed, mass and response
         time will not be able to follow the line."
      category Kind.Assumption
      val Minimum_Curvature_Radius = 100.0 mm
```

```
    value predicate #Physics_Properties::Curvature_Radius >=
        Minimum_Curvature_Radius
    ]
]
```

Listing 21: Environmental assumptions for the robot CPS

Value predicates are set to the assumptions for their automated verification. For example, the predicate of *EA_1* (line 8) compares the value of the *Illuminance* property of the *light_source* feature of the warehouse to a minimum value of 100.0 lx declared in the requirement (line 7).

Note that if an entity constrained by the assumption were to be developed by another organisation, then the organisation would provide an extension of the abstract component representing the entity and the assumption would then be converted to an equivalent requirement for the entity.

8.6.4 Functional Architecture

The next step of the followed development process is to develop the functional architecture. This step assumes that the operational concepts have been developed as mentioned in section 8.6.2 through use case modelling. Developing the functional architecture then consists of providing a first set of system functions identified from the use cases scenarios.

Following the architecture-led ALISA approach, high level requirements are first modeled for the system functions. Then, their corresponding architecture model elements are specified. Finally, the requirements are assigned to these architecture elements. This is an iterative process where design decisions are taken on the decomposition of the architecture of the system into subcomponents and the initial high level requirements are decomposed accordingly.

System Requirements

Constructs

Like for assumptions, requirements are modelled with the ReqSpec notation. Requirements modelling has already been introduced during the modelling of environmental assumptions. Additional properties needed for this modelling are presented below:

- *decomposes*: refers to one or more requirements that this requirement decomposes or is derived from. The decomposing requirement is assigned to an architecture component contained in a component to which the decomposed requirement is assigned to.
- *see goal*: refers to one or more stakeholder goals that the requirement represents, typically when the referred goal is transformed into this verifiable requirement.

Running Example Specification

The high level stakeholder goals of Listing 15 transformed into verifiable requirements are shown as requirements *R_Behav_1* and *R_Behav_2* of Listing 22. The *see goal* clause of line 6 refers to the stakeholder goals of Listing 15.

```
system requirements Line_Follower_Robot_Behavior for
    Line_Follower_Functions::Cary_Object.basic [

    requirement R_Behav_1 : "Carry_Object_Function" [
        description
                "The robot shall carry an object between two specified
                    points by following a predefined trajectory in the
                    warehouse."
```

```
        see goal Line_Follower_Robot_Behavior.G_Behav_1
        category Quality.Behavior
]

        requirement R_Behav_1_1 : "Pick_Up_Object_Function" for
            pick_up_object [
            description "At the beginning of the path, the robot shall
                pick up an object on the floor."
            category Quality.Behavior
            decomposes R_Behav_1
        ]

        requirement R_Behav_1_2 : "Follow_Line_Function" for follow_line
            [
            description "The robot shall follow a line on the floor of
                the warehouse."
            category Quality.Behavior
            decomposes R_Behav_1
        ]

        requirement R_Behav_1_3 : "Drop_Off_Object_Function" for
            drop_off_object [
            description "At the end of the path, the robot shall drop off
                the carried object on the floor."
            category Quality.Behavior
            decomposes R_Behav_1
        ]

    requirement R_Behav_2 : "Avoid_Obstacles_Function" for
        detect_obstacle [
        description
                "The robot shall avoid colliding with obstacles along the
                    path."
        see goal Line_Follower_Robot_Behavior.G_Behav_2
        category Quality.Behavior
    ]
]
```

Listing 22: High level system requirements

Those two high level requirements are further decomposed into sub-requirements that each must be verified for the high level requirement to be verified. For instance, *R_Behav_1* is decomposed into *R_Behav_1_1*, *R_Behav_1_2* and *R_Behav_1_3* (Listing 22), which state that carrying an object between two points is achieved by picking up the object, moving it along a path and dropping it at destination. This is specified by the *decomposes* keyword referring to the decomposed requirement.

Architecture

Following the architecture-lead modelling approach, the system functions must be represented in the system architecture. This is performed again using the *abstract* component category, which was also used for modelling the components of the environment. Since the abstract category has already been introduced for modelling the system in its environment, we can proceed directly to the running example specification below.

Running Example Specification

System functions are shown in Listing 23. An abstract component type is first declared for the *Carry_Object* function, and a following implementation defines subcomponents for the *pick_up_object, follow_line, drop_off_object* and *detect_obstacles* sub-functions. Abstract component types are declared for each of these sub-functions with their features representing function variables. An abstract component is declared for the type of a robot state variable determining if an obstacle has been detected. The *detect_obstacles* function will output such variable and pass it to the *follow_line* function that will stop the robot in case an obstacle is detected. This is declared by the connection of line 15 in Listing 23.

```
package Line_Follower_Functions
public
    with Physics;

    abstract Cary_Object
    end Cary_Object;


    abstract implementation Cary_Object.basic
        subcomponents
            pick_up_object : abstract Pick_Up_Object.basic;
            follow_line: abstract Follow_Line.basic;
            drop_off_object: abstract Drop_Off_Object.basic;
            detect_obstacle: abstract Detect_Obstacle;
        connections
            obstacle_detection_state: feature detect_obstacle.state ->
                follow_line.state;
    end Cary_Object.basic;
```

Listing 23: High level architecture system functions

Now that a component type has been provided for the global carry object system function, the stakeholder goals of Listing 15 can be assigned to this component with the *for* clause of the goals package. Note that the component type is used for the assignment instead of the component implementation because of the visibility of requirements and goals in ALISA. Similar to property visibility (see section 8.6.1), a goal assigned to a component type is automatically assigned to all implementations and subcomponents of the type and to the extending classifiers and their subcomponents, and so on.

Similarly, the top level requirements created from the goals are assigned to the *Carry_Object.basic* component implementation (Listing 22) and each decomposing requirement is assigned to the corresponding subcomponent decomposing the main component.

Decomposing the Follow Line Function

As an example of further requirements and architecture decomposition, the abstract component of the *follow_line* function of Listing 23 is detailed in Listing 24. The *Follow_Line* component type declares an input feature for the light intensity and output features for the left and right wheel power. Such features are typed by the abstract component types of the Physics package (Listing 18). A data access to a robot state internal variable is further added. This variable will be set by the obstacle detection subprogram to stop the robot when an obstacle is detected.

```
abstract Follow_Line
    features
        light_intensity: in feature Physics::Light;
        left_motor_power: out feature Physics::Power;
        right_motor_power: out feature Physics::Power;
        state: feature Robot_State;
end Follow_Line;
```

Listing 24: Follow Line architecture system functions

The line following sub function is further decomposed into two sub functions as illustrated by the the *Follow_Line.basic* component implementation shown in Listing 25, and by its corresponding diagram of figure 8.13. The implementation specifies subcomponents for two functions for computing the turn angle of the robot as a function of the measured light intensity and to compute the wheels motor power from the desired angle.

```
abstract implementation Follow_Line.basic
    subcomponents
        compute_turn_angle: abstract Compute_Turn_Angle.pid;
        compute_wheels_motors_power: abstract Compute_Wheels_Motors_Power
            .basic;
    connections
        light_intensity_compute_turn_angle: feature light_intensity ->
            compute_turn_angle.light_intensity;
        turn_angle_compute_wheels_power: feature compute_turn_angle.
            turn_angle -> compute_wheels_motors_power.turn_angle;
        compute_wheels_motors_power_left_motor_power: feature
            compute_wheels_motors_power.left_motor_power ->
            left_motor_power;
        compute_wheels_motors_power_right_motor_power: feature
            compute_wheels_motors_power.right_motor_power ->
            right_motor_power;
        state_compute_wheels_motors_power: feature state ->
            compute_wheels_motors_power.state;
end Follow_Line.basic;
```

Listing 25: Line following architecture system function implementation

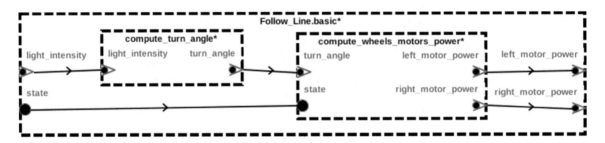

Fig. 8.13: The decomposition of the line following function into sub functions

Feature connections connect the subcomponent functions to the features of the enclosing component and the features between each contained subcomponent. The graphical representation of figure 8.13 clearly illustrates the dependency between the two sub functions, which is a practice recommended by the REMH.

On the requirements side, a new requirements set is created for the *Follow_Line.basic* component implementation as shown in Listing 26. It contains requirements decomposing *R_Behav_1_2* into requirements *R_Behav_1_2_1*, *R_Behav_1_2_2* and *R_Behav_1_2_3*. These requirements require setting the values of the turn angle internal variable and the left and right wheel power controlled variables. These requirements are allocated via their *for* clauses to the *compute_turn_angle* and the *compute_wheels_motors_power* subcomponents having the aforementioned variables as output features.

```
system requirements Follower_Line_Behavior for Line_Follower_Functions::
    Follow_Line.basic [

    requirement R_Behav_1_2_1 : "Set_Turn_Angle_Function" for
        compute_turn_angle [
        description "The controller shall set the value of the turn angle
            variable."
        category Quality.Behavior
        decomposes Line_Follower_Robot_Behavior.R_Behav_1_2
    ]
```

```
    requirement R_Behav_1_2_2 : "Set_Left_Wheel_Power_Function" for
        compute_wheels_motors_power [
          description "The left wheel controller shall set the value of the
              left wheel power variable."
          category Quality.Behavior
          decomposes Line_Follower_Robot_Behavior.R_Behav_1_2
    ]

    requirement R_Behav_1_2_3 : "Set_Right_Wheel_Power_Function" for
        compute_wheels_motors_power [
          description "The right wheel controller shall set value of the
              right wheel power variable."
          category Quality.Behavior
          decomposes Line_Follower_Robot_Behavior.R_Behav_1_2
    ]
]
```

Listing 26: Sub-requirements for the sub-functions of the line following system function

8.6.5 Physical Plant Model

A functional architecture model has been provided with associated requirements created from the original stakeholder goals of the system overview. Such model remains quite abstract and needs to be refined by providing a software model and an execution platform model for execution of the software. These models will allow more accurate analyses and verification since many analyses strongly depend on the plant (hardware) model providing execution resources for the system functions. Therefore, the next step is to provide a detailed model of the robot hardware plant that will be considered to execute the system functions in order to meet the system requirements.

In the following, the modelling of a simple Lego NXT Mindstorm robot is first introduced. A Lego robot is a highly configurable system and the first step is to model a specific configuration suitable for achieving the system goals. The modelling of this configuration will allow briefly introducing how component families can be modelled in AADL, for which more information can be found in [99].

Constructs

The composite component categories, which consists of the *system* and *abstract* sub-categories have been introduced in section 8.6.1 and were used to model the system overview. Abstract components were also used to model a first set of system functions. The execution platform (hardware) component sub-categories are now presented in order to use them to model the plant. They are decomposed into six sub-categories:

- *Processor*: A processor is an abstraction of hardware and software responsible for scheduling and executing threads, as well as virtual processors representing processor partitions such as ARINC653. Processors may contain memories to model a processor's internal structure such as caches. A processor can access memories and devices via bus access features.
- *Virtual Processor*: A virtual processor represents a logical resource that is capable of scheduling and executing threads and other virtual processors bound to it. Such component is useful to model partitioned processors such as ARINC653 for integrated modular avionics. Virtual processors can be declared as subcomponents of a processor or of another virtual processor. They are implicitly bound to the processor or virtual processor that contains them. They can also be used to represent hierarchical schedulers.
- *Memory*: A memory is a component for storing code and data. Memories can represent randomly accessible physical storage such as RAM, ROM, or more permanent storage such as disks or logical storage. Subprograms, data and processes are bound to memory components for being accessed by processors executing threads.
- *Bus*: A bus is a component that can exchange control and data between memories, processors and devices. It is an abstraction of a communication channel and associated communication protocols. Communication protocols can also be explicitly modelled with virtual buses.

- *Virtual Bus*: A virtual bus is a logical bus abstraction such as a virtual channel or communication protocol.
- *Device*: A device is a dedicated hardware performing built-in function(s) and acting as an interface to the physical world of CPSs. Devices can be used for both hardware and software parts of systems. Hardware sensors and actuators are modelled as devices, which can physically be connected to processors via buses. For software, devices can be logically connected to application software components thus representing the software part of corresponding hardware devices, such as drivers residing in a memory and executed on an external processor.

Running Example Specification

The NXT Lego robot includes plastic blocks that can be assembled to create the robot physical structure. It also includes two motors and wheels assemblies, a light sensor, a sonar sensor and a so-called brick. The brick itself contains several components such as a battery, an ARM7 processor containing a RAM memory, a Bluetooth module, input and output circuits, etc. An informal diagram of the NXT brick obtained from the NXT hardware developer kit [2] is shown in figure 8.14.

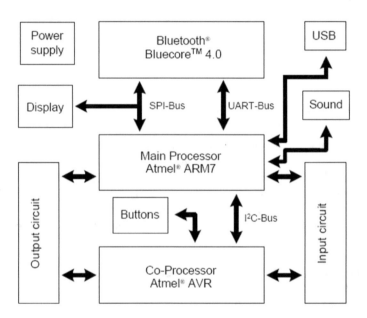

Fig. 8.14: An informal diagram of the NXT brick (reproduced from [2])

In order to model these elements in AADL, a *Robots_Library* package (Listing 27) is first created to contain classifiers for being instantiated to specify the configuration of the Lego NXT robot. A system type (line 5) and a system implementation (line 10) are declared for the robot. At this level of abstraction, only a single subcomponent for the brick is contained in the robot's system implementation (line 12). This is because it can be can assumed that any robot will always contain a brick while the presence of other components will depend on the specific hardware configuration selected for the robot according to its mission.

System classifiers are then declared for the brick subcomponent (lines 15 and 20). For the purpose of this chapter, only a simplified version of the brick will be modeled, which is sufficient for the analyses that will be presented in this chapter. Such version is limited to the modelling of the processor, the power supply, the input and output circuit and buses to connect these subcomponents.

The basic implementation of the generic brick is illustrated in line 20 of Listing 27 and in the diagram of figure 8.15. Power consumption is a concern for robots as it determines their autonomy. Therefore, the electrical power supply is modelled and represented as a device component connected to a power bus also connected to the processor. Note that for this hardware model, bus access and bus access connections are used instead of abstract features such as those of the system overview. Having explicit components for buses allows representing concurrency on communication resources. In addition, logical connections modelled in the software application can be mapped to hardware bus components to estimate resources consumption.

```
package Robots_Library
public
    with OSEK, Physics, Physics_Properties, SEI;

    system Robot
        properties
            Classifier_Substitution_Rule => Type_Extension;
    end Robot;

    system implementation Robot.basic
        subcomponents
            brick: system Brick.basic;
    end Robot.basic;

    system Brick
        properties
            Classifier_Substitution_Rule => Type_Extension;
    end Brick;

    system implementation Brick.basic
        subcomponents
            main_processor: processor Generic_Processor;
            power_supply: device Generic_Battery;
            power_bus: bus Physics::Power_Bus;
        connections
            power_supply_power_bus: bus access power_supply.power ->
                power_bus;
            power_bus_main_processor: bus access power_bus ->
                main_processor.power_in;
    end Brick.basic;

    processor Generic_Processor extends Physics::Power_Consuming_Object
        features
            data_bus: requires bus access Data_Bus;
    end Generic_Processor;

    bus Data_Bus
        properties
            Transmission_Time => [Fixed => 10 ns .. 100 ns; PerByte => 10
                ns .. 40 ns;];
    end Data_Bus;

    device Generic_Battery
        features
            power: requires bus access Physics::Power_Bus;
    end Generic_Battery;
```

Listing 27: AADL package for a robot components library

This component for the brick is generic in the sense that it has nothing specific to the NXT version. In order to capture these specificities, an extension of the brick system implementation is provided as illustrated by Listing 28 and the corresponding diagram of figure 8.16. The system type *Nxt_Brick* extends the previously defined *Brick* system type and adds bus access features (lines 3 to 9) for the brick to be connected to the sensors and actuators of the robot. The corresponding implementation *Nxt_Brick.basic* (line 12) extends the *Brick.basic* implementation and adds subcomponents specific to the NXT brick for the input and output circuits and a data bus to connect them to the processor. The input and output circuits take data from the processor as input and

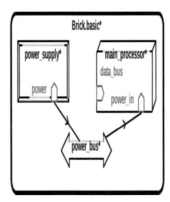

Fig. 8.15: A basic system implementation for a brick of a Lego Mindstorm robot

read / send data accordingly to the connected sensors and actuators. Device component types are provided for the input and output circuits (lines 40 and 49), which inherit their *power_in* bus access feature from the extended *Power_Consuming_Object* provided by the Physics package.

```
system Nxt_Brick extends Brick
    features
        in_1: requires bus access Sensor_Bus;
        in_2: requires bus access Sensor_Bus;
        in_3: requires bus access Sensor_Bus;
        in_4: requires bus access Sensor_Bus;
        out_1: requires bus access Actuator_Bus;
        out_2: requires bus access Actuator_Bus;
        out_3: requires bus access Actuator_Bus;
end Nxt_Brick;

system implementation Nxt_Brick.basic extends Brick.basic
    subcomponents
        main_processor: refined to processor Arm_Processor.nxt;
        input_circuit: device Input_Circuit;
        output_circuit: device Output_Circuit;
        data_bus: bus Data_Bus;
    connections
        brick_input_circuit_1: bus access in_1 -> input_circuit.in_1;
        brick_input_circuit_2: bus access in_2 -> input_circuit.in_2;
        brick_input_circuit_3: bus access in_3 -> input_circuit.in_3;
        brick_input_circuit_4: bus access in_4 -> input_circuit.in_4;
        brick_output_circuit_1: bus access output_circuit.out_1 -> out_1;
        brick_output_circuit_2: bus access output_circuit.out_2 -> out_2;
        brick_output_circuit_3: bus access output_circuit.out_3 -> out_3;
        power_bus_input_circuit: bus access power_bus -> input_circuit.
            power_in;
        power_bus_output_circuit: bus access power_bus -> output_circuit.
            power_in;
        main_processor_data_bus: bus access main_processor.data_bus ->
            data_bus;
        data_bus_input_circuit: bus access data_bus -> input_circuit.
            control;
        data_bus_output_circuit: bus access data_bus -> output_circuit.
            control;
    properties
```

```
        SEI::PowerCapacity => 9.0 W applies to power_supply; -- 6 AA
            batteries in series at 1 A
        Physics_Properties::Mass => 200.0 g;
        Scheduling_Protocol => (RMS) applies to main_processor;
        OSEK::SystemCounter_MaxAllowedValue => 2000 applies to
            main_processor;
        OSEK::SystemCounter_TicksPerBase => 1 applies to main_processor;
        OSEK::SystemCounter_MinCycle => 1 applies to main_processor;
end Nxt_Brick.basic;

device Input_Circuit extends Physics::Power_Consuming_Object
    features
        in_1: requires bus access Sensor_Bus;
        in_2: requires bus access Sensor_Bus;
        in_3: requires bus access Sensor_Bus;
        in_4: requires bus access Sensor_Bus;
        control: requires bus access Data_Bus;
end Input_Circuit;

device Output_Circuit extends Physics::Power_Consuming_Object
    features
        out_1: requires bus access Actuator_Bus;
        out_2: requires bus access Actuator_Bus;
        out_3: requires bus access Actuator_Bus;
        control: requires bus access Data_Bus;
end Output_Circuit;
```

Listing 28: NXT brick extension

The generic *main_processor* subcomponent of the extended generic brick system implementation is *refined to* an *ARM_Processor.nxt* type as shown in Listing 28 (line 14). The type of this refined subcomponent is an extension of the generic processor component type as shown in Listing 29 below (line 1). It contains a property association setting the MIPS capacity of the processor (line 3). The processor implementation provides a RAM memory subcomponent (line 8) with a memory capacity property value of 64 KByte (line 10) as given by the Lego hardware specification of the NXT brick.

```
processor Arm_Processor extends Generic_Processor
    properties
        SEI::MIPSCapacity => 43.2 MIPS;
end Arm_Processor;

processor implementation Arm_Processor.nxt
    subcomponents
        ram: memory RAM_Memory;
    properties
        SEI::RAMCapacity => 64.0 KByte applies to ram;
end Arm_Processor.nxt;

memory RAM_Memory
    features
        data_bus: requires bus access Data_Bus;
end RAM_Memory;
```

Listing 29: ARM processor for the NXT brick extension

Such combination of classifier extensions and subcomponent and feature refinements is extremely useful in modelling component families as further explained in [99]. In this modelling, two different ways are used to provide the NXT-specific information. One of them is to add property associations within the extending

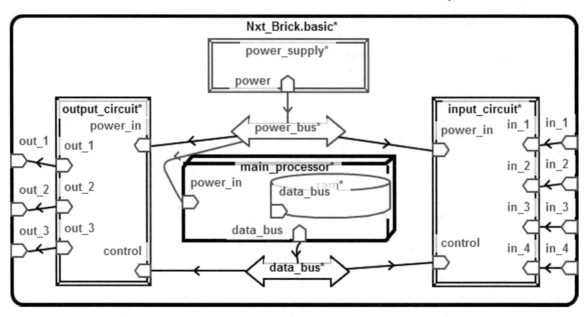

Fig. 8.16: The internal composition of a NXT Lego Mindstrom brick

component implementation of the brick with values pertaining to this specific model, as shown for the power capacity value set for the power supply subcomponent (line 32 of Listing 28).

The second way is to refine the classifier of a subcomponent as is done for the ARM processor extending the generic processor. Having this new classifier then allows providing more details such as the internal composition of the processor, which includes a RAM memory subcomponent. MIPS and RAM capacity property values can be declared in theses processor extensions (lines 3 and 10 of Listing 29). Deciding on which mechanism should be used depends on the modelling context. For example, the advantage of creating such extension classifiers for the processor is that they allow encapsulating the NXT specific information and makes it easy to reuse in other specifications.

Now that component classifiers are provided for the brick, other classifiers must be declared for the sensors and actuators of the robot. Those are presented in Listing 30 where classifiers of the *device* category are declared for the light sensor, the sonar, the wheel assemblies and the gripper, which is used to lift the carried object. Note that such gripper does not actually exist for the Mindstorm kit but is added for illustration purposes only. The component types for the sensors and actuators declare interaction features whose types for the corresponding physical resources are provided by the *Physics* package of Listing 18. They also declare bus access features typed with classifiers for the sensor and actuator buses. Those will represent the standard RJ12 connectors provided with the NXT kit. Note that no component implementation is provided for sensors and actuators since their internal composition is irrelevant for this modelling.

```
device Light_Sensor extends Physics::Power_Consuming_Object
    features
        light_in: in feature Physics::Light;
        light_data: requires bus access Sensor_Bus;
    properties
        Physics_Properties::Mass => 20.0 g;
end Light_Sensor;

device Sonar extends Physics::Power_Consuming_Object
    features
        sound_in: in feature Physics::Sound;
        sound_out: out feature Physics::Sound;
        sound_data: requires bus access Sensor_Bus;
    properties
        Physics_Properties::Mass => 20.0 g;
```

```
end Sonar;

device Wheel_Assembly extends Physics::Power_Consuming_Object
    features
        power_in: refined to requires bus access Actuator_Bus;
        force: out feature Physics::Force;
    properties
        Physics_Properties::Mass => 100.0 g;
end Wheel_Assembly;

device Gripper extends Physics::Power_Consuming_Object
    features
        power_in: refined to requires bus access Actuator_Bus;
        force: out feature Physics::Force;
    properties
        Physics_Properties::Mass => 20.0 g;
end Gripper;
```

Listing 30: Classifiers for robot sensors and actuators

Now that classifiers have been declared for all the components of the NXT Mindstorm robot kit, they can be instantiated and assembled into a robot plant system configured for object transportation in the warehouse. This is illustrated by Listing 31 and its corresponding diagram of figure 8.17. New *Line_Follower_Robot* classifiers are provided extending the partially-configured basic robot classifiers previously defined, which only contains a brick subcomponent (Listing 27 line 10). The extending *Line_Follower_Robot* system type adds features specific to the line following configuration for using the wheel assemblies, the gripper and the light and sonar sensors (line 2).

```
system Line_Follower_Robot extends Robots_Library::Robot
    features
        light_sensor_in: in feature Physics::Light;
        sonar_in: in feature Physics::Sound;
        sonar_out: out feature Physics::Sound;
        force_left_wheel: out feature Physics::Force;
        force_right_wheel: out feature Physics::Force;
        force_gripper: out feature Physics::Force;
end Line_Follower_Robot;

system implementation Line_Follower_Robot.nxt extends Robots_Library::
    Robot.basic
    subcomponents
        left_wheel_assembly: device Robots_Library::Wheel_Assembly;
        left_wheel_cable: bus Robots_Library::Actuator_Bus;
        righ_wheel_assembly: device Robots_Library::Wheel_Assembly;
        rght_wheel_cable: bus Robots_Library::Actuator_Bus;
        light_sensor: device Robots_Library::Light_Sensor;
        light_sensor_cable: bus Robots_Library::Sensor_Bus;
        sonar: device Robots_Library::Sonar;
        sonar_cable: bus Robots_Library::Sensor_Bus;
        gripper: device Robots_Library::Gripper;
        gripper_cable: bus Robots_Library::Actuator_Bus;
        brick: refined to system Robots_Library::NXT_Brick.basic;
    connections
        brick_left_wheel_cable: bus access brick.out_1 ->
            left_wheel_cable;
        left_wheel_cable_left_wheel_assembly: bus access left_wheel_cable
            -> left_wheel_assembly.power_in;
```

```
        left_wheel_force_robot: feature left_wheel_assembly.force ->
            force_left_wheel;
        brick_right_wheel_cable: bus access brick.out_2 ->
            rght_wheel_cable;
        right_wheel_cable_right_wheel_assembly: bus access
            rght_wheel_cable -> righ_wheel_assembly.power_in;
        right_wheel_force_robot: feature righ_wheel_assembly.force ->
            force_right_wheel;
        brick_gripper_cable: bus access brick.out_3 -> gripper_cable;
        gripper_cable_gripper: bus access gripper_cable -> gripper.
            power_in;
        gripper_force_robot: feature gripper.force -> force_gripper;
        robot_light_sensor: feature light_sensor_in -> light_sensor.
            light_in;
        light_sensor_light_sensor_cable: bus access light_sensor.
            light_data -> light_sensor_cable;
        light_sensor_cable_brick: bus access light_sensor_cable -> brick.
            in_1;
        sonar_out_robot: feature sonar.sound_out -> sonar_out;
        robot_sonar_in: feature sonar_in -> sonar.sound_in;
        sonar_sonar_cable: bus access sonar.sound_data -> sonar_cable;
        sonar_cable_brick: bus access sonar_cable -> brick.in_2;
end Line_Follower_Robot.nxt;
```

Listing 31: Classifiers for the line follower robot plant

The configured robot system implementation (line 11) instantiates sensor and actuator subcomponents including buses for the RJ12 connectors. The generic brick subcomponent (line 23) of the extended generic robot is refined to the NXT brick of Listing 28 (line 12) and figure 8.16.

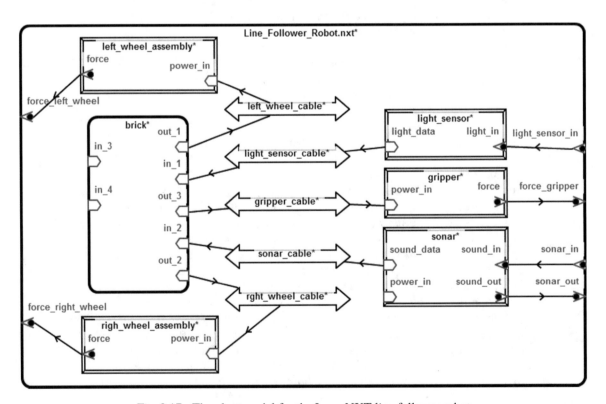

Fig. 8.17: The plant model for the Lego NXT line follower robot

This constitutes the physical plant model of our robot carrier system. Such hardware plant will be used to deploy a software application (the cyber part of the robot CPS) to control the speed of the wheels according to the data sensed by the light and sonar sensors.

8.6.6 Deployment

This section considers concrete implementations in software of the system functions introduced in section 8.6.4 and the deployment of the software onto the plant model introduced in section 8.6.5: the NXT Mindstorm Lego robot configured for carrying objects in the warehouse. In order to deploy the system, the system functions first need to be refined into a software application executable by the NXT brick of the plant model.

Software Application

Constructs

Software applications can be modelled in AADL using the six software component sub-categories introduced below:

- *Data*: A data component represents hierarchical data structures such as instance variables of a class in object-oriented programming languages or the fields of a record. Operations on a data type can also be modelled by declaring a *provides subprogram access* feature on the data component type. The feature is typed with classifiers of the accessed subprogram.
- *Subprogram*: A subprogram represents an atomic operation with parameters. It cannot have any state persisting after the call (static data), but can have local variables represented by data subcomponents declared in the subprogram implementation. Subprogram parameters and required access to data must be explicitly declared in a subprogram type. Events raised within a subprogram can also be specified as event or event data ports in a subprogram type.
- *Subprogram group*: A subprogram group represents subprogram libraries, whose content is declared through a subprogram group type. Subprogram groups are made accessible to other components using *subprogram group access* features that can be connected with *subprogram group access connections*. This allows for reducing the number of connections by providing a single connection for the whole group.
- *Thread*: A thread represents a scheduleable unit that can be executed concurrently with other threads. It represents sequential flows of control executing instructions. Threads can be dispatched periodically or upon the arrival of data or events on ports, or upon arrival of subprogram calls from other threads.
- *Thread group*: A thread group represents logically grouped threads. A thread group type specifies the features and required subcomponent accesses for thread subcomponents declared in a thread group implementation also declaring the connections between threads.
- *Process*: A process represents a virtual memory address space. This address space contains the program formed by the process's subcomponents, which can be threads, thread groups, subprogram, subprogram group or data subcomponents.

The AADL standard includes several annexes supporting software modelling:

- The Behaviour Annex (BA) consists of a state-machine based sublanguage allowing modelling the behaviour of AADL components. From this neutral behavioural description, executable code can be generated. However presenting the BA is beyond the scope of this chapter.
- The data modelling annex, which consists of predefined properties and classifiers provides guidance to represent data types and structures in AADL to be used for code generation.
- The ARINC653 annex, which consists of predefined properties and classifiers, provides a standard way of representing ARINC653 compliant partitioned embedded system architectures in AADL. Again, this annex is beyond the scope of this chapter.
- The code generation annex defines language-specific rules for generation of runtime systems from AADL models.

The refined software model presented in this section will provide subprogram classifiers extending the system function abstract classifiers. Thread classifiers will be provided for specifying scheduling properties for these software subprograms. Software-specific properties for supporting timing and scheduling analyses and code generation will be set to these components.

The execution semantics of a thread is specified using predefined properties, which are found under the standard *Thread_Properties* property set. Those properties are:

- *Dispatch_Protocol*, which can have the values of:

    - *Periodic*: the thread is activated periodically
    - *Aperiodic*: the thread is activated upon receiving messages
    - *Sporadic*: the thread is activated upon receiving messages with a minimum delay between two activations
    - *Timed*: the thread is activated *either* upon receiving a message or a given deadline (the timer is reset upon message reception)
    - *Hybrid*: the thread is activated *both* upon receiving a message and a given deadline

- *Scheduling_Protocol*: specifies the scheduling policy
- *Compute_Execution_Time*: represents the execution time of a thread or a subprogram
- *Stack_Size*: Defines the size of the stack for the thread

Threads can communicate with each other using three predefined communication mechanisms represented by their port and connection types:

- *Data ports* represent unenqueued state data
- *Event data ports* represent queued message data
- *Event ports* represent asynchronous events

Port connections can have different timing semantics:

- *sampled*: the receiving thread samples at dispatch or during execution
- *immediate*: the data is communicated upon execution completion of the sending thread
- *delayed*: the data is communicated upon deadline of the sending thread

The standard *Programming_Prperties* property set defines properties for code generation:

- *Source_Language* specifies the programming language. Predefined values are Ada95, Ada2005, C, Java, Simulink_6_5 and SCADE
- *Source_Text* specifies a source code file name
- *Source_Name* specifies the name of a data structure or function in the code file specified by *Source_Text*.
- *Compute_Entrypoint_Call_Sequence* specifies the name of a call sequence in a thread that will execute after the thread has been dispatched.

Running Example Specification

A package is created to contain declarations for the software classifiers as shown in Listing 32. Concrete data types are declared for the software variables, which are derived from the system function variables by extending them and adding a *SW* suffix to their names. To these classifiers, properties specifying their concrete representation in code are set using the *Data_Representation* property of the data model annex (line 7). The *Robot_State_SW* software variable is also refined by providing properties for an equivalent data structure *Robot_state* declared in a C code file *data_types.h* and to be used for code generation (lines 24 and 25).

```
package Line_Follower_Software
public
    with Line_Follower_Functions, Code_Generation_Properties, Base_Types,
        Data_Model;

    data Light_Intensity_SW extends Physics::Light
        properties
```

```
            Data_Model::Data_Representation => integer;
            Data_Size => 4 bytes;
      end Light_Intensity_SW;


      data Turn_Angle_SW extends Line_Follower_Functions::Turn_Angle
         properties
            Data_Model::Data_Representation => integer;
      end Turn_Angle_SW;


      data Power_SW extends Line_Follower_Functions::Power
         properties
            Data_Model::Data_Representation => integer;
      end Power_SW;


      data Robot_State_SW extends Line_Follower_Functions::Robot_State
         properties
            Data_Model::Data_Representation => Enum;
            Data_Model::Enumerators => ( "FORWARD", "STOP" );
            Source_Name => "Robot_state";
            Source_Text => ("data_types.h");
      end Robot_State_SW;
```

Listing 32: Data classifiers for the software application

Concrete software subprogram implementations for the *compute turn angle* and the *compute wheels motor power* functions are declared as shown in Listing 33. Like for the data types, those subprograms extend the system function abstract components and refine the abstract features to subprogram parameters typed with appropriate SW data types previously introduced.

```
subprogram Compute_Turn_Angle_SW extends Line_Follower_Functions::
   Compute_Turn_Angle
   features
      light_intensity: refined to in parameter Light_Intensity_SW;
      turn_angle:  refined to out parameter Turn_Angle_SW;
   properties
      Classifier_Substitution_Rule => Type_Extension;
      Source_Language => (C);
end Compute_Turn_Angle_SW;


subprogram implementation Compute_Turn_Angle_SW.pid extends
   Line_Follower_Functions::Compute_Turn_Angle.pid
   properties
      Source_Name => "computePID";
      Source_Text => ("line_follower.h", "line_follower.c");
end Compute_Turn_Angle_SW.pid;


subprogram Compute_Wheels_Motors_Power_SW extends Line_Follower_Functions
   ::Compute_Wheels_Motors_Power
   features
      turn_angle: refined to in parameter Turn_Angle_SW;
      left_motor_power: refined to out parameter Power_SW;
      right_motor_power: refined to out parameter Power_SW;
      state: refined to requires data access Robot_State_SW;
   properties
      Classifier_Substitution_Rule => Type_Extension;
      Source_Language => (C);
end Compute_Wheels_Motors_Power_SW;
```

```
subprogram implementation Compute_Wheels_Motors_Power_SW.basic extends
    Line_Follower_Functions::Compute_Wheels_Motors_Power.basic
    properties
        Source_Name => "computeWheelsPower";
        Source_Text => ("line_follower.h", "line_follower.c");
end Compute_Wheels_Motors_Power_SW.basic;
```

Listing 33: Subprogram classifiers for the compute turn angle and the compute wheels motor power software functions

The implementation of the *compute_turn_angle* subprogram makes use of a PID control mechanism, which computes a turn angle that is proportional to the observed light intensity as illustrated on the RHS of figure 8.8. The C code for such subprogram can actually be derived from control simulations with tools such as Matlab Simulink. This code can then be associated with an AADL subprogram classifier via the *Source_Language*, *Source_Text* and *Source_Name* properties as shown in Listing 33 (lines 7, 12 and 13). Note that the *Source_Language* property is set on the subprogram types (lines 7 and 24) while the other source code properties are set on the subprogram implementations. This ease design space exploration by capturing the variability of different subprogram code implementations of a given programming language.

In order to read the value from the light sensor and to actuate the wheel motors, subprograms provided by the NXT OSEK ECRobot C library [219] are used. A new subprogram implementation is created to encapsulate calls to the *get_light_intensity*, *follow_line*, *set_left_motor_power* and *set_right_motor_power* subprograms as shown in Listing 34 and in figure 8.18. Implementation details are added consisting of data subcomponents for setting the value of required input parameters for the ECRobot subprograms such as the port number and a braking coefficient and initial values on data subcomponents (lines 8 to 19) for the corresponding data structures in code.

```
subprogram Follow_Line_SW_NXT
    features
        state: requires data access Robot_State_SW;
end Follow_Line_SW_NXT;

subprogram implementation Follow_Line_SW_NXT.basic
    subcomponents
        left_wheel_port: data Base_Types::Integer {
            Data_Model::Initial_Value => ( "NXT_PORT_B" );
        };
        left_wheel_brake: data Base_Types::Integer {
            Data_Model::Initial_Value => ( "0" );
        };
        right_wheel_port: data Base_Types::Integer {
            Data_Model::Initial_Value => ( "NXT_PORT_A" );
        };
        right_wheel_brake: data Base_Types::Integer {
            Data_Model::Initial_Value => ( "0" );
        };
    calls
        main_call: {
            get_light_intensity: subprogram ECRobot_Get_Light_Intensity;
            follow_line: subprogram Follow_Line_SW.basic;
            set_left_motor_power: subprogram NXT_Motor_Set_Power;
            set_right_motor_power: subprogram NXT_Motor_Set_Power;
        };
    connections
        light_intensity_follow_line: parameter get_light_intensity.
            light_intensity -> follow_line.light_intensity;
```

```
        left_motor_power: parameter follow_line.left_motor_power ->
            set_left_motor_power.power;
        right_motor_power: parameter follow_line.right_motor_power ->
            set_right_motor_power.power;
        state_follow_line: data access state -> follow_line.state;
        left_wheel_port_set_left_motor_power: parameter left_wheel_port
            -> set_left_motor_power.portNb;
        left_wheel_brake_set_left_motor_power: parameter left_wheel_brake
            -> set_left_motor_power.brake;
        right_wheel_port_set_right_motor_power: parameter
            right_wheel_port -> set_right_motor_power.portNb;
        right_wheel_brake_set_right_motor_power: parameter
            right_wheel_brake -> set_right_motor_power.brake;
    properties
        Compute_Execution_Time => 3 ms .. 5 ms;
end Follow_Line_SW_NXT.basic;
```

Listing 34: Subprogram classifiers for the follow line software function

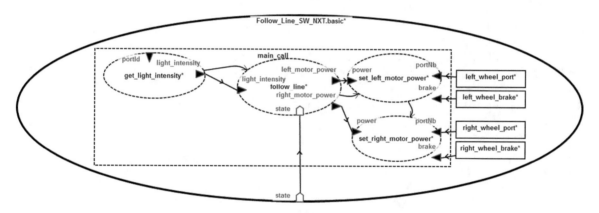

Fig. 8.18: The NXT line following subprogram

Software subprograms extending the *Pick_Up_Object* and *Drop_Off_Object* system function components must also be declared. Those are not presented here but can be found in the complete set of models from [63].

The next step is to specify how the subprograms of the software application will be executed. The *Pick_Up_Object_SW.basic* subprogram should be executed once the robot has reached the start of the path. The *Follow_Line_SW_NXT.basic* subprogram should be executed once the object has been picked up and periodically at a frequency high enough to follow the line, given a required tolerance, the minimum line curvature radius and the robot speed. The *Drop_Off_Object_SW.basic* subprogram should be executed once the end of the path is reached. The *Detect_Obstacles_SW* subprogram should be executed periodically while the line is being followed at a frequency high enough to detect obstacle on time in order to avoid collisions. Finally, the *Logging_SW* subprogram should be always executed periodically, with a period given by logging requirements.

The modelling of these threads is shown in Listing 35. The *Pick_Up_Object_Thread* thread is set with an *at_beginning_path* input event port for its activation when the robot reaches the beginning of the path and an *object_picked* output event port for notification when the object has been picked by the gripper. An *aperiodic* dispatch protocol is set meaning that the thread will be dispatched upon events on its input event port.

```
thread Pick_Up_Object_Thread
    features
        at_beginning_path: in event port;
        object_picked: out event port;
    properties
        Dispatch_Protocol => aperiodic;
```

```
        Deadline => 5 ms;
        Priority => 2;
        Stack_Size => 512 Bytes;
end Pick_Up_Object_Thread;

thread Follow_Line_Thread
    features
        state: requires data access Robot_State_SW;
        object_picked: in event port;
        object_dropped: in event port;
        at_beginning_path: out event port;
        at_end_path: out event port;
        right_wheel: out data port Power_SW;
        left_wheel: out data port Power_SW;
    flows
        light_to_left: flow path light_intensity -> left_wheel;
        light_to_right: flow path light_intensity -> right_wheel;
    properties
        Dispatch_Protocol => periodic;
        Period => 15 ms;
        Deadline => 15 ms;
        Priority => 3;
        Stack_Size => 512 Bytes;
end Follow_Line_Thread;

thread implementation Follow_Line_Thread.basic
    calls
        main_call: {
                follow_line: subprogram Follow_Line_SW_NXT.basic;
        };
    connections
        state_follow_line: data access state -> follow_line.state;
    properties
        Compute_Entrypoint_Call_Sequence => reference (main_call);
        Latency => 3 ms .. 5 ms applies to light_to_left, light_to_right;
end Follow_Line_Thread.basic;

thread Detect_Obstacles_Thread
    features
        state: requires data access Robot_State_SW;
    properties
        Dispatch_Protocol => periodic;
        Period => 10 ms;
        Deadline => 5 ms;
        Priority => 2;
        Stack_Size => 512 Bytes;
end Detect_Obstacles_Thread;

thread Log_Thread
    properties
        Dispatch_Protocol => periodic;
        Period => 20 ms;
        Deadline => 5 ms;
        Priority => 5;
        Stack_Size => 512 Bytes;
```

```
end Log_Thread;
```
Listing 35: Thread classifiers for the tasks of the software application

The *Follow_Line_Thread* thread is set with an *object_picked* input event port for its activation when the object has been picked by the gripper and an *object_arrived* output event port for its deactivation and a periodic dispatch protocol. The determination of the period of this thread will be detailed in the analysis section 8.6.7. A thread implementation is provided specifying a call to the follow line subprogram (line 21) and a data access connection for the robot state variable, which will be valued by the obstacle detection subprogram. Finally, the *Compute_Entrypoint_Call_Sequence* property (line 35) specifies which call sequence (there could be many) should be executed upon dispatch of the thread.

A *Drop_Off_Object_Thread* thread classifier (not shown in the listing due to space constraints) is provided for executing the *Drop_Off_Object_SW* subprogram with features and properties similar to the ones of the *Pick_Up_Object_Thread* thread. Another thread classifier is provided for the obstacle detection subprogram with a data access feature for the robot state and a periodic dispatch protocol. Finally, the logging function is encapsulated within a periodic thread. More details on other properties of these threads can be found in the analysis section 8.6.7.

Now that these thread classifiers have been declared, they can be instantiated as subcomponents of a *process* implementation, which represents a memory address space for the global software application. This is shown in Listing 36 and figure 8.19, where a data subcomponent is declared for the robot state data with a *FORWARD* initial value. Proper port connections are declared between the subcomponents.

```
process Carry_Object_Process
    features
        light_intensity: in data port Light_Intensity_SW;
        left_wheel_power: out data port Power_SW;
        right_wheel_power: out data port Power_SW;
    flows
        light_to_left: flow path light_intensity -> left_wheel_power;
        light_to_right: flow path light_intensity -> right_wheel_power;
end Carry_Object_Process;

process implementation Carry_Object_Process.basic
    subcomponents
        state: data Robot_State_SW {
            Data_Model::Initial_Value => ( "FORWARD" );
        };
        pick_up_object: thread Pick_Up_Object_Thread;
        follow_line: thread Follow_Line_Thread.basic;
        drop_off_object: thread Drop_Off_Object_Thread;
        detect_obstacle: thread Detect_Obstacles_Thread;
        logging: thread Log_Thread;
    connections
        pick_up_object_folow_line: port pick_up_object.object_picked ->
            follow_line.object_picked;
        follow_line_drop_object: port follow_line.at_end_path ->
            drop_off_object.object_arrived;
        drop_object_follow_line: port drop_off_object.object_dropped ->
            follow_line.object_dropped;
        obstacle_detection_state: data access detect_obstacle.state ->
            state;
        state_follow_line: data access state -> follow_line.state;
        follow_line_begin_path_pick_up_object: port follow_line.
            at_beginning_path -> pick_up_object.at_beginning_path;
        light_intensity_follow_line: port light_intensity -> follow_line.
            light_intensity;
```

```
        follow_line_right_wheel: port follow_line.right_wheel ->
            right_wheel_power;
        follow_line_left_wheel: port follow_line.left_wheel ->
            left_wheel_power;
    properties
        Timing => Immediate applies to follow_line_right_wheel,
            follow_line_left_wheel;
end Carry_Object_Process.basic;
```

Listing 36: Process classifiers for the software application

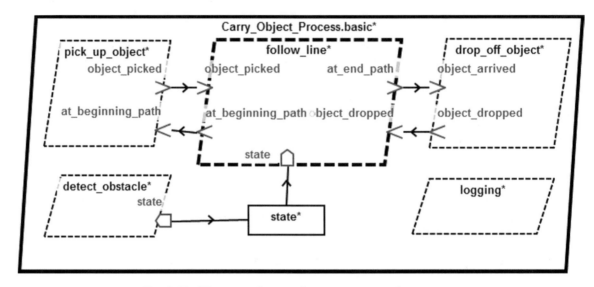

Fig. 8.19: The carry object software process implementation

This constitutes a complete software application that can be deployed on the brick of the NXT robot plant model. However, as will be presented in the analysis section, we are interested in determining the latency of steering the robot upon a change in the value of the observed path light intensity. For this purpose, we extend the software model by modeling the drivers of the input and output circuits of the brick driving the light sensor and the two motor assemblies.

In order to achieve this, we first create a system component for the software application as illustrated in figure 8.20. To this system we add a subcomponent of the software application process previously defined and device subcomponents with data ports typed with the software power and light intensity variable types. Such data ports are connected to the software process containing the threads. As explained in section 8.6.5, devices can be used for both hardware and software parts of systems. For software, they represent the software part of corresponding hardware devices, such as drivers residing in a memory and executed on an external processor. Such is the case for the input and output circuit devices of the robot brick (figure 8.16), which include built-in drivers to control sensors and actuators.

Binding of the Software Application on the Execution Platform

The last step in modeling the robot CPS is to deploy the software application on the plat model. This is presented in the following.

Concepts

In order to specify the binding of the software application on the plant model, first a system implementation must be provided for the *Line_Follower_Robot_CPS*, for the system type of the system overview of Listing 19. This implementation should have as subcomponent the software application previously defined and the robot plant model.

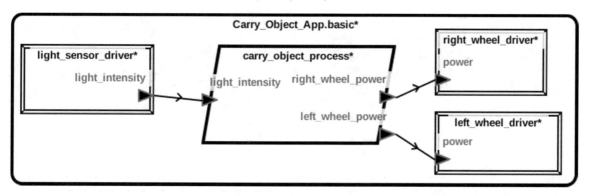

Fig. 8.20: A system component for the software application including devices for modeling drivers of sensors and actuators

The binding of the thread and data subcomponents of the process on the execution platform components can then be specified. Properties from the standard *Deployment_Properties* property set are used for that:

- *Actual_Processor_Binding* specifies the processor that will execute threads or the hardware device that will execute software devices (driver)
- *Actual_Memory_Binding* specifies the memory that will store processes and data.
- *Actual_Connection_Binding* specifies the bus that will transmit the data of port connections.

Running Example Specification

The system implementation for the line follower CPS component type is shown in Listing 37 and figure 8.21. A system subcomponent for the NXT software application and a system subcomponent for the configured NXT line following robot plant model are instantiated. The features of the robot plant model are connected to the features of the CPS system type of the system overview (Listing 19).

```
system implementation Line_Follower_Robot_Cps.nxt
    subcomponents
        app: system Line_Follower_Software::Carry_Object_App.basic;
        robot: system Line_Follower_Robot.nxt;
    connections
        force_left_wheel_conn: feature robot.force_left_wheel ->
            force_left_wheel;
        force_right_wheel_conn: feature robot.force_right_wheel ->
            force_right_wheel;
        light_sensor_conn: feature light_sensor_in -> robot.
            light_sensor_in;
        force_gripper_conn: feature robot.force_gripper -> force_gripper;
        sonar_in_conn: feature sonar_in -> robot.sonar_in;
        sonar_out_conn: feature robot.sonar_out -> sonar_out;
    properties
        Actual_Processor_Binding => (reference (robot.brick.
            main_processor)) applies to app.carry_object_process;
        Actual_Memory_Binding => (reference (robot.brick.main_processor.
            ram)) applies to app.carry_object_process;
        Actual_Processor_Binding => (reference (robot.brick.input_circuit
            )) applies to app.light_sensor_driver;
        Actual_Connection_Binding => (reference (robot.brick.data_bus))
            applies to app.light_sensor_process_conn;
        Actual_Processor_Binding => (reference (robot.brick.
            output_circuit)) applies to app.left_wheel_driver;
```

```
        Actual_Connection_Binding => (reference (robot.brick.data_bus))
            applies to app.process_left_wheel_conn;
        Actual_Processor_Binding => (reference (robot.brick.
            output_circuit)) applies to app.right_wheel_driver;
        Actual_Connection_Binding => (reference (robot.brick.data_bus))
            applies to app.process_right_wheel_conn;
end Line_Follower_Robot_Cps.nxt;
```

Listing 37: System implementation for the Robot CPS

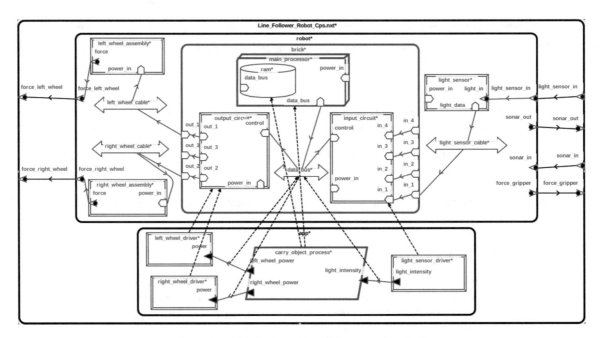

Fig. 8.21: The deployed line follower robot CPS

The binding properties *Actual_Processor_Binding* and *Actual_Memory_Binding* are set to respectively the ARM7 processor and its contained memory as shown in Listing 37 under the properties section (lines 13 and 14). This is also illustrated in figure 8.21 with the dashed arrows pointing from the process subcomponent to the NXT brick ARM processor and its included RAM memory. Since there is only one processor and one memory, it was chosen to apply the binding properties to the enclosing process, which means that all contained threads and data subcomponents are bound to the same processor and memory. If several processors were available, a binding property could have been be set for each individual thread for specifying their execution by different processors.

The device drivers for the light sensor and the wheel motors are also bound to the input and output circuits on the NXT brick. The port connections between the devices and the process are bound to the data bus connecting the input and output circuits to the ARM processor.

8.6.7 Analyses

The modelling of the object carrier robot CPS has been completed. The next step is to verify some of its properties to ensure that the system will operate properly and will meet its requirements. Furthermore, it would be interesting to estimate its performances. A first question to ask is if the ARM 7 processor can schedule all the threads given their timing properties. This strongly depends on the computation time of the subprograms themselves. Several methods exist to answer this question. For the case of the NXT brick, assume that the times are given by executing the programs alone on the brick and by logging the times to the display of the robot so that they can be memorized.

Scheduling

Scheduling analysis can be performed from the software application of section 8.6.6 and using tools such as Cheddar or its commercial implementation in AADL Inspector. This analysis will rely on the following properties attached to software components:

In Listing 35, property values are given to AADL threads in order to define their dispatch protocol, period, priority, and their sequence of subprogram calls. From the call sequence of a thread, we can compute an upper bound on the thread execution time by summing the upper bound of the called subprograms execution time. The latter is given for subprogram *Follow_Line_SW_NXT* in listing 34 using the *Compute_Execution_Time* property. Last but not least, the scheduling protocol is attached to the AADL processor component on which threads are going to run. This is done in Listing 28 using the *Scheduling_Protocol* property (line 34). Here, the scheduling protocol is a well known protocol named Rate Monotonic Scheduling (RMS). RMS is a fixed priority scheduling technique (tasks are executed according to a predefined priority) and tasks priorities are ordered as follows: the most frequent tasks are given a higher priority.

Semantics

The AADL modeling elements listed above define a periodic task set model (where an AADL thread is a periodic task) which can be summarized on figure 8.22.

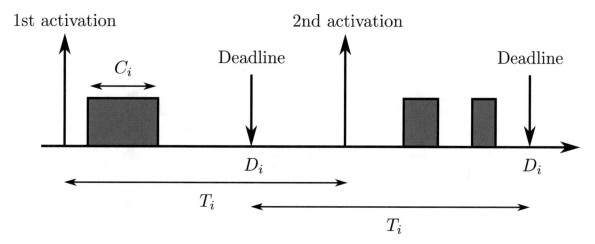

Fig. 8.22: Periodic task set

On this figure, a task τ_i is characterized by its period T_i, its deadline D_i, its capacity C_i. The capacity of a task corresponds to the upper bound on its execution time, and we explained above how this can be extracted from the execution time of AADL subprograms called in sequence by an AADL thread. In practice, we measure the execution time of subrpograms running on the NXT brick to estimate tasks capacity.

From this task model, the CPU utilization of U_i of a task τ_i can be computed as follows: $U_i = \frac{C_i}{T_i}$ and the CPU utilization of the task is simply the sum of every task utilization.

Given a periodic task set model, schedulability analysis can be performed to ensure all the tasks of the object carrier robot can finish their execution within their deadlines. A first option to perform this analysis is to compare the worst-case processor utilization with a safe upper bound: it has been proven by Liu and Layland in 1973 that a necessary condition for the schedulability of a task set scheduled with RMS, on a mono-core processor, is that the CPU utilisation of the task set remains under 69.3147%. Another option, for task sets scheduled with a fixed priority scheduling, is to compute tasks Worst Case Response Time (WCRT) and check that they remain inferior to the tasks deadline. The WCRT of a task τ_i, on a mono-core processor, can be computed using the following recursive formula:

$$WCRT_i = \sum_{\tau_j \in HP_i} \lceil \frac{WCRT_i}{T_j} \rceil \cdot C_j$$

where HP_i is the set of tasks having a higher priority than τ_i.

Analysis results

Figures 8.23 and 8.24 show the different results provided by AADL Inspector after performing the scheduling analysis of an AADL model.

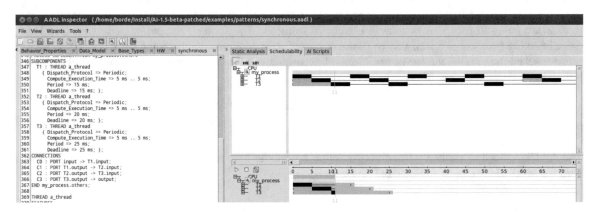

Fig. 8.23: Tasks execution gantt chart

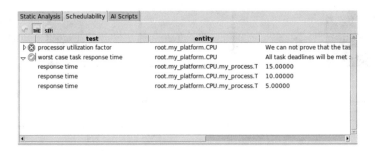

Fig. 8.24: Response time analysis result

This analysis can help us configure the task set either by modifying the period of AADL threads, reducing the number of tasks executed on the platform, or changing the execution platform itself. In the case of the object carrier robot, given the number of tasks, their execution time, and the physics of the system, the scheduling analysis helped to define tasks period.

Latency

Another question to ask about our object carrying robot is whether it will be able to follow the line given a minimum speed requirement. For this, the latency of the system in steering the robot must be known. In particular, we are interested in the latency of the data flows involved in the control loop, i.e. the amount of time separating the production of a data from the light sensor, to the production of commands on the engine. AADL is well suited for such analysis using its *flow* constructs.

Constructs

AADL flows are used to annotate AADL elements in order to specify paths of information flow across components. As such, flows do not represent any concrete system architecture elements but facilitate analyses and allow characterizing information flows with additional properties. By defining connections among directed ports, possible data flows are already defined by the structure of an AADL model. However, they are **possible** data flows and potentially some of them are not present in the software application. Besides, only a subset of data flows require latency analysis.

In AADL, a flow consists of:

- *Flow source*: describes the origin of an information flow
- *Flow path*: describes a path of information flow between components
- *Flow sink*: describes the end of an information flow
- *End to end flow*: describes a complete flow including a source, a path and a sink

Component types can only contain flow specifications (sources, sinks and paths) annotating the declared features.

Component implementations can only contain flow implementations (sources, sinks and paths) annotating features, connections and subcomponent flows. They can also contain end to end flows beginning by a flow source and ending by a flow sink.

Latency upper bounds can either be attached to end-to-end flows or set in ReqSpec requirements assigned to end-to-end flows. Latency contributions are attached to model elements implementing these flows. In practice, the latency of an end-to-end flow is the sum of latencies contributed by components and connections traversed by the flow.

Figure 8.25 represents the architecture of the robot carrier application along with devices representing the light sensor and the right and left motors. They are represented with an AADL device, representing the logical execution of the device driver. In addition, the end-to-end flow from the light sensor output (flow source) to the left motor input (flow sink) passing by the software application (flow path in *carry_object_process*) is coloured in yellow.

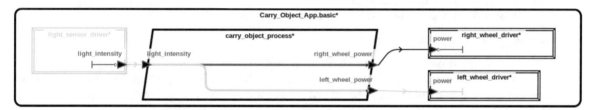

Fig. 8.25: Light sensor to left motor end-to-end flow

There exists three types of contributions to a flow latency in AADL:

1. Latency property value can be attached to a flow path in order to decompose the end-to-end latency into latencies attached to subcomponents.
2. Physical communications (i.e. connections bound to a bus component) contribute to the latency of flows passing by these links.
3. Logical communications (i.e. communications among software components) also contribute to the latency of flows passing by these links.

We explain further how to use these modeling capabilities in the remainder of the following paragraph, illustrating these types.

Running Example Specification

Contribution type 1: Latency property value attached to a flow path. In the early phase of the development process, or in order to decompose the end-to-end latency requirement into latency requirements attached to subcomponents, AADL users can attach flow latency values to flow path as illustrated in Listing 35 (property *Latency* defined in *Follow_Line_Thread.basic*). Note that without this property, the deadline of the thread would be used as an upper bound (and 0 as a lower bound).

Contribution type 2: Physical communications contribution. When a flow passes by a logical connection (i.e. connection between software components) bound to a bus component, the communication itself takes time. AADL offers the possibility to model this contribution as a linear function of the number of transmitted bytes.

Property *Transmission_Time* is meant for this usage, which is illustrated in Listing 38. In this listing we can see that the *Transmission_Time* property consists of two fields: *Fixed* and *PerByte* which respectively represent a communication offset and a communication time per byte. Each of these fields is a time range, representing the lower and upper bound of the communication offset and time per byte. Note that when this property is used, it is necessary to define the size of the transmitted data. This is feasible with the *Data_Size* property, as illustrated in Listing 32 for the *Light_Intensity_SW* data type.

```
bus Data_Bus
    properties
        Transmission_Time => [Fixed => 10 ns .. 100 ns; PerByte => 10 ns
            .. 40 ns;];
end Data_Bus;
```

Listing 38: Data bus component transmission time

Note that the example we took is rather simple: AADL architectures may be much more complex, for instance by binding logical connections to virtual busses (representing communication routes) themselves bound to busses and devices representing a network.

Contribution type 3: Local communications contribution. This type of contribution is probably the less intuitive and though the main contribution to flow latency from a software architecture viewpoint. To explain how local communications contribute to flow latency, it is necessary to understand further the model of computation and communications entailed by AADL software components, in particular communications among threads and/or devices.

We decompose this explanation in two parts, answering the following questions:

1. When are data consumed/produced by threads?
2. What are the additional constraints on threads communications?

By default, the answers to these questions are:

1. Threads consume data on their input ports when they are dispatched, and produce their output at completion time (this can be refined using the *Input_Time* and *Output_Time* AADL properties). One data is consumed when a thread is dispatched (this can be refined using the *Dequeue_Protocol* AADL property), and one data is produced per activation of a thread (this can be refined using the *Output_Rate* AADL property). In the remainder of this paragraph, we keep this default configuration as this is the one we used in the AADL model of the object carrier robot, except for the *Dequeue_Protocol*: we used the *AllItem* property value meaning that the queued is emptied by the activation of a receiving thread.
2. Communications are not really constrained, meaning that threads consume the latest (set of) data arrived at dispatch time. This can be refined using the *Timing* AADL property (default value being *Sampled*). We explain how this property value impacts the flow latency analysis in the case of the object carrier robot.

Listing 36 represents the configuration of connections between process *carry_object_process* and device *light_sensor_driver* on the one hand, and between process *carry_object_process* and *left_wheel_driver* and *right_wheel_driver*.

For the latter, the value *Immediate* is assigned for the *Timing* property, which means the recipient can only start its execution when new outputs are produced by the process. As a consequence, the activation of the device always receive the latest data produced by the process.

Overall, the end-to-end latency time can be computed in OSATE: with the configuration described in this book, the result obtained is in a range of 3 to 30 milliseconds.

Other Analyses The two analyses presented in the previous sections are very common and essential for embedded systems design. However several other analyses can be conducted from an AADL model such as memory usage, power consumption [17], mass, costs, etc. On the behavior side, verification and validation can also be performed via tools such as the BLESS proof checker [179] and model checking [42, 223]. More information on analysis tools for AADL can be found at [6].

8.6.8 Code Generation

The ultimate activity to be performed from an AADL specification is automatic code generation. Once the analyses have shown a correct design, a large part of the code can be generated automatically, thus reducing the risk of errors introduced by humans. There are currently two main code generators for AADL: the Ocarina tool [222] and RAMSES (Refinement of AADL Models for the Synthesis of Embedded Systems) [235]. In this chapter, RAMSES is presented due to its refinement-based approach providing more accurate analyses and supporting automated design optimizations.

Concepts The process flow of the RAMSES approach is presented in figure 8.26. RAMSES takes as input an AADL model and transforms it into another AADL model into which details specific to an operating system specification selected by the user has been added. The POSIX, ARINC 653 and NXT OSEK are currently supported. This step is called model refinement since the generated AADL model includes details specific to the operating system and therefore is of a lower level of abstraction than the original model. The advantage of this approach as opposed to code generated directly from the initial model is a reduced semantic gap between the refined model and the code. In addition, the refined model can be further analyzed giving more accurate analysis results due to this reduced semantic gap.

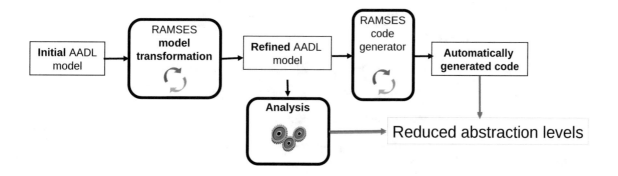

Fig. 8.26: The RAMSES process flow

The result of such analyses can be used to select of appropriate model transformations for applying design patterns to the initial AADL model for optimizing non-functional properties. The model can be refined iteratively until the desired quality attributes are achieved. Then, a code generator for the selected operating system generates platform-specific code including the required configuration files for the operating system such as thread scheduling and memory partitions.

There are many ways code can be generated from an AADL model. A first approach consists of generating skeletons from AADL components by assembling the source code set by the programming properties to subprograms and / or threads as presented in section 8.6.6. An alternative is to translate into code the behavior specification of components as specified using the AADL behavior annex. This later approach is not presented in this chapter, since the behavior annex has not been introduced. In any case, RAMSES supports these two approaches.

Running Example Specification

Refinement

RAMSES applies several refinement rules to produce the refined model for code generation. The set of applied rules varies depending on the selected operating system platform. We illustrate here one of the simplest RAMSES refinement rule for transforming event data port communications between threads into communication via shared data access. This is illustrated by the diagram of figure 8.27 where the port connection between two periodic threads is transformed into a shared data (queue) and two shared data access connections between the threads and the shared data. For this, the event data ports of the threads have also been converted into data access features. In addition, a call to a subprogram to manage the access of the queued data has been added to each thread.

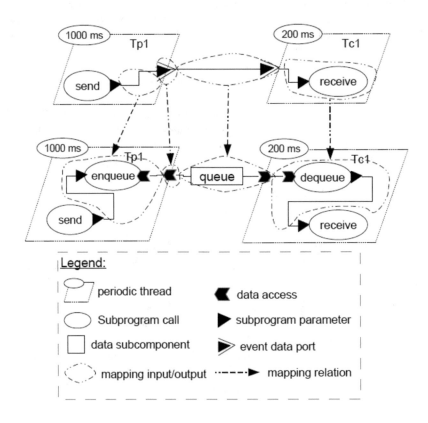

Fig. 8.27: The RAMSES local communication refinement rule (reproduced from [234])

Code Generation step

During the modelling of the software application of the line follower robot, various properties have been set to subprograms and data components for their mapping to existing C code artifacts. We briefly describe the essence of these properties using our software application example.

As shown in Listing 32, we created new data types for our software variables that map to the system variable abstract types that they extend. To these types we added the *Data_Representation* property from the data modelling annex to specify that our variables will be represented as integers in code. For the robot state internal variable, we have set the *Data_Representation* as an enumeration with values *FORWARD* and *STOP*. The *Source_Name* property specifies the name of a corresponding data structure in C code located in a C file specified by the *Source_Text* property. An initial value of *FORWARD* is set on the data subcomponent of the *Carry_Object_Process* process of Listing 36.

For subprograms such as *Compute_Turn_Angle_SW* (Listing 33), the *Source _Language* property has been set to "C" and similar to the Robot_State data component type, properties *Source_Name* and *Source_Text* are used to identify corresponding subprograms in C code.

From the AADL threads of our software application and their properties such as their priorities, as well as other properties set on the NXT ARM processor of the NXT brick (Listing 28) onto which the NXT OSEK firmware has been installed, the code generator also generates the configuration file for the NXT OSEK operating system specifying which threads should be scheduled and how they should be scheduled.

8.7 Summary

This chapter introduced the AADL language with a focus on its use for the development of CPSs. The modelling and analysis of a simple object carryer robot was used to introduce AADL by following a typical V-cycle development process (figure 8.3) augmented with some of the requirements engineering management practices promoted by the REMH (figure 8.9) and supported by the ALISA set of notations and workbench. The required AADL notions have been introduced gradually as they were required by the modelling steps of the followed development process.

This allowed presenting how AADL can be used to design a CPS with a top down approach starting from modelling the overview of the system to be built, followed by the capture of stakeholder and goals, verifiable requirements, system functions, physical plant model, software application model and deployment model. From such complete model including the software (cyber) part deployed on the hardware part, scheduling and latency analysis have been illustrated followed by model refinement, refined analyses and finally automatic C code generation with the RAMSES approach and tool.

We hope that introducing AADL this way allowed a better understanding of how AADL can be used and what it can achieve in termes of analyses, design space exploration and system synthesis via automatic code generation.

8.8 Literature and Further Reading

The AADL language and its annexes are quite rich and it was not possible to present the concepts thoroughly in this chapter due to space constraints. However, many documents are readily available for more in depth learning.

- The official AADL website [6];
- "Model-Based Engineering with AADL: An Introduction to the SAE Architecture Analysis & Design Language" [100] is the reference book about the core AADL language written by Peter Feiler and Dave Gluch. Peter Feiler is the main architect of the AADL;
- An introduction to the modelling of system families / product lines with AADL is given at [99];
- More information on ALISA can be found at [205, 82];
- More information on RAMSES can be found at [235].

8.9 Self Assessment

1. Write a non-functional requirement in ReqSpec for the latency in steering the robot that will constrain the results of analysis in section 8.6.7.
2. Add a predicate to verify that the latency is below the limit of 20 ms.
3. Complete the Lego Mindstorm robot product line family by adding AADL system classifiers to the robots library package of Listing 27 for the EV3 version of the Lego Mindstorm brick and robot.[2].
4. Model software requirem ents in ReqSpec derived from the example functional requirements for the system functions of section 8.6.4.
5. Perform a latency analysis for the obstacle detection function similar to the one presented in section 8.6.7 for steering the robot. First add the missing device driver to the software application system for the sonar sensor and add the required flow annotations and properties. Perform the analysis with OSATE.

---

[2] The EV3 Hardware Developer Kit document can be found at https://www.mikrocontroller.net/attachment/338591/hardware_developer_kit.pdf

Acknowledgements

The authors are grateful to the ISC chair (Ingénierie des Systèmes Complexes) on cyber-physical systems and distributed control systems for supporting this work. ISC is sponsored by renowned industries such as Thalès, Dassault Aviation, DCNS and DGA and operated by the École polytechnique, ENSTA ParisTech and Télécom ParisTech engineering schools.

Chapter 9
FTG+PM: Describing Engineering Processes in Multi-Paradigm Modelling

Moharram Challenger, Ken Vanherpen, Joachim Denil, and Hans Vangheluwe

Abstract Model-based System Engineering (MBSE) is a methodology that uses models throughout the engineering to replace the paper-based approach of communication among stakeholders. Multi-Paradigm Modelling (MPM) is at the core of this engineering approach as for each phase in the engineering process the most appropriate models at the most appropriate levels of abstraction is used. A design process includes a set of activities in which the design decisions or evaluations of the (sub-) system properties are done. Furthermore, the design artifacts are transformed by the design activities. We can define transformations as the manipulation of a model with a specific purpose. MPM approaches do not have a standard way of representing processes. A process model for MPM should focus on the languages, model instances and transformations between these models at different levels of abstraction. In this chapter, we propose the Formalism Transformation Graph and Process Model (FTG+PM) as a standard representation of MPM processes. The described process can be simulated for analysis and orchestration, as a set of (automatic) transformations.

Learning Objectives

After reading this chapter, we expect you to be able to:

- Understand why modelling the design process is of importance.
- Represent the MPM processes using the Formalism Transformation Graph and Process Model (FTG+PM).
- Reason on the orchestration of a modelled design process to support the designers.

9.1 Introduction

To tackle the increasing complexity of today's systems, engineers already practice a Model-Based Systems Engineering (MBSE) methodology [95]. In MBSE, models are used to support requirements engineering,

Moharram Challenger
University of Antwerp, Belgium
e-mail: moharram.challenger@uantwerpen.be

Ken Vanherpen
University of Antwerp, Belgium
e-mail: ken.vanherpen@uantwerpen.be

Joachim Denil
Flanders Make, Belgium
e-mail: joachim.denil@uantwerp.be

Hans Vangheluwe
McGill University, Canada
e-mail: hans.vangheluwe@uantwerp.be

© The Author(s) 2020   259
P. Carreira et al. (eds.), *Foundations of Multi-Paradigm Modelling for Cyber-Physical Systems*,
https://doi.org/10.1007/978-3-030-43946-0_9

design, verification, and validation activities of a system beginning in the conceptual design phase and continuing throughout development and later life cycle phases [153]. While these models often operate at different abstraction levels, model transformations are applied to manipulate models between different appropriate representations. They are typically used for code synthesis, integration, analysis, simulation, and optimization purposes. Multi-Paradigm Modelling (MPM) as a method consolidates these modelling methods and techniques, encouraging engineers to model each aspect of the system explicitly at the most appropriate level(s) of abstraction using the most appropriate formalism(s) [212].

While engineering a system, a design process is followed, in which a set of design activities are executed in a well-defined order. In each design activity a set of design artifacts are consumed and/or produced. These design artifacts explicitly model the informed design decisions taken by the engineers or the evaluations done by the engineer. As such, they can also be regarded as model transformations where information in a set of input models is consumed to produce a set of output models.

Although this series of design activities should result in an operational system that meets the predefined set of requirements, MPM approaches do not have a standard way of representing the process. This is in spite of the many advantages that the explicit modelling of the development process can offer, such as optimization, consistency analysis, time and risk analysis, etc. Generally, having well defined processes reduces the risk of failing a design project, as we can explicitly reason on problems that have to be overcome in the design of a system [143].

In this chapter, we propose the Formalism Transformation Graph and Process Model (FTG+PM) as a standard representation of MPM processes. As its name implies, it consists of two (related) parts: (a) the FTG where the focus is on the languages and transformations used throughout the process and (b) the flow of design activities and design artifacts. As tool-support is crucial in the MBSE (and MPM) life-cycle, we demonstrate how the orchestration of a modelled design process can support designers by scheduling transformations.

9.2 Model-based Systems Engineering

Document-centric approaches used to be at the core of communicating design choices between different engineers. However, as systems steadily enlarge in size, number of components, and features offered to the user, this is no longer appropriate. The added systems' complexity leads to an increase in the development time, number of errors, and development cost. As such, model-based systems engineering (MBSE) approaches have gained momentum to mend these shortcomings [95]. In MBSE, models are used as an integral part of the technical baseline of the system under design. This includes models in different development phases of a capability, system, and/or product [76].

MBSE is an engineering methodology in which the domain models and modeling techniques are used as the primary means of information exchange between engineers, rather than on document-based information exchange. The MBSE approach was outlined and popularized by INCOSE when it kicked off its MBSE initiative in January 2007 [154]. It focuses on distributed but integrated model management and the main goal is to increase the productivity and reducing the risk, by minimizing unnecessary manual transcription of concepts when coordinating the work of large teams. MBSE methodology provides several advantages:

- System Engineers focus on the technicalities of the problem rather than document structure
- Diagrammatic descriptions are often less ambiguous than textural descriptions
- Greater consistency across related documents
- Dependencies are explicitly captured across stovepipes resulting in less duplication and inconsistency

The MBSE methodology supports system requirement engineering, analysis, design, implementation, verification, and validation activities of a system beginning in the conceptual design phase and continuing throughout development and later life cycle phases [154]. There are a multitude of modelling techniques and approaches that fall within MBSE. Some of them include:

- Structured Analysis and Design
- Data Flow Diagramming
- State Transition Diagramming
- Behavioural Modelling
- Entity Relationship Modelling

- Process Modelling

Recently, the focus of MBSE is also the model execution in computer simulation experiment, to overcome the gap between the system model and the respective simulation software. As a result, the term Modeling and Simulation-based Systems Engineering (M&SBSE) has also been used along with MBSE [123].

The models used in an MBSE approach pave the way to tackle system complexity by providing proper levels of abstraction where engineers make design choices or analyze the systems' capabilities or performance. Models operate at a similar or different levels of abstraction called horizontal and vertical abstractions, respectively [124][254]. The level of abstraction can be regarded as which properties of the system under design are taken into account in the models. Certain design approaches, such as the Model Driven Architecture (MDA) approach, explicitly introduce such levels of abstractions: Platform-specific Modeling (PSM) level or Platform Independent Modeling (PIM) level [166]. In this sense, the horizontal abstraction can be either among PIM models or PSM models. While the vertical abstraction can be considered among PIM and PSM models.

Model transformations can be applied to manipulate models from one abstraction to another. They are typically used for code synthesis, integration, analysis, simulation, and optimization purposes. By using model transformations, models in one formalism are automatically transformed to another formalism. This allows the possibility to benefit from the power of the target formalism, e.g. to simulate the system under design. Two types of transformations exist, Model-to-Model (M2M) and Model-to-Text (M2T) transformations.

Multi-paradigm Modelling (MPM) consolidates the modeling methods and techniques, such as MBSE, enabling engineers to model each aspect of the system explicitly at the most appropriate level(s) of abstraction using the most appropriate formalism(s), while modeling the development process(es) explicitly [213].

Cyber-physical Systems (CPS) are one of the inter-disciplinary systems which require different formalisms, levels of abstraction, tools, and viewpoints in their design and development. Therefore, they need for MPM and MBSE techniques to deal with the complexity and available risks. Also, the evolution in the requirements of different views during the CPS design and development needs for a MPM process management. The next sections deals with the life-cycle and design process in MPM for complex systems such as CPSs and IoT systems [307].

9.3 Development Lifecycle Models

As designing complex systems involves engineers from various disciplines, engineers follow a set of guidelines to ensure that the product implements the given system requirements. The order in which the guidelines are followed is called the design process. Inspired by the software engineering community, that has defined Software Development Life Cycle (SDLC) models over the past decades [246]. We can distinguish four models that are commonly used when designing complex systems: the waterfall model, the V model, the spiral model, and the agile model. They are conceptually shown in Figure 9.1.

9.3.1 Waterfall Model

The waterfall model is typed by a sequence of design steps/activities in which a step/activity can only be started if one deliverable from the previous step/activity is arrived, as such it is a gated process. The gate is a decision point to define if a next phase of the project can start. The process starts by defining a set of requirements, which are then refined to a set of technical requirements called specifications. Given these specification, engineers reason about a high-level architecture in which the structure of the system is defined. In the detailed design and implementation phase, the behaviour of the architectural elements is defined after which the system under design is tested. Finally, the designed system is deployed in its intended environment. There is often feed-back between steps/activities.

Note that the waterfall suits projects that are not subject to changes in one of the previous phases during the process e.g. the requirement should not change during architecture design. Otherwise, the design process must start over which may be costly in large engineering projects.

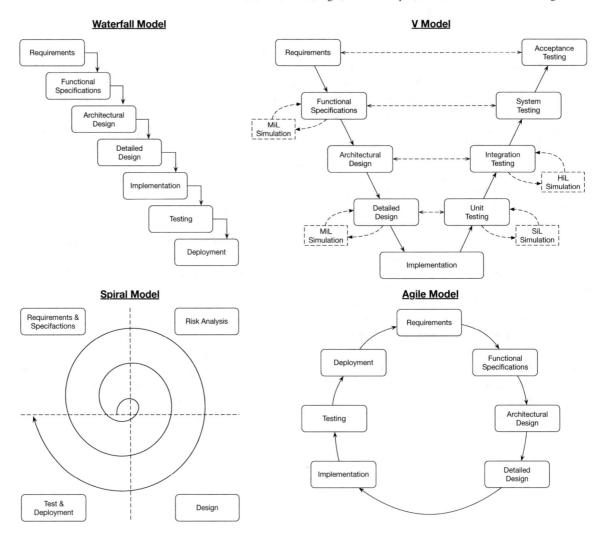

Fig. 9.1: Conceptual overview of commonly used MBSE design processes

9.3.2 V Model

Similar to the waterfall model, the V model is typed by a sequential design process. However, the V model distinguishes between *design* and *verification & validation* steps, respectively on the left-hand side and right-hand side of the V model, while keeping them tightly coupled as illustrated by the horizontal dashed lines, see Figure 9.1. While going from high-level requirements to a low-level implementation, the specifications of the design are verified by executing acceptance tests, ranging from low-level unit tests to high-level system tests. Ultimately, at the top of the V model, it is validated whether the system satisfies the requirements. Model-based techniques allow engineers to front-load a large set of the V&V activities.

Although the verification steps are executed while designing the system so that late detection of implementation errors can be avoided, a changing product requirement still requires one to redesign (parts of) the system which, again, may be costly. Nevertheless, in the context of designing (large) complex systems using a MBSE approach, the V model is widely accepted as the industry standard for designing and testing complex systems [95].

9.3.3 Spiral Model

Using an iterative design process, spiral model avoid these (costly) redesign by repeatedly going through four design phases: Requirements and specification election, risk analysis, design, and testing and deployment. In the first iteration, preliminary requirements and specifications are elected so that the first prototype of the system can be designed, tested and deployed.

In subsequent iterations, requirements and specifications are added and/or become more detailed so that the initial prototype further evolves towards the final product. During each iteration, the process is monitored such that possible project risks (e.g. technical feasibility and project cost) can be better assessed. Although this design model better mitigates changing requirements and design errors, one should note that an increasing number of iterations can lead to an increasing (cumulative) project cost.

9.3.4 Agile Model

Agile design process models are typed by so called sprints (i.e. design iterations) in which a small part of the overall system is designed, tested, and deployed. At the end of each sprint, a finalized, working, system is delivered to the stakeholders. This differs from a spiral design approach in which a prototype might be delivered. Moreover, the design iterations of the agile design approach are typically shorter than the ones of the spiral design approach. Agile design processes are therefore suitable when requirements are subject to change during the design and implementation phases, and when multidisciplinary teams need to cooperate. On the downside, the lack of a thorough system analysis may result in a more complex and more expensive design process.

9.4 Modelling the Design Process

A process for a complex system (such as CPS) should cover different aspects of the system and should make the different levels of abstraction clear. These levels of abstraction could be at the domain-specific design of the computational or physical components, the verification of the system at a specific abstraction level, and/or during deployment where different approximations can be used to obtain a better understanding of the parameters involved. At the defined level of abstraction, it has to be clear which languages and transformations are involved and how they are used together in the design and development process. Though, from a tool building perspective, this definition could be too restrictive, as some activities can require manual intervention before the activity are completed. Since tool-support is crucial in the MDE (and MPM) life-cycle, it is important that the described process can be executed automatically. One approach to do this is by means of a set of transformations scheduled after each other.

The process model has to act as a guide for the design, verification and deployment of the system. This means that the design of a large set of applications has to be described using the modelling language. The language should focus on the constructs of MPM, mainly formalisms and transformations and how these interact with each other. There should also be an explicit representation of control- and data-flow since the output of a single phase in the process does not necessarily means that the produced models are the input of the next phase. This also includes the use of control structures that allow parallel design, iterations, etc.

9.4.1 Rationale

Over the years, the process engineering community has proposed various process modelling means for software development. Process modelling has a large research community, resulted in many modelling languages. Rolland [241] gives a definition of process modelling: *Process models are processes of the same nature that are classified together into a model. Thus, a process model is a description of a process at the type level. Since the process model is at the type level, a process is an instantiation of it. The same process model is used repeatedly for the development of many applications and thus, has many instantiations.* [241]

Generally, a process model is used for three different goals:

- **Descriptive:** The models are used by an external observer to look at what happens during a process. It can be used to improve the process.
- **Prescriptive:** The models define a set of rules to prescribe a desired process. If this process is followed, it would lead to a desired outcome.
- **Explanatory:** The goal of an explanatory model is to explain the rationale behind the process.

9.4.2 What to model?

The process models can describe different aspects of the system development. These aspects can be divided in four groups (see Figure 9.2), namely: Functional view, Dynamic view, Informational view and Organization view [78].

- **Functional view:** This view of a process model discusses the dependencies in the functions (i.e. the transformations) of the system under development. A well-known example of a modeling language employing this perspective is data flow diagrams.
- **Dynamic/Behavior view:** This view focuses on the activities, actions, and tasks of the system and their sequences. This aspect deals with the control-flow and the timing of the activities in the system.
- **Informational view:** This view deals with the informational entities (i.e. the data for process, activities, artifacts, and products) which are produced or manipulated by the process. This view includes both the structure of the informational entities and the relationships among them.
- **Organisational view:** This view presents the stakeholders of the system as well as who performs the process elements in what part of the organization.

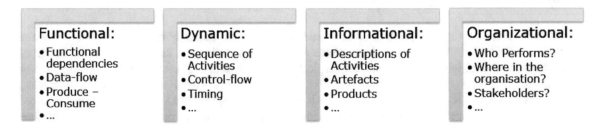

Fig. 9.2: Describing Processes

9.4.3 Reasoning about maturity

When processes are repeated multiple times, it is important to extend, evolve and mature the process models as well. Adding more information to process models can ultimately lead to better (e.g. faster, more robust, etc.) processes. The term *maturity* relates to the degree of formality and optimization of processes, from ad-hoc practices, to formally defined steps, to managed result metrics, to active optimization of the processes.

The Capability Maturity Model (CMM[1]) is a methodology used to develop and refine a development process. The model's aim is to improve existing development processes. CMM was developed and is promoted by the Software Engineering Institute (SEI), a research and development center sponsored by the U.S. Department of Defense (DoD). CMM was originally developed as a tool for objectively assessing the ability of government contractors' processes to implement a contracted software project [303].

---

[1] CMM is a registered service mark of Carnegie Mellon University (CMU).

The CMM is similar to ISO 9001, one of the ISO 9000 series of standards specified by the International Organization for Standardization (ISO). The ISO 9000 standards specify an effective quality system for manufacturing and service industries; ISO 9001 deals specifically with software development and maintenance. The main difference between the two systems lies in their respective purposes: ISO 9001 specifies a minimal acceptable quality level for software processes, while the CMM establishes a framework for continuous process improvement and is more explicit than the ISO standard in defining the means to be employed to that end [198].

CMM describes a five-level evolutionary path, see Figure 9.3, of increasingly organized and systematically more mature processes. These maturity levels of processes are:

- At the *ad-hoc/initial* level, the processes are disorganized, even chaotic. Success is likely to depend on individual efforts, and is not considered to be repeatable, because processes would not be sufficiently defined and documented to allow them to be replicated.
- At the *repeatable* level, basic project management techniques are established, and successes could be repeated, because the requisite processes would have been made established, defined, and documented.
- At the *defined* level, an organization has developed its own standard process model through greater attention to documentation, standardization, and integration.
- At the *managed/capable* level, an organization monitors and controls its own processes through data collection and analysis.
- At the *optimized/efficient* level, processes are constantly being improved through monitoring feedback from current processes and introducing innovative processes to better serve the organization's particular needs.

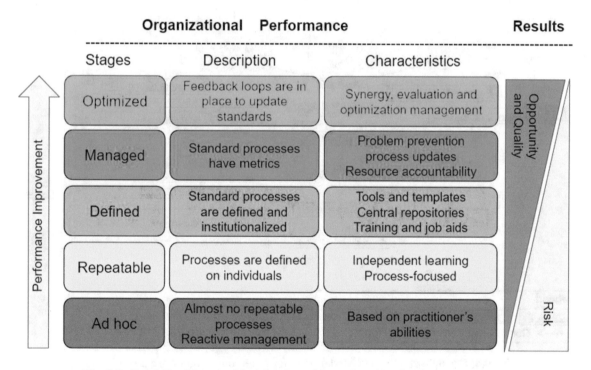

Fig. 9.3: Capability Maturity Model (adopted from: http://performancexpress.org/)

9.5 Activities 2.0 for modelling processes

When engineering complex system using a MBSE approach, design artifacts will be created at different levels of abstraction within the same or between different engineering domains. A model of the design process should make the relation between different design artifacts, and their respective abstraction level, clear. This process model has to act as a guide for the design, verification & validation, and implementation of the system. There should also be an explicit representation of control- and data-flow since the output of a single phase in the process does not necessarily means that the produced models are the input of the next phase. This also includes the use of control structures that allow parallel design, iterations, etc.

In [194, 196, 216] a Process Model (PM) language is proposed to describe the control and data flow between design activities. To this end, a subset of the Unified Modelling Language (UML) 2.0 activity diagrams are used. It enables to define the process as a descriptive and prescriptive model.

An example of a PM, describing the (partial) process of generating code from a control model, is shown in Figure 9.4.

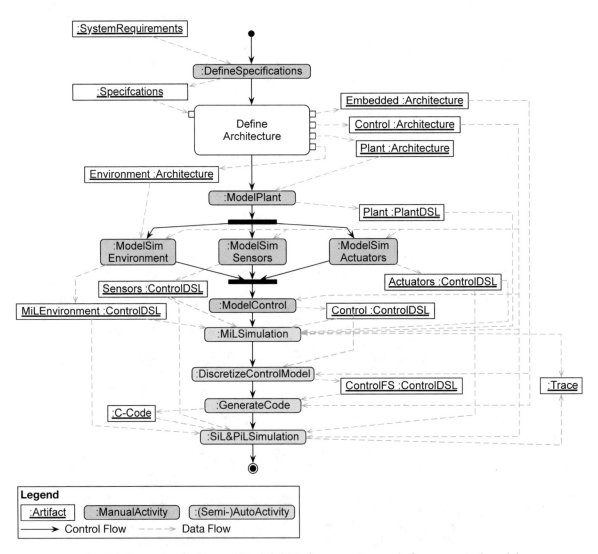

Fig. 9.4: Example of a Process Model (PM) for generating code from a control model

The labelled square edged rectangles represent the data objects (design artifacts) that flow throughout the design process. In a MBSE approach, these objects corresponds to models that are consumed or produced by design activities, represented by the labelled rectangles with rounded corners. For example, to verify the closed loop behaviour of a control model, a model of the physical system (i.e., a plant model) is required such that a Model-in-the-Loop (MiL) simulation can be executed. Note there the process is typed by a control flow and a data flow represented by solid and dashed edges, respectively. Furthermore, a distinction is made between manual and (semi-)automated activities. Finally, the join and fork Activity Diagram flow constructs, represented in Figure 9.4 as horizontal bars, allow one to represent concurrent design activities.

9.6 The tool perspective: Formalism Transformation Graph

In each phase of the Process Model, it has to be clear what formalisms and transformations are involved. Corresponding to mega-model principles [19, 31, 142], the Formalism Transformation Graph (FTG) [194, 196, 216] is a language that allows one to declare those formalisms and transformations. It models the relations between the different languages using transformations.

As can be seen in Figure 9.5, domain-specific formalisms are represented as labelled rectangles in an FTG model.

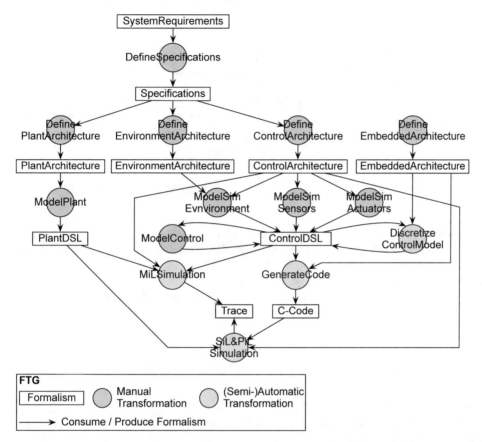

Fig. 9.5: Example of a Formalism Transformation Graph (FTG) for generating code from a control model

Transformations between those formalisms are depicted as labelled circles. Similar to the Process Model, these transformations can be either manual or (semi-) automated. The arrows from formalisms into transformations describe the inputs to the transformations and the arrows from the transformations into formalisms describe the outputs of the transformation. The FTG model is thus a graph describing the modelling languages and the transformations statically available to the engineers of a given domain. In the example of generating code from a control model, one can observe that a MiL Simulation is a (semi-)automated transformation from models expressed in a control, plant, and architectural formalism to a trace model represented in its appropriate formalism.

9.7 FTG+PM: Formalism Transformation Graph and Process Model

In the previous sections we have introduced models that enable engineers to describe their design activities, the formalisms that are used, and the transformations that exist between formalisms. One may have noticed that there exists a relation between the Formalism Transformation Graph and the Process Model. Indeed, data objects (i.e., models) and design activities in the PM are instances of the formalisms and transformations defined in the FTG, respectively. These typing relations are made clear by the colon defined in the rectangles of the PM.

As such, we define the unified metamodel of the Formalism Transformation Graph and Process Model as depicted in Figure 9.6.

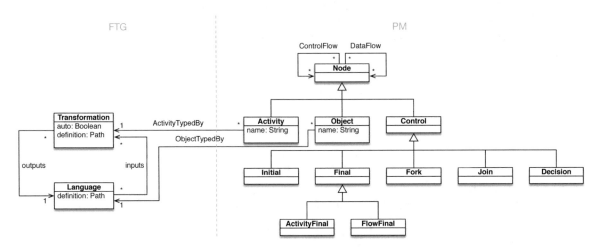

Fig. 9.6: Formalism Transformation Graph and Process Model (FTG+PM) metamodel

On the left-hand side the FTG metamodel can be observed in which the concepts *Language* and *Transformation* are depicted. Both concepts have a *definition* attribute that refers to the definition of the language or transformation. Additionally, an *auto* attribute defines whether the transformation is automatic or not. On the right-hand side the metamodel of the PM language can be seen with its *Activity*, *Object*, and *Control* nodes. Note the typing relation between an *Activity* and an *Object* in the PM, and an FTG *Transformation* and *Language*, respectively.

9.7.1 Reasoning about appropriateness of formalisms and heterogeneous modelling

The formalism transformation graph can be used for much more than typing alone. The mega-model describes the relations between different formalisms. As such, we can attach properties to formalisms and transformations. We can query the extended FTG to look for appropriate formalisms that satisfy a set of these properties.

The same applies for giving semantics to languages. The relations between the different formalisms can be used to define a path towards a certain analysis goal. For example, if we are interested in getting behavioral traces, we might use a simulator that creates behavioural traces from the model. However, you could also leverage the properties (and capabilities) of other formalisms and define another path in the FTG to create the state trajectories. For example, if a certain formalism can apply symbolic transformations to create a more run-time performant model, you can leverage these transformation and define a path in the FTG that goes over this formalism. As such the different translational semantics of the formalisms can be reused. This also has drawbacks as it is necessary to keep enough traceability information available to allow the engineer to e.g. debug the models. Figure 9.7 shows the FTG from [292]. Here two different properties are attached to the relations: (a) translational semantics (full line) versus simulation (dashed line) and (b) no approximation (blue) versus allowing approximation e.g. sampling (green). The vertical dashed line in the middle shows the split between the continuous-time domain (left) and the discrete-time domain (right).

The FTG also allows us to reason on combining different formalisms together. Multi-paradigm modelling advocates to use the most appropriate levels of abstraction using the most appropriate formalisms. When applying this principle to the design of complex cyber-physical systems, we have to combine both discrete and continuous-time formalisms. The cyber-part is much more naturally described in the discrete-time domain, while models-of-the-physics are much more naturally described in the continuous-time domain. However, to

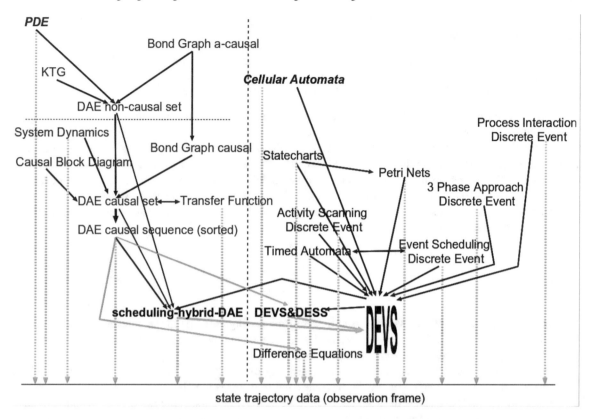

Fig. 9.7: Formalism Transformation Graph from [292]

evaluate the full system behaviour, we have to combine these models. Therefore, we need to reason on what it means to combine formalisms together, and how we can achieve this.

There are three techniques available to simulate the behaviour of hybrid systems.

- Hybrid modelling: A new modelling language is created based on the formalisms and the connections between the formalisms. The modelling language is operationalized with a simulator.
- Co-simulation: Co-simulation reasons on the combination of models by combining the state trajectories. For this, the simulator of the different formalisms is used and orchestrated explicitly.
- Mapping to a common formalism: Finally, the different formalisms and their connection are mapped to a common formalism. For example, In Figure 9.7 you can see that DEVS can be used as a common formalisms to map both discrete-time and continuous-time models to. However, it also shows that there are other possibilities, for example, Bond graphs and Causal-block diagrams can be combined using DAE causal equations.

9.7.2 Orchestrating Processes

Orchestration of a process means that the engineers are supported in their design process. This support can take various forms, from showing the engineers a dashboard with various metrics about the process and the steps to follow (e.g. in [193]) to automatically setting up modelling environments, opening the correct models and automatically saving and transforming models when possible. We assume the latter and reason about the orchestration of a process model, more specifically an FTG+PM, in a multi-paradigm modelling environment.

The orchestration of the FTG+PM language requires that the action nodes, denoting a transformation, are properly executed. The order of execution of these nodes is based on the mapping of the activity diagram to coloured Petri nets. The mapping of the activity diagram to the Petri net is described in [168]. When an action node is encountered the action has to be executed. Depending on the state of the auto attribute in the transformation, the framework has to:

- false: open a modelling environment containing the input model(s) in the specified language(s) and the modelling environment of the output language
- true: automatically execute the transformation with the desired input models

For this end, the tool should be able to capture commands from the engineers, e.g. to indicate when the designer is done with her/his task and the process can continue. Furthermore, the FTG part is very important as it gives the information to the MPM environment to open a certain which modelling environment should be opened to aid the designer. Automating the model transformations that are already defined, results in the consecutive execution of different transformations. A multi-paradigm modelling environment, such as AToMPM [259], has dedicated languages to execute model transformations. We can use these languages to enact the FTG+PM by transforming the process model to such a language. Other modelling environments have similar tools available that can be used to setup such an orchestration, e.g. [240, 145, 131].

9.8 Summary

This section discusses the use of Model-based System Engineering (MBSE) in the development life-cycle of Cyber-Physical Systems (CPS). It focuses specifically on the design processes and transformation of paradigm in the Multi-Paradigm Modelling (MPM) of CPSs.

Generally, MPM approaches do not have a standard way of representing processes. A process model for MPM should focus on the languages, model instances and transformations between these models at different levels of abstraction. These levels of abstraction could be at the domain-specific design of the computational or physical components, the verification of the system at a high abstraction level, but also during deployment where different approximations can be used to obtain a better understanding of the parameters involved.

In this regard, a standard representation for MPM processes of CPS is introduced called Formalism Transformation Graph and Process Model (FTG+PM). The automation and tooling is crucial in MPM lifecycle and it is key that the described process can be simulated for analysis and orchestrated to support the designers. To support the designers, the reasoning on the orchestration of a modelled design process is elaborated.

9.9 Literature and Further Reading

The interested readers can study the following papers and books for more details. To elaborate on the development/design processes, you can read [246] for software development life cycle models, [95] for the V model which is widely used for MBSE, and the book in [276] for agile development process.

To investigate more on FTG+PM, you can study [194] and [196] for the use of FTG in MDE, as well as the paper in [216] for a case study which uses FTM-PM in automotive. The background studies for FTG+PM can be also found in multiple references [81][41][292][295][291][293][290][178][288][289].

Finally, to read more on the reasoning and orchestration using FTG+PM, you can study tthe work of István et. al [79].

9.10 Self-Assessment

1. Create an FTG-PM for a design process or development process of a system of your choice.
2. For the development of the system indicated in exercise (1), discuss which design process is the most appropriate?
3. Formulate the difference between waterfall and agile development processes?
4. Considering your design process in exercise (1), where is it (regarding its maturity) according to the levels of CMM?
5. How you can combine partial differential equations with a a casual bond graphs models?

Acknowledgements

The authors would like to thank the followings: (1) The COST Action networking mechanisms and support of IC1404 Multi-Paradigm Modeling for Cyber-Physical Systems (MPM4CPS). COST is supported by the EU Framework Programme Horizon 2020. (2) Flanders Make, Belgium. Flanders Make is the strategic research centre for the manufacturing industry in the Flanders area, Belgium.

Bibliography

[1] jUCMNav User Requirements Notation Tool.
http://jucmnav.softwareengineering.ca/jucmnav/.

[2] Lego MINDSTORMS NXT Hardware Developer Kit. Technical report, Lego, 2006.

[3] Complexity-Reducing Design Patterns for Cyber-Physical Systems, Steven P. Miller (presentation at the AADL standards committee meeting). https://wiki.sei.cmu.edu/aadl/images/8/85/Rockwell-META-CPS--Jan-2011.pdf, 2010.

[4] Competition on software verification, 2014. http://sv-comp.sosy-lab.org/2014/.

[5] Airbus 380 Cables Bug.
http://calleam.com/WTPF/?p=4700.

[6] AADL Website. http://www.aadl.info/aadl/.

[7] AADL Inspector Homepage.
https://www.ellidiss.com/products/aadl-inspector/.

[8] E. Allen, R. Cartwright, and B. Stoler. *DrJava: A Lightweight Pedagogic Environment for Java*. ACM, 2002.

[9] M. Andersson. *Object-Oriented Modeling and Simulation of Hybrid Systems*. PhD thesis, 1994.

[10] Ariane 5 Bug.
https://blog.bugsnag.com/bug-day-ariane-5-disaster/.

[11] A. Asghar, S. Tariq, M. Torabzadeh-Tari, P. Fritzson, A. Pop, M. Sjölund, P. Vasaiely, and W. Schamai. An open source modelica graphic editor integrated with electronic notebooks and interactive simulation. In *Proceedings of the 8th International Modelica Conference - Modelica'2011, Dresden, Germany*, 2011.

[12] U. Assmann. *Invasive Software Composition*. Spinger Verlag, 2002.

[13] M. Association. *Modelica - A Unified Object-Oriented Language for System Modelling and Simulation*. 2018.

[14] K. J. Astrom, H. Elmqvist, D. Ab, and S. E. Mattsson. Evolution of continuous-time modeling and simulation, 1998.

[15] K. J. Åström, H. Elmqvist, and S. E. Mattsson. Evolution of Continuous-Time Modeling and Simulation. In *ESM*, pages 9–18, 1998.

[16] K. J. Aström and R. M. Murray. *Feedback systems: an introduction for scientists and engineers*. Princeton university press, 2010.

[17] R. B. Atitallah, E. Senn, D. Chillet, M. Lanoe, and D. Blouin. An Efficient Framework for Power-Aware Design of Heterogeneous MPSoC. *IEEE Transactions on Industrial Informatics*, 9(1):487–501, Feb 2013.

[18] B. Bachmann. *Proceedings of the 6th International Modelica Conference*. Bielefeld University, Bielefeld, Germany, 2008.

[19] M. Barbero, M. D. Del Fabro, and J. Bézivin. Traceability and provenance issues in global model management. In *ECMDA-TW: Proceedings of the 3rd Workshop on Traceability*, 2007.

[20] J. Barnat, L. Brim, and P. Ročkai. Scalable multi-core ltl model-checking. In *Model Checking Software*, volume 4595 of *LNCS*, pages 187–203. Springer, 2007.

[21] J. Barnat, L. Brim, I. Černá, and P. Šimeček. DiVinE – The Distributed Verification Environment. In M. Leucker and J. van de Pol, editors, *Proceedings of 4th International Workshop on Parallel and Distributed Methods in verifiCation*, pages 89–94, July 2005.

© The Author(s) 2020
P. Carreira et al. (eds.), *Foundations of Multi-Paradigm Modelling for Cyber-Physical Systems*,
https://doi.org/10.1007/978-3-030-43946-0

[22] J. Barnat, V. Forejt, M. Leucker, and M. Weber. DivSPIN – A SPIN compatible distributed model checker. In M. Leucker and J. van de Pol, editors, *Proceedings of 4th International Workshop on Parallel and Distributed Methods in verifiCation*, pages 95–100, July 2005.

[23] F. J. Barros. Dynamic structure discrete event system specification: a new formalism for dynamic structure modeling and simulation. In *Proceedings of the 1995 Winter Simulation Multiconference*, pages 781–785, 1995.

[24] F. J. Barros. Modeling formalisms for dynamic structure systems. *ACM Transactions on Modeling and Computer Simulation*, 7:501–515, 1997.

[25] F. J. Barros. Abstract simulators for the DSDE formalism. In *Proceedings of the 1998 Winter simulation Multiconference*, pages 407–412, 1998.

[26] F. Bause and P. S. Kritzinger. Stochastic petri nets - an introduction to the theory (2. ed.). In *Advanced studies of computer science*, 1996.

[27] G. Behrmann, A. David, K. G. Larsen, P. Pettersson, and W. Yi. Developing UPPAAL over 15 years. *Softw., Pract. Exper.*, 41(2):133–142, 2011.

[28] L. A. Belady, M. W. Blasgen, C. J. Evangelisti, and R. D. Tennison. A computer graphics system for block diagram problems. *IBM Systems Journal*, 10(2):143–161, 1971.

[29] B. Berthomieu and F. Vernadat. Time petri nets analysis with TINA. In *Third International Conference on the Quantitative Evaluation of Systems (QEST 2006), 11-14 September 2006, Riverside, California, USA*, pages 123–124. IEEE Computer Society, 2006.

[30] Y. Bertot and P. Castéran. *Interactive Theorem Proving and Program Development - Coq'Art: The Calculus of Inductive Constructions*. Texts in Theoretical Computer Science. An EATCS Series. Springer, 2004.

[31] J. Bézivin and B. Jean. On the unification power of models. *Vital And Health Statistics. Series 20 Data From The National Vitalstatistics System Vital Health Stat 20 Data Natl Vital Sta*, 4(2):171–188, 2005.

[32] O. Biberstein, D. Buchs, and N. Guelfi. Object-Oriented Nets with Algebraic Specifications: The {CO-OPN/2} Formalism. In G. Agha, F. D. Cindio, and G. Rozenberg, editors, *Advances in Petri Nets on Object-Orientation*, LNCS, pages 70–127. Springer, 2001.

[33] A. Biere, A. Cimatti, E. M. Clarke, and Y. Zhu. Symbolic model checking without bdds. In *TACAS '99: Proceedings of the 5th International Conference on Tools and Algorithms for Construction and Analysis of Systems*, pages 193–207, London, UK, 1999. Springer.

[34] A. Biere and K. Heljanko. Hardware model checking contest, 2013. http://fmv.jku.at/hwmcc13/.

[35] G. Birtwistle, O. J. Dahl, B. Myhrhaug, and K. Nygaard. *Speed-Up in Chemical Engineering Design*. Auerbach Publishers, Inc., 1973.

[36] T. Blochwitz, M. Otter, J. Åkesson, M. Arnold, C. Clauss, H. Elmqvist, M. Friedrich, A. Junghanns, J. Mauss, D. Neumerkel, H. Olsson, and A. Viel. Functional mockup interface 2.0: The standard for tool independent exchange of simulation models. In *Proceedings of the 9th International Modelica Conference*, pages 173–184. The Modelica Association, 2012.

[37] D. Blouin. *Modeling Languages for Requirements Engineering and Quantitative Analysis of Embedded Systems*. PhD thesis, University of South Britanny, France, 2013.

[38] D. Blouin and H. Giese. Combining Requirements, Use Case Maps and AADL Models for Safety-Critical Systems Design. In *2016 42th Euromicro Conference on Software Engineering and Advanced Applications (SEAA)*, pages 266–274, Aug 2016.

[39] E. L. Bóbeda, M. Colange, and D. Buchs. Building a symbolic model checker from formal language description. In *ACSD*, pages 50–59. IEEE Computer Society, 2015.

[40] B. Boehm. A spiral model of software development and enhancement. *SIGSOFT Softw. Eng. Notes*, 11(4):14–24, Aug. 1986.

[41] J.-S. Bolduc and H. Vangheluwe. Expressing ODE models as DEVS: Quantization approaches. In *Proceedings of the AIS'2002 Conference (AI, Simulation and Planning in High Autonomy Systems), April 2002, Lisboa, Portugal/F. Barros and N. Giambiasi (eds.)*, pages 163–169, 2002.

[42] M. Bozzano, R. Cavada, A. Cimatti, J.-P. Katoen, V. Y. Nguyen, T. Noll, and X. Olive. Formal verification and validation of AADL models. *Proc. ERTS*, 2010.

[43] P. C. Breedveld. *Physical Systems Theory in Terms of Bond Graphs*. Phd thesis, University of Twente, 1984.

[44] P. C. Breedveld. Multibond-graph elements in physical systems theory. *Journal of the Franklin Institute*, 319:1–36, 1985.

[45] R. D. Brennan. Digital simulation for control system design. In *Proceedings of the SHARE design automation project on - DAC '66*, DAC '66, pages 1.1–1.12, New York, New York, USA, 1966. ACM Press.

[46] J. F. Broenink. *Computer-Aided Physical-Systems Modeling and Simulation: A Bond-Graph Approach.* Phd thesis, University of Twente, Enschede, The Netherlands, mar 1990.

[47] J. F. Broenink. Bond-graph modeling in modelica. pages 137–141. 9th European Simulation Symposium (ESS97), Passau, Germany, 1997.

[48] J. F. Broenink. Modelling, simulation and analysis with 20-sim. *Journal A*, 38:22 –25, sep 1997.

[49] J. F. Broenink. 20-sim software for hierarchical bond-graph/block-diagram models. *Simulation Practice and Theory, nr. 7*, pages 481–492, 1999.

[50] J. F. Broenink and C. Kleijn. Computer-aided design of mechatronic systems using 20-sim 3.0. In *Proc WESIC, Workshop on Scientific and Industrial Collaboration*, Newport, UK, 1999. IEE.

[51] D. Broman, C. Brooks, L. Greenberg, E. A. Lee, M. Masin, S. Tripakis, and M. Wetter. Determinate composition of fmus for co-simulation. In *Proceedings of the Eleventh ACM International Conference on Embedded Software*, EMSOFT '13, pages 2:1–2:12, Piscataway, NJ, USA, 2013. IEEE Press.

[52] J. Brooks, F.P. No silver bullet essence and accidents of software engineering. *Computer*, 20(4):10 –19, april 1987.

[53] F. Brown. Direct application of the loop rule to bond graphs. *Journal of Dynamic Systems, Measurement and Control*, 94:253–261, 1972.

[54] R. Bruttomesso, M. Deters, and A. Griggio. Satisfiability modulo theories competition, 2014. `http://www.smtcomp.org/`.

[55] R. E. Bryant. Symbolic Boolean Manipulation with Ordered Binary-Decision Diagrams. *ACM Comput. Surv.*, 24(3):293–318, 1992.

[56] D. Buchs, S. Hostettler, A. Marechal, and M. Risoldi. AlPiNA: A symbolic model checker. In J. Lilius and W. Penczek, editors, *Applications and Theory of Petri Nets*, volume 6128 of *Lecture Notes in Computer Science*, pages 287–296. Springer, 2010.

[57] D. Buchs, S. Hostettler, A. Marechal, and M. Risoldi. Alpina: An algebraic petri net analyzer. In J. Esparza and R. Majumdar, editors, *Tools and Algorithms for the Construction and Analysis of Systems*, volume 6015 of *Lecture Notes in Computer Science*, pages 349–352. Springer, 2010.

[58] J. R. Burch, E. M. Clarke, D. E. Long, K. L. McMillan, and D. L. Dill. Symbolic model checking for sequential circuit verification. *IEEE Transactions on Computer-Aided Design of Integrated Circuits and Systems*, 13:401–424, 1994.

[59] J. R. Burch, E. M. Clarke, K. L. McMillan, D. L. Dill, and L. J. Hwang. Symbolic model checking: 10^{20} states and beyond. *Information and Computation*, 98(2):142–170, 1992.

[60] R. L. Burden and J. D. Faires. *Numerical Analysis*. Cengage Learning, 9 edition, 2010.

[61] J. Butcher. *Numerical Methods for Ordinary Differential Equations*. John Wiley & Sons, Ltd, Chichester, UK, jun 2003.

[62] L. Cabac, M. Haustermann, and D. Mosteller. Renew 2.5 - towards a comprehensive integrated development environment for petri net-based applications. In F. Kordon and D. Moldt, editors, *Application and Theory of Petri Nets and Concurrency - 37th International Conference, PETRI NETS 2016, Toruń, Poland, June 19-24, 2016. Proceedings*, volume 9698 of *Lecture Notes in Computer Science*, pages 101–112. Springer, 2016.

[63] Case Study AADL Models. `https://gitlab.telecom-paristech.fr/mbe-tools/RAMSES/tree/develop/examples/line_follower_mindstorm_robot`.

[64] F. Casella. *Proceedings of the 7th International Modelica Conference, Como, Italy*. 2009.

[65] F. Celier. *Combined Continuous/Discrete System Simulation by Use of Digital Computers. Techniques and Tools*. ETH No 6483 Swiss Federal Institute of Technology Zurich, Switzerland, 1979.

[66] F. Cellier, P. Fritzson, E. Lee, and D. Broman. *Proceedings of the 4th International Workshop on Equation-Based Object-Oriented Languages and Tools, (EOOLT'2011), Zürich, Switzerland*. Linköping University Electronic Press, 2011.

[67] F. E. Cellier. *Continuous System Modeling*. Springer Verlag, 1991.

[68] F. E. Cellier and E. Kofman. *Continuous System Simulation*. Springer-Verlag, Berlin, Heidelberg, 2006.

[69] F. E. Cellier and E. Kofman. *Continuous System Simulation*. Springer Science & Business Media, 2006.

[70] B. Chen and H. Vangheluwe. Symbolic flattening of DEVS models. In *Proceedings of the 2010 Summer Simulation Multiconference*, pages 209–218, 2010.

[71] G. Chiola. Characterization of timed well-formed petri nets behavior by means of occurrence equations. In *Proceedings of the Sixth International Workshop on Petri Nets and Performance Models, PNPM 1995, Durham, NC, USA, October 3-6, 1995*, pages 127–136. IEEE Computer Society, 1995.

[72] A. C. H. Chow and B. P. Zeigler. Parallel DEVS: a parallel, hierarchical, modular, modeling formalism. In *Proceedings of the 1994 Winter Simulation Multiconference*, pages 716–722, 1994.

[73] A. C. H. Chow, B. P. Zeigler, and D. H. Kim. Abstract simulator for the parallel DEVS formalism. In *AI, Simulation, and Planning in High Autonomy Systems*, pages 157–163, 1994.

[74] E. M. Clarke, Jr., O. Grumberg, and D. A. Peled. *Model checking*. MIT Press, Cambridge, MA, USA, 1999.

[75] C. Clauss. *Proceedings of the 8th International Modelica Conference, Dresden, Germany*. 2011.

[76] N. S. E. D. M. Committee et al. Final report of the Model Based Engineering (MBE) subcommittee. *National Defense Industrial Association (NDIA)*, 2011.

[77] O. Coudert, C. Berthet, and J. C. Madre. Formal boolean manipulations for the verification of sequential machines. In *EURO-DAC '90: Proceedings of the conference on European design automation*, pages 57–61, Los Alamitos, CA, USA, 1990. IEEE Computer Society Press.

[78] B. Curtis, M. I. Kellner, and J. Over. Process Modeling. *Communications of the ACM*, 35(9):75–90, 1992.

[79] I. Dávid, H. Vangheluwe, and Y. Van Tendeloo. Translating engineering workflow models to DEVS for performance evaluation. In *Proceedings of the 2018 Winter Simulation Conference*, pages 616–627. IEEE Press, 2018.

[80] B. Davis, H. Porta, and J. Uhl. *Calculus & Mathematica Vector Calculus: Measuring in Two and Three Dimensions*. Addison-Wesley, 1994.

[81] J. De Lara and H. Vangheluwe. Computer Aided Multi-paradigm Modelling to Process Petri-nets and Statecharts. In *International Conference on Graph Transformation*, pages 239–253. Springer, 2002.

[82] J. Delange, P. Feiler, and E. Neil. Incremental Life Cycle Assurance of Safety-Critical Systems. In *8th European Congress on Embedded Real Time Software and Systems (ERTS 2016)*, Toulouse, France, Jan. 2016.

[83] J. Denil, B. Meyers, P. D. Meulenaere, and H. Vangheluwe. Explicit semantic adaptation of hybrid formalisms for FMI co-simulation. In *Proceedings of the Symposium on Theory of Modeling & Simulation: DEVS Integrative M&S Symposium, part of the 2015 Spring Simulation Multiconference, SpringSim '15, Alexandria, VA, USA, April 12-15, 2015*, pages 99–106, 2015.

[84] P. Derler, E. A. Lee, M. Törngren, and S. Tripakis. Cyber-physical system design contracts. In *2013 ACM/IEEE International Conference on Cyber-Physical Systems (ICCPS)*, pages 109–118, April 2013.

[85] M. Diaz. *Petri Nets: Fundamental Models, Verification and Applications*. ISTE. Wiley, 2013.

[86] J. Dixhoorn, van. Bond graphs and the challenge of a unified modelling theory of physical systems. In F. E. Cellier, editor, *Progress in Modelling and Simulation*, pages 207 – 245. Academic Press, 1982.

[87] H. Ehrig and B. Mahr. *Fundamentals of Algebraic Specification 1: Equations und Initial Semantics*, volume 6 of *EATCS Monographs on Theoretical Computer Science*. Springer, 1985.

[88] H. Elmqvist. *A Structured Model Language for Large Continuous Systems*. PhD thesis, 1978.

[89] H. Elmqvist, D. Bruck, and M. Otter. *Dymola—User's Manual*. Dynasim AB, Research Park Ideon, SE-223 70, Lund, Sweden, 1996.

[90] H. Elmqvist, S.-E. Mattsson, and M. Otter. Modelica—a language for physical system modeling, visualization and interaction. In *Proceedings of the 1999 IEEE Symposium on Computer-Aided Control System Design (CACSD'99), Hawaii*, 1999.

[91] H. Elmqvist, M. Otter, S.-E. Mattsson, and H. Olsson. *Modeling, Simulation, and Optimization with Modelica and Dymola,Book draft*. Dynasim AB, Research Park Ideon, SE-223 70, Lund, Sweden, 2002.

[92] E. A. Emerson, S. Jha, and D. Peled. Combining partial order and symmetry reductions. In *TACAS*, pages 19–34, 1997.

[93] T. Ernst, S. Jahnichen, M. Klose, and R. Chaussee. The architecture of the smile/m simulation environment. In *Proceedings 15th IMACS World Congress on Scientific Computation, Modelling and Applied Mathematics*, 01 1998.

[94] S. Esmaeilsabzali, N. A. Day, J. M. Atlee, and J. Niu. Deconstructing the semantics of big-step modelling languages. *Requirements Engineering*, 15(2):235–265, 2010.

[95] J. A. Estefan. Survey of Model-Based Systems Engineering (MBSE) Methodologies. Technical Report Rev. B, INCOSE MBSE Initiative, May 2008.

[96] S. Evangelista and J.-F. Pradat-Peyre. Memory efficient state space storage in explicit software model checking. In *in Proceedings of the 12th International SPIN Workshop on Model Checking of Software*, volume 3639, pages 43–57. Springer, 2005.

[97] J. Farkas. Theorie der einfachen ungleichungen. *Journal für die reine und angewandte Mathematik*, 124:1–27, 1902.

[98] P. Feiler, J. Hansson, D. de Niz, and L. Wrage. System Architecture Virtual Integration: An Industrial Case Study. Technical Report CMU/SEI-2009-TR-017, Software Engineering Institute, Carnegie Mellon University, Pittsburgh, PA, 2009.

[99] P. H. Feiler. Modeling of System Families. Technical report, Software Engineering Institute (SEI), 2007.

[100] P. H. Feiler and D. P. Gluch. *Model-Based Engineering with AADL: An Introduction to the SAE Architecture Analysis & Design Language*. Addison-Wesley Professional, 1st edition, 2012.

[101] Design and Analysis of Cyber-Physical Systems: AADL and Avionics Systems. https://resources.sei.cmu.edu/library/asset-view.cfm?assetid=48553.

[102] M. Felleisen, R. Findler, M. Flatt, and S. Krishnamurthi. The drscheme project: An overview. In *Proceedings of the ACM SIGPLAN 1998 Conference on Programming Language Design and Implementation (PLDI'98), Montreal, Canada*. ACM, 1998.

[103] A. Fernström, I. Axelsson, P. Fritzson, A. Sandholm, and A. Pop. *OMNotebook – Interactive WYSIWYG Book Software for Teaching Programming*. Proc. of the Workshop on Developing Computer Science Education – How Can It Be Done? Linköping University, Dept. Computer & Inf. Science, Linköping, Sweden, 2006.

[104] P. Fritzson. *Proceedings of SIMS'99—The 1999 Conference of the Scandinavian Simulation Society*. 1999.

[105] P. Fritzson. *Proceedings of Modelica 2000 Workshop*. Lund University, Lund, Sweden, 2000.

[106] P. Fritzson. *Proceedings of the 3rd International Modelica Conference*. Linköping University, Linköping, Sweden, 2003.

[107] P. Fritzson. *Principles of Object Oriented Modeling and Simulation with Modelica 2.1*. Wiley-IEEE Press, 2004.

[108] P. Fritzson. *Proceedings of the 2nd Conference on Modeling and Simulation for Safety and Security (SimSafe'2005)*. Linköping University, Linköping, Sweden, 2005.

[109] P. Fritzson. *Proceedings of the 4rd International Conference on Modeling and Simulation in Biology, Medicine, and Biomedical Engineering (BioMedSim'2005)*. Linköping University, Linköping, Sweden, 2005.

[110] P. Fritzson. Mathmodelica - an object oriented mathematical modeling and simulation environment. 10(1), 2006.

[111] P. Fritzson. *Introduction to Modeling and Simulation of Technical and Physical Systems with Modelica*. Wiley-IEEE Press, 2011.

[112] P. Fritzson. *Principles of Object Oriented Modeling and Simulation with Modelica 3.3: A Cyber-Physical Approach*. Wiley IEEE Press, 2014.

[113] P. Fritzson and K.-F. Berggren. A pseudopotential calculation of the density of states of expanded crystalline mercury. *Solid State Communications*, 19(4):385 – 387, 1976.

[114] P. Fritzson and P. Bunus. Modelica—a general object-oriented language for continuous and discrete-event system modeling and simulation. In *Proceedings of the 35th Annual Simulation Symposium, San Diego, California*. IEEE, 2002.

[115] P. Fritzson, F. Cellier, and D. Broman. *Proceedings of the 2nd International Workshop on Equation-Based Object-Oriented Languages and Tools, (EOOLT'2008), in conjunction with ECOOP, Paphos, Cyprus*. Linköping University Electronic Press, 2008.

[116] P. Fritzson, F. Cellier, and C. Nytsch Geusen. *Proceedings of the 1st International Workshop on Equation-Based Object-Oriented Languages and Tools, (EOOLT'2007), in conjunction with ECOOP, Berlin*. Linköping University Electronic Press, 2007.

[117] P. Fritzson, V. Engelson, and J. Gunnarsson. An integrated modelica environment for modeling, documentation and simulation. In *Proceedings of Summer Computer Simulation Conference '98, Reno, Nevada, USA*, 1998.

[118] P. Fritzson, E. Lee, F. Cellier, and D. Broman. *Proceedings of the 3rd International Workshop on Equation-Based Object-Oriented Languages and Tools, (EOOLT'2010), in conjunction with MODELS, Oslo, Norway*. Linköping University Electronic Press, 2010.

[119] P. Fritzson, L. Viklund, J. Herber, and D. Fritzson. Industrial application of object-oriented mathematical modeling and computer algebra in mechanical analysis. 1992.

[120] P. Fritzson, L. Viklund, O. J. Herber, and D. Fritzson. High-level mathematical modeling and programming. *IEEE Software*, 12(4):77–87, July 1995.

[121] H. Garavel, R. Mateescu, and I. Smarandache. Parallel state space construction for model-checking. In *Proceedings of the 8th international SPIN workshop on Model checking of software*, SPIN '01, pages 217–234, New York, NY, USA, 2001. Springer-Verlag New York, Inc.

[122] W. Gellert and V. N. Reinhold. *The VNR concise encyclopedia of mathematics / W. Gellert ... [et al.], editors ; K.A. Hirsch, H. Reichardt, scientific advisors*. Van Nostrand Reinhold New York, 2nd ed. edition, 1989.

[123] D. Gianni, A. D'Ambrogio, and A. Tolk. Introduction to the modeling and simulation-based systems engineering handbook., 2014.

[124] H. Giese, S. Neumann, O. Niggemann, and B. Schätz. Model-based integration. In *Proceedings of the 2007 International Dagstuhl conference on Model-based engineering of embedded real-time systems*, pages 17–54. Springer-Verlag, 2007.

[125] P. Godefroid. Using partial orders to improve automatic verification methods. In E. Clarke and R. Kurshan, editors, *Computer-Aided Verification*, volume 531 of *Lecture Notes in Computer Science*, pages 176–185. Springer, 1991.

[126] C. Gomes, C. Thule, D. Broman, P. G. Larsen, and H. Vangheluwe. Co-simulation: State of the art. *CoRR*, abs/1702.00686, 2017.

[127] R. Goud, K. M. van Hee, R. D. J. Post, and J. M. E. M. van der Werf. Petriweb: A repository for petri nets. In S. Donatelli and P. S. Thiagarajan, editors, *Petri Nets and Other Models of Concurrency - ICATPN 2006, 27th International Conference on Applications and Theory of Petri Nets and Other Models of Concurrency, Turku, Finland, June 26-30, 2006, Proceedings*, volume 4024 of *Lecture Notes in Computer Science*, pages 411–420. Springer, 2006.

[128] F. Gracer and R. A. Myers. Graphic Computer-assisted Design of Optical Filters. *IBM Journal of Research and Development*, 13(2):172–178, mar 1969.

[129] J. Granda. Computer generation of physical system differential equations using bond graphs. *Journal of the Franklin Institute*, 319:243–256, 1985.

[130] C. Greer, M. Burns, D. Wollman, and E. Griffor. Cyber-physical systems and internet of things. *NIST Special Publication 1900-202*.

[131] E. Guerra, J. De Lara, D. S. Kolovos, R. F. Paige, and O. M. Dos Santos. transML: A family of languages to model model transformations. In *International Conference on Model Driven Engineering Languages and Systems*, pages 106–120. Springer, 2010.

[132] A. Hamez, L. Hillah, F. Kordon, A. Linard, E. Paviot-Adet, X. Renault, and Y. Thierry-Mieg. New features in CPN-AMI 3: focusing on the analysis of complex distributed systems. In *Sixth International Conference on Application of Concurrency to System Design (ACSD 2006), 28-30 June 2006, Turku, Finland*, pages 273–275. IEEE Computer Society, 2006.

[133] J. Hansson, S. Helton, and P. Feiler. ROI Analysis of the System Architecture Virtual Integration Initiative. Technical Report CMU/SEI-2018-TR-002, Software Engineering Institute, Carnegie Mellon University, Pittsburgh, PA, 2018.

[134] D. Harel. Statecharts: A visual formalism for complex systems. *Science of Computer Programming*, 8(3):231–274, 1987.

[135] D. Harel. Statecharts: A Visual Formulation for Complex Systems. *Sci. Comput. Program.*, 8(3):231–274, 1987.

[136] D. Harel. On visual formalisms. *Commun. ACM*, 31(5):514–530, May 1988.

[137] D. Harel and E. Gery. Executable object modeling with statecharts. *IEEE Computer*, 30(7):31–42, 1997.

[138] D. Harel and H. Kugler. The rhapsody semantics of statecharts (or, on the executable core of the UML). In *Integration of Software Specification Techniques for Applications in Engineering*, volume 3147 of *Lecture Notes in Computer Science*, pages 325–354. Springer Berlin Heidelberg, 2004.

[139] D. Harel and H. Kugler. The Rhapsody semantics of Statecharts (or, on the executable core of the UML). In H. Ehrig, W. Damm, J. Desel, M. Große-Rhode, W. Reif, E. Schnieder, and E. Westkämper,

editors, *Integration of Software Specification Techniques for Applications in Engineering*, pages 325–354. Springer, Berlin Heidelberg, 2004.

[140] D. Harel and A. Naamad. The STATEMATE semantics of Statecharts. *ACM Transactions on Software Engineering Methodology*, 5(4):293–333, 1996.

[141] D. Harel and A. Naamad. The STATEMATE Semantics of Statecharts. *ACM Transactions on Software Engineering and Methodology*, 5(4):293–333, 1996.

[142] R. Hebig, A. Seibel, and H. Giese. On the Unification of Megamodels. In *Electronic Communications of the EASST: Proceedings of the 4th International Workshop on Multi-Paradigm Modeling*, volume 42, 2011.

[143] F. J. Heemstra and R. J. Kusters. Dealing with risk: a practical approach. *Journal of Information Technology*, 11(4):333–346, 1996.

[144] D. C. Heggie. J. fauvel, r. flood, m. shortland and r. wilson, let newton be! (clarendon press, oxford, 1989). *Proceedings of the Edinburgh Mathematical Society*, 33(1):166–168, 1990.

[145] F. Heidenreich, J. Kopcsek, and U. Aßmann. Safe Composition of Transformations. In *International Conference on Theory and Practice of Model Transformations*, pages 108–122. Springer, 2010.

[146] L. Hillah and F. Kordon. Petri nets repository: A tool to benchmark and debug petri net tools. In W. M. P. van der Aalst and E. Best, editors, *Application and Theory of Petri Nets and Concurrency - 38th International Conference, PETRI NETS 2017, Zaragoza, Spain, June 25-30, 2017, Proceedings*, volume 10258 of *Lecture Notes in Computer Science*, pages 125–135. Springer, 2017.

[147] L. Hillah, F. Kordon, L. Petrucci, and N. Trèves. PNML framework: An extendable reference implementation of the petri net markup language. In J. Lilius and W. Penczek, editors, *Applications and Theory of Petri Nets, 31st International Conference, PETRI NETS 2010, Braga, Portugal, June 21-25, 2010. Proceedings*, volume 6128 of *Lecture Notes in Computer Science*, pages 318–327. Springer, 2010.

[148] J. Himmelspach and A. M. Uhrmacher. Sequential processing of PDEVS models. In *Proceedings of the 3rd European Modeling & Simulation Symposium*, pages 239–244, 2006.

[149] G. J. Holzmann. *The SPIN Model Checker*. Addison Wesley, 2003.

[150] J. E. Hopcroft, R. Motwani, and J. D. Ullman. *Introduction to Automata Theory, Languages, and Computation*. Addison-Wesley Longman Publishing Co., Inc., Boston, MA, USA, 3rd edition, 2006.

[151] S. Hostettler, A. Marechal, A. Linard, M. Risoldi, and D. Buchs. High-Level Petri Net Model Checking with AlPiNA. *Fundamenta Informaticae*, 113, 2011.

[152] M.-H. Hwang. Generating finite-state global behavior of reconfigurable automation systems: DEVS approach. In *Proceedings of the International Conference on Automation Science and Engineering*, pages 254 – 260, 2005.

[153] INCOSE. Systems Engineering Vision 2020. Technical Report INCOSE-TP-2004-004-02, International Council on Systems Engineering (INCOSE), 2007.

[154] T. INCOSE. Systems engineering vision 2020. *INCOSE, San Diego, CA, accessed Jan*, 26:2019, 2007.

[155] ITU URN Specification. http://www.itu.int/rec/T-REC-Z.150/en/.

[156] iTunes Crashes when Ripping. https://discussions.apple.com/thread/247969.

[157] J. F. Jensen, T. Nielsen, L. K. Oestergaard, and J. Srba. TAPAAL and reachability analysis of P/T nets. *T. Petri Nets and Other Models of Concurrency*, 11:307–318, 2016.

[158] K. Jensen. *Coloured Petri Nets - Basic Concepts, Analysis Methods and Practical Use - Volume 1, Second Edition*. Monographs in Theoretical Computer Science. An EATCS Series. Springer, 1996.

[159] M. Järvisalo, D. Le Berre, and O. Roussel. Sat competitions, 2011. http://www.satcompetition.org/.

[160] D. C. Karnopp, D. L. Margolis, and R. C. Rosenberg. *System Dynamics: Modeling and Simulation of Mechatronic Systems*. Wiley, 2006.

[161] D. C. Karnopp and R. C. Rosenberg. *Analysis and Simulation of Multiport Systems - The Bond Graph Approach to Physical System Dynamics*. MIT Press, 1968.

[162] D. C. Karnopp and R. C. Rosenberg. *System Dynamics, a Unified Approach*. Wiley, 1975.

[163] S. Katz and D. Peled. An efficient verification method for parallel and distributed programs. In *Linear Time, Branching Time and Partial Order in Logics and Models for Concurrency, School/Workshop*, pages 489–507, London, UK, 1989. Springer.

[164] E. Kindler. The epnk: An extensible petri net tool for PNML. In L. M. Kristensen and L. Petrucci, editors, *Applications and Theory of Petri Nets - 32nd International Conference, PETRI NETS 2011, Newcastle,*

UK, June 20-24, 2011. Proceedings, volume 6709 of *Lecture Notes in Computer Science*, pages 318–327. Springer, 2011.

[165] V. Klebanov, P. Müller, N. Shankar, G. T. Leavens, V. Wüstholz, E. Alkassar, R. Arthan, D. Bronish, R. Chapman, E. Cohen, M. Hillebrand, B. Jacobs, K. R. M. Leino, R. Monahan, F. Piessens, N. Polikarpova, T. Ridge, J. Smans, S. Tobies, T. Tuerk, M. Ulbrich, and B. Weiß. The 1st Verified Software Competition: Experience report. In M. Butler and W. Schulte, editors, *FM'2011: 17th International Symposium on Formal Methods*, volume 6664 of *LNCS*. Springer, 2011. http://www.vscomp.org.

[166] A. G. Kleppe, J. Warmer, J. B. Warmer, and W. Bast. *MDA explained: the model driven architecture: practice and promise*. Addison-Wesley Professional, 2003.

[167] S. Klikovits, R. Al-Ali, M. Amrani, A. Barisic, F. Barros, D. Blouin, E. Borde, D. Buchs, H. Giese, M. Goulao, M. Iacono, F. Leon, E. Navarro, P. Pelliccione, and K. Vanherpen. COST IC1404 WG1 Deliverable WG1.1: State-of-the- art on Current Formalisms used in Cyber-Physical Systems Development, Jan. 2019.

[168] C. Knieke and U. Goltz. An Executable Semantics for UML 2 Activity Diagrams. In *Proceedings of the International Workshop on Formalization of Modeling Languages*, page 3. ACM, 2010.

[169] F. Kordon and D. Buchs. Model Checking Contest @ Petri Nets, 2014. http://mcc.lip6.fr.

[170] F. Kordon, H. Garavel, L. Hillah, E. Paviot-Adet, L. Jezequel, C. Rodríguez, and F. Hulin-Hubard. Mcc'2015 - the fifth model checking contest. *T. Petri Nets and Other Models of Concurrency*, 11:262–273, 2016.

[171] F. Kordon, A. Linard, D. Buchs, M. Colange, S. Evangelista, J.-F. Jensen, K. Lampka, N. Lohmann, E. Paviot-Adet, Y. Thierry-Mieg, and H. Wimmel. Report on the Model Checking Contest at Petri Nets 2011. *Transactions on Petri Nets and Other Models of Concurrency*, 2011.

[172] C. Kral and A. Haumer. *Proceedings of the 6th International Modelica Conference, Vienna, Austria.* 2006.

[173] G. Kron. *Diakoptics: The Piecewise Solution of Large-scale Systems*. McDonald & Co. Ltd., 1963.

[174] P. Kruchten. *The Rational Unified Process An Introduction*. NY, 1980.

[175] T. S. Kuhn, O. Neurath, J. Dewey, and T. S. Kuhn. *The Structure of scientific revolutions*. Number ed.-in-chief: Otto Neurath ; Vol. 2 No. 2 in International encyclopedia of unified science Foundations of the unity of science. Chicago Univ. Press, Chicago, Ill, 2nd ed., enlarged edition, 1994. OCLC: 258260085.

[176] T. Kühne. Matters of (meta-)modeling. *Software and System Modeling*, 5(4):369–385, 2006.

[177] Y. W. Kwon, H. C. Park, S. H. Jung, and T. G. Kim. Fuzzy-DEVS formalism: concepts, realization and application. In *Proceedings of AI, Simulation and Planning in High Autonomy Systems*, pages 227 – 234, 1996.

[178] J. d. Lara, H. Vangheluwe, and P. J. Mosterman. Modelling and analysis of traffic networks based on graph transformation. *Formal Methods for Automation and Safety in Railway and Automotive Systems*, 11, 2004.

[179] B. R. Larson, P. Chalin, and J. Hatcliff. BLESS: Formal Specification and Verification of Behaviors for Embedded Systems with Software. In *NASA Formal Methods, 5th International Symposium, NFM 2013, Moffett Field, CA, USA, May 14-16, 2013. Proceedings*, pages 276–290, 2013.

[180] D. P. Y. Lawrence, C. Gomes, J. Denil, H. Vangheluwe, and D. Buchs. Coupling petri nets with deterministic formalisms using co-simulation. In *Proceedings of the Symposium on Theory of Modeling & Simulation, TMS/DEVS 2016, part of the 2016 Spring Simulation Multiconference, SpringSim '16, Pasadena, CA, USA, April 3-6, 2016*, page 6, 2016.

[181] E. A. Lee. The problem with threads. *Computer*, 39(5):33–42, 2006.

[182] Lego-Mindstorm Website. https://www.lego.com/mindstorms/.

[183] D. Lempia and S. Miller. Requirements Engineering Management Findings Report. Technical report, Federal Aviation Administration, 2009.

[184] D. Lempia and S. Miller. Requirements Engineering Management Handbook. Technical report, Federal Aviation Administration, 2009.

[185] E.-L. Lengquist Sandelin and S. Monemar. *DrModelica–An Experimental Computer-Based Teaching Material for Modelica*. Master's thesis, LITH-IDA-Ex03/3, Department of Computer and Information Science, Linköping University, Linköping, Sweden, 2003.

[186] E.-L. Lengquist-Sandelin, S. Monemar, P. Fritzson, and P. Bunus. Drmodelica - an interactive tutoring environment for modelica. In *Proceedings of the 3rd International Modelica Conference, Nov. 3-4, Linköping, Sweden*, 2003.

[187] C. Lin, A. Chandhury, A. B. Whinston, and D. C. Marinescu. Logical inference of horn clauses in petri net models. In *CSD-TR-91-031*. Purdue University, 1991.

[188] G. Liu and K. Barkaoui. A survey of siphons in Petri nets. *Information Sciences*, 363:198–220, 10 2016.

[189] S. Liu, V. Stavridou, B. Dutertre, and S. L. V. Stavridou. The Practice of Formal Methods in Safety Critical Systems. *Journal of Systems and Software*, 28:28–77, 1995.

[190] E. López Bóbeda, M. Colange, and D. Buchs. StrataGEM: A Generic Petri Net Verification Framework. In G. Ciardo and E. Kindler, editors, *Proceedings of the 35th International Conference on Application and Theory of Petri Nets and Concurrency*, June 2014.

[191] F. Lorenz. *Modelling System 1, Users Manual*. iege, Belgium, 1997.

[192] LRDE. Spot home page. http://spot.lip6.fr/.

[193] L. Lúcio, S. bin Abid, S. Rahman, V. Aravantinos, R. Kuestner, and E. Harwardt. Process-Aware Model-driven Development Environments. In *MODELS (Satellite Events)*, pages 405–411, 2017.

[194] L. Lúcio, S. Mustafiz, J. Denil, B. Meyers, and H. Vangheluwe. The Formalism Transformation Graph as a Guide to Model Driven Engineering. SOCS-TR-2012.1, McGill University, 2012.

[195] L. Lúcio, S. Mustafiz, J. Denil, H. Vangheluwe, and M. Jukss. Ftg+pm: An integrated framework for investigating model transformation chains. In F. Khendek, M. Toeroe, A. Gherbi, and R. Reed, editors, *Proceedings of SDL 2013: Model-Driven Dependability Engineering: 16th International SDL Forum*, pages 182–202, Berlin, Heidelberg, 2013. Springer.

[196] L. Lúcio, S. Mustafiz, J. Denil, H. Vangheluwe, and M. Jukss. FTG+PM: An Integrated Framework for Investigating Model Transformation Chains. In *SDL 2013: Model-Driven Dependability Engineering*, volume 7916 of *Lecture Notes in Computer Science*, pages 182–202, Berlin, Heidelberg, 2013. Springer Berlin Heidelberg.

[197] A. Marechal and D. Buchs. Generalizing the Compositions of Petri Nets Modules. *Fundamenta Informaticae*, 137(1), 2015.

[198] R. Margaret. Capability Maturity Model (CMM). https://searchsoftwarequality.techtarget.com/definition/Capability-Maturity-Model. Last access: April 2019.

[199] D. C. Marinescu, M. Beaven, and R. Stansifer. A parallel algorithm for computing invariants of petri net models. In *Petri Nets and Performance Models, 1991. PNPM91., Proceedings of the Fourth International Workshop on*, pages 136–143. IEEE, 1991.

[200] MASIW Homepage.
https://forge.ispras.ru/projects/masiw-oss.

[201] Mathworks. *Simulink User's Guide*. MathWorks Inc., 2001.

[202] Mathworks. *MATLAB User's Guide*. MathWorks Inc., 2002.

[203] S. Mattsson, M. Andersson, and K. Åström. *Object-Oriented Modelling and Simulation*. Marcel Dekker, 1993.

[204] S. E. Mattsson, H. Elmqvist, and J. F. Broenink. Modelica: The next generation modeling language - an international design effort. In *Proceedings of First World Congress of System Simulation, Singapore*, 1997.

[205] J. D. McGregor, D. P. Gluch, and P. H. Feiler. Analysis and Design of Safety-critical, Cyber-Physical Systems. *Ada Lett.*, 36(2):31–38, May 2017.

[206] W. Michael and L. M. Kristensen. The Access-CPN Framework: A Tool for Interacting with the CPN Tools Simulator. In F. Giuliana and K. Wolf, editors, *Applications and Theory of Petri Nets*, pages 313–322, Berlin, Heidelberg, 2009. Springer Berlin Heidelberg.

[207] R. Milner. *Communication and Concurrency*. Prentice-Hall, Inc., Upper Saddle River, NJ, USA, 1989.

[208] A PID Controller for Lego Mindstorms Robots. http://www.inpharmix.com/jps/PID_Controller_For_Lego_Mindstorms_Robots.html.

[209] E. Mitchell and J. S.Gauthier. *ACSL: Advanced Continuous Simulation Language-User Guide and Reference Manual*. Mitchell & Gauthier Assoc., Concord, Mass, 1986.

[210] T.-T. Mohsen, J. Remala, M. Sjolund, A. Pop, and P. Fritzson. Omsketch - graphical sketching in the openmodelica interactive book, omnotebook. In *Proceedings of the 52th Scandinavian Conference on Simulation and Modeling (SIMS 2011), Vasteras, Sweden*, 2011.

[211] P. J. Mosterman and G. Biswas. A theory of discontinuities in physical system models. *Journal of the Franklin Institute*, 335(3):401–439, apr 1998.

[212] P. J. Mosterman and H. Vangheluwe. Computer Automated Multi-Paradigm Modeling: An Introduction. *Simulation*, 80(9):433–450, sep 2004.

[213] P. J. Mosterman and H. Vangheluwe. Computer automated multi-paradigm modeling: An introduction. *Simulation*, 80(9):433–450, 2004.

[214] T. Murata, B. Shenker, and S. M. Shatz. Detection of ada static deadlocks using petri net invariants. *IEEE Transactions on Software engineering*, 15(3):314–326, 1989.

[215] T. Murata and D. Zhang. A predicate-transition net model for parallel interpretation of logic programs. *IEEE Transactions on Software Engineering*, 14(4):481–497, 1988.

[216] S. Mustafiz, J. Denil, L. Lúcio, and H. Vangheluwe. The FTG+PM framework for multi-paradigm modelling: An automotive case study. In *Proceedings of the 6th International Workshop on Multi-Paradigm Modeling*, pages 13–18. ACM, 2012.

[217] H. Nilsson. *Proceedings of the 5th International Workshop on Equation-Based Object-Oriented Languages and Tools, (EOOLT'2013), Nottingham.* Linköping University Electronic Press, 2013.

[218] J. J. Nutaro. *Building Software for Simulation: Theory and Algorithms, with Applications in C++.* Wiley, 1st edition, 2010.

[219] NXT-OSEK Homepage.
http://lejos-osek.sourceforge.net/.

[220] NXT Programs Website. http://www.nxtprograms.com/line_follower/steps.html.

[221] Object Management Group. Precise Semantics of UML State Machines Specification, May 2019.

[222] Ocarina Website. http://www.openaadl.org/ocarina.html.

[223] P. C. Ölveczky, A. Boronat, and J. Meseguer. Formal Semantics and Analysis of Behavioral AADL Models in Real-Time Maude. In *Formal Techniques for Distributed Systems, Joint 12th IFIP WG 6.1 International Conference, FMOODS 2010 and 30th IFIP WG 6.1 International Conference, FORTE 2010, Amsterdam, The Netherlands, June 7-9, 2010. Proceedings*, pages 47–62, 2010.

[224] Open Source AADL Tool Environment (OSATE).
http://osate.org/.

[225] M. Otter. Objektorientierte modellierung physikalischer systeme, teil 1: Übersicht. *at Automatisierungstechnik*, 47(1):A1–A4, 1999. LIDO-Berichtsjahr=1999,.

[226] M. Otter. *Proceedings of the 2nd International Modelica Conference,Oberpfaffenhofen, Germany.* 2002.

[227] Papyrus-RT Homepage.
https://www.eclipse.org/papyrus-rt/.

[228] H. M. Paynter. *Analysis and Design of Engineering Systems.* MIT Press, 1961.

[229] F. Peter, M. Jirstrand, and J. Gunnarsson. Mathmodelica—an extensible modeling and simulation environment with integrated graphics and literate programming. In *Proceedings of the 2nd International Modelica Conference, Oberpfaffenhofen, Germany*, 2002.

[230] F. Pommereau. SNAKES: A flexible high-level petri nets library (tool paper). In R. R. Devillers and A. Valmari, editors, *Application and Theory of Petri Nets and Concurrency - 36th International Conference, PETRI NETS 2015, Brussels, Belgium, June 21-26, 2015, Proceedings*, volume 9115 of *Lecture Notes in Computer Science*, pages 254–265. Springer, 2015.

[231] L. Popova-Zeugmann. *Time and Petri Nets.* Springer, 2013.

[232] E. Posse, J. de Lara, and H. Vangheluwe. Processing causal block diagrams with graphgrammars in atom3. In *Workshop on Applied Graph Transformation (AGT)*, pages 23–34, Grenoble, France, apr 2002. Springer, Berlin, Heidelberg.

[233] A. Pritsker and B. Alan. *The GASP IV simulation language [by] A. Alan B. Pritsker.* Wiley New York, 1974.

[234] S. Rahmoun, E. Borde, and L. Pautet. multi-objectives refinement of aadl models for the synthesis embedded systems (mu-ramses). In *2015 20th International Conference on Engineering of Complex Computer Systems (ICECCS)*.

[235] RAMSES Homepage.
https://mem4csd.telecom-paristech.fr/blog/index.php/ramses/.

[236] RDAL Homepage.
https://mem4csd.telecom-paristech.fr/blog/index.php/rdal/.

[237] W. Reisig. *Petri Nets: An Introduction.* Springer-Verlag New York, Inc., New York, NY, USA, 1985.

[238] W. Reisig. Petri Nets and Algebraic Specifications. In *Theoretical Computer Science*, volume 80, pages 1–34. Elsevier, 1991.

[239] W. Reisig. *Understanding Petri Nets: Modeling Techniques, Analysis Methods, Case Studies.* Springer Berlin Heidelberg, 2013.

[240] J. E. Rivera, D. Ruiz-Gonzalez, F. Lopez-Romero, J. Bautista, and A. Vallecillo. Orchestrating ATL Model Transformations. *Proc. of MtATL*, 9:34–46, 2009.

[241] C. Rolland. A Comprehensive View of Process Engineering. In *International Conference on Advanced Information Systems Engineering*, pages 1–24. Springer, 1998.

[242] R. C. Rosenberg. *A User's Guide to ENPORT 4*. Wiley, New York, 1974.

[243] J. Rothenberg. *The nature of modeling*. The Rand Corporation Santa Monica, 1989.

[244] W. W. Royce. Managing the development of large software systems: concepts and techniques. *Proc. IEEE WESTCON, Los Angeles*, pages 1–9, August 1970. Reprinted in *Proceedings* of the Ninth International Conference on Software Engineering, March 1987, pp. 328–338.

[245] K. Y. Rozier and M. Y. Vardi. Ltl satisfiability checking. In *Model Checking Software*, pages 149–167. Springer, 2007.

[246] N. B. Ruparelia. Software Development Lifecycle Models. *SIGSOFT Softw. Eng. Notes*, 35(3):8–13, May 2010.

[247] SAE AADL Specification. `http://standards.sae.org/as5506b/`.

[248] P. Sahlin, A. Bring, and E. Sowell. *The Neutral Model Format for Building Simulation, Version 3.02*. Royal Institute of Technology, Stockholm, Sweden, 1996.

[249] O. Salo and P. Abrahamsson. Agile methods in european embedded software development organisations: a survey on the actual use and usefulness of extreme programming and scrum. *Software, IET*, 2(1):58–64, February 2008.

[250] A. Sandholm, P. Fritzson, V. Arora, S. Delp, G. Petersson, and J. Rose. The gait e-book - development of effective participatory learning using simulation and active electronic books. In *Proceedings of the 11th Mediterranean Conference on Medical and Biological Engineering and Computing (Medicon'2007), Ljubljana, Slovenia*, 2007.

[251] H. Sargent and A. Westerberg. *Speed-Up in Chemical Engineering Design*. Transaction Institute in Chemical Engineering, 1964.

[252] H. S. Sarjoughian and Y. Chen. Standardizing DEVS models: an endogenous standpoint. In *Proceedings of the 2011 Spring Simulation Multiconference*, pages 266–273, 2011.

[253] SAVI Homepage.
`https://savi.avsi.aero/`.

[254] B. Schätz, A. Pretschner, F. Huber, and J. Philipps. Model-based development of embedded systems. In *International Conference on Object-Oriented Information Systems*, pages 298–311. Springer, 2002.

[255] G. Schmitz. *Proceedings of the 4th International Modelica Conference*. Technical University Hamburg-Harburg, 2005.

[256] H. Stachowiak. *Allgemeine Modelltheorie*. 1973.

[257] U. Stern and D. L. Dill. Parallelizing the Murφ verifier. In *Computer Aided Verification. 9th International Conference*, pages 256–267. Springer, 1997.

[258] J. Strauss, D. Augustin, M. Fineberg, B. Johnson, R. Linebarger, and F. Sansom. The sci continuous system simulation language (cssl). *SIMULATION*, 9(6):281–303, 1967.

[259] E. Syriani, H. Vangheluwe, R. Mannadiar, C. Hansen, S. Van Mierlo, and H. Ergin. AToMPM: A web-based modeling environment. In *Joint proceedings of MODELS'13 Invited Talks, Demonstration Session, Poster Session, and ACM Student Research Competition co-located with the 16th International Conference on Model Driven Engineering Languages and Systems (MODELS 2013): September 29-October 4, 2013, Miami, USA*, pages 21–25, 2013.

[260] C. Szyperski. *Component Software- Beyond Object-Oriented Programming*. Addison-Wesley, 1997.

[261] H. Takeuchi and I. Nonaka. The new new product development game. *Harvard Business Review*, 1986.

[262] R. Tarjan. Depth-first search and linear graph algorithms. In *12th Annual Symposium on Switching and Automata Theory (swat 1971)*, volume 1, East Lansing, MI, USA, oct 1971.

[263] R. Tarjan. Depth-first search and linear graph algorithms. *SIAM journal on computing*, 1(2):146–160, 1972.

[264] A. Tarski. A lattice-theoretical fixpoint theorem and its applications. *Pacific J. Math.*, 5(2):285–309, 1955.

[265] E. Technologies. SCADE Suite, 2012. `http://www.esterel-technologies.com/products/scade-suite/`.

[266] Y. V. Tendeloo. *A Foundation for Multi-Paradigm Modelling*. PhD thesis, University of Antwerp, 2018.

[267] Y. Thierry-Mieg. Symbolic model-checking using its-tools. In C. Baier and C. Tinelli, editors, *Tools and Algorithms for the Construction and Analysis of Systems - 21st International Conference, TACAS 2015, Held as Part of the European Joint Conferences on Theory and Practice of Software, ETAPS 2015, London, UK, April 11-18, 2015. Proceedings*, volume 9035 of *Lecture Notes in Computer Science*, pages 231–237. Springer, 2015.

[268] J. Thoma. *Introduction to Bond Graphs and Their Applications*. Pergamon Press, 1975.

[269] J. Thoma. *Simulation by Bond Graphs - Introduction to a Graphical Mathod*. Springer Verlag, 1989.

[270] M. Tiller. *Introduction to Physical Modeling with Modelica*. Kluwer Academic Publishers, 2011.

[271] M. Torabzadeh-Tari, Z. Hossain, P. Fritzson, and T. Richter. Omweb - virtual web-based remote laboratory for modelica in engineering courses. In *Proceedings of the 8th International Modelica Conference (Modelica'2011), Dresden, Germany, March*, 2011.

[272] S. Tripakis. Bridging the semantic gap between heterogeneous modeling formalisms and FMI. In *2015 International Conference on Embedded Computer Systems: Architectures, Modeling, and Simulation (SAMOS)*, pages 60–69. IEEE, 7 2015.

[273] A. Troccoli and G. Wainer. Implementing Parallel Cell-DEVS. In *Proceedings of the 2003 Spring Simulation Symposium*, pages 273–280, 2003.

[274] TTool Homepage.
https://ttool.telecom-paristech.fr/.

[275] A. M. Uhrmacher. Dynamic structures in modeling and simulation: a reflective approach. *ACM Transactions on Modeling and Computer Simulation*, 11:206–232, 2001.

[276] B. Unhelkar. *The art of agile practice: A composite approach for projects and organizations*. Auerbach Publications, 2016.

[277] A. University. UPPAAL homepage.

[278] A. Valmari. A stubborn attack on state explosion. *Form. Methods Syst. Des.*, 1(4):297–322, 1992.

[279] J. van Amerongen. *Dynamical Systems for Creative Technology*. Controllab Products B.V., Enschede, 2010.

[280] P. P. J. Van den Bosch and P. Bruijn. The directed digital computer as a teaching tool in control engineering; interactive instruction and design. In *Proceedings of the IFAC Symposium on Trends in Automatic Control Education*, pages 260–271, 1977.

[281] S. Van Mierlo, Y. Van Tendeloo, B. Meyers, J. Exelmans, and H. Vangheluwe. SCCD: SCXML extended with class diagrams. In *3rd Workshop on Engineering Interactive Systems with SCXML, part of EICS 2016*, 2016.

[282] Y. Van Tendeloo and H. Vangheluwe. The modular architecture of the Python(P)DEVS simulation kernel. In *Proceedings of the 2014 Spring Simulation Multiconference*, pages 387–392, 2014.

[283] Y. Van Tendeloo and H. Vangheluwe. An overview of PythonPDEVS. In C. W. RED, editor, *JDF 2016 – Les Journées DEVS Francophones – Théorie et Applications*, pages 59–66, 2016.

[284] Y. Van Tendeloo and H. Vangheluwe. An evaluation of DEVS simulation tools. *SIMULATION*, 93(2):103–121, 2017.

[285] Y. Van Tendeloo and H. Vangheluwe. Extending the DEVS formalism with initialization information. *ArXiv e-prints*, 2018.

[286] G. C. V. Vangheluwe, Hans L. and E. J. J. Kerckhoffs. Simulation for the future: Progress of the esprit basic research working group 8467. *Proceedings of the 8th European Simulation Symposium*, 1996.

[287] H. Vangheluwe. DEVS as a common denominator for multi-formalism hybrid systems modelling. In *IEEE International Symposium on Computer-Aided Control System Design*, pages 129–134, 2000.

[288] H. Vangheluwe and J. De Lara. Computer automated multi-paradigm modelling for analysis and design of traffic networks. In *Proceedings of the 36th conference on Winter simulation*, pages 249–258. Winter Simulation Conference, 2004.

[289] H. Vangheluwe, J. De Lara, and P. J. Mosterman. An introduction to multi-paradigm modelling and simulation. In *Proceedings of the AIS'2002 conference (AI, Simulation and Planning in High Autonomy Systems), Lisboa, Portugal*, pages 9–20, 2002.

[290] H. Vangheluwe, B. H. Li, Y. Reddy, and G. C. Vansteenkiste. A framework for concurrent simulation engineering. *Simulation in Concurrent Engineering*, pages 50–55, 1993.

[291] H. Vangheluwe, G. Vansteenkiste, and R. Reddy. Simformatics: Meaningful Model Storage, Re-use, Exchange and Simulation. In *Summer Computer Simulation Conference*, pages 936–944. SOCIETY FOR COMPUTER SIMULATION, ETC, 1997.

[292] H. L. Vangheluwe. DEVS as a common denominator for multi-formalism hybrid systems modelling. In *Cacsd. conference proceedings. IEEE international symposium on computer-aided control system design (cat. no. 00th8537)*, pages 129–134. IEEE, 2000.

[293] H. L. Vangheluwe and G. C. Vansteenkiste. A multi-paradigm modelling and simulation methodology: Formalisms and languages. In *Simulation in Industry, Proceedings 8th European Simulation Symposium (Genoa Italy*, pages 168–172, 1996.

[294] G. C. Vansteenkiste and H. L. Vangheluwe. European thoughts, actions, and plans for more effective modelling and simulation simulation in europe: A forum for basic research in modelling and simulation. *SIMULATION*, 66(5):331–335, 1996.

[295] G. C. Vansteenkiste and H. L. Vangheluwe. European Thoughts, Actions, and Plans for More Effective Modelling and Simulation Simulation in Europe: A Forum for Basic Research in Modelling and Simulation. *Simulation*, 66(5):331–335, 1996.

[296] M. Y. Vardi and P. Wolper. An Automata-Theoretic Approach to Automatic Program Verification (Preliminary Report). In *Proceedings 1st Annual IEEE Symp.\ on Logic in Computer Science, LICS'86, Cambridge, MA, USA, 16–18 June 1986*, pages 332–344. IEEE, Washington, DC, 1986.

[297] J. Vautherin. Parallel systems specitications with coloured petri nets and algebraic specifications. In G. Rozenberg, editor, *Advances in Petri Nets 1987, covers the 7th European Workshop on Applications and Theory of Petri Nets, Oxford, UK, June 1986*, volume 266 of *Lecture Notes in Computer Science*, pages 293–308. Springer, 1986.

[298] Verimag. Openkronos -a model-checker for timed (buchi) automata. `http://www-verimag.imag.fr/~tripakis/openkronos.html`.

[299] L. Viklund and P. Fritzson. Objectmath–an object-oriented language and environment for symbolic and numerical processing in scientific computing. *Scientific Programming*, 4(4):229–250, 1995.

[300] J. Voron, C. Demoulins, and F. Kordon. Adaptable intrusion detection systems dedicated to concurrent programs: A petri net-based approach. In L. Gomes, V. Khomenko, and J. M. Fernandes, editors, *10th International Conference on Application of Concurrency to System Design, ACSD 2010, Braga, Portugal, 21-25 June 2010*, pages 57–66. IEEE Computer Society, 2010.

[301] G. A. Wainer. *Discrete-Event Modeling and Simulation: A Practitioner's Approach*. CRC Press, 1st edition, 2009.

[302] G. Wanner and E. Hairer. *Solving ordinary differential equations I: Nonstiff Problems*, volume 1. Springer-Verlag, springer s edition, 1991.

[303] Wikipedia. Capability Maturity Model. https://en.wikipedia.org/wiki/Capability_ Maturity_Model. Last access: April 2019.

[304] J. C. Willems. The behavioral appraoch to open interconnected systems: Modelling by tearing, zooming, and linking. *IEEE Control Systems Magazine*, 27:46 – 99, 2007.

[305] S. Wolfram. Statistical mechanics of cellular automata. *Reviews of Modern Physics*, 55(3):601 – 644, 1983.

[306] S. Wolfram. *The Mathematica Book*. Wolfram Media Inc., 1997.

[307] T. Zafeer Asici, B. Karaduman, R. Eslampanah, M. Challenger, J. Denil, and H. Vangheluwe. Applying Model Driven Engineering Techniques to the Development of Contiki-based IoT Systems. In *1st International Workshop on Software Engineering Research & Practices for the Internet of Things (SERP4IoT 2019) held at 41st ACM/IEEE International Conference on Software Engineering (ICSE 2019)*, Accepted.

[308] B. P. Zeigler, H. Praehofer, and T. G. Kim. Academic Press, 2000.

[309] B. P. Zeigler, H. Praehofer, and T. G. Kim. *Theory of Modeling and Simulation*. Academic Press, 2nd edition, 2000.

[310] F. Zhang, M. Yeddanapudi, and P. J. Mosterman. Zero-Crossing Location and Detection Algorithms For Hybrid System Simulation. In *IFAC Proceedings Volumes*, volume 41, pages 7967–7972, Seoul, Korea, jul 2008. Elsevier Ltd.

Printed in the United States
by Baker & Taylor Publisher Services